INTERNATIONAL SERIES OF MONOGRAPHS IN
NATURAL PHILOSOPHY

GENERAL EDITOR: D. TER HAAR

VOLUME 41

EMISSION, ABSORPTION AND TRANSFER OF RADIATION IN HEATED ATMOSPHERES

OTHER TITLES IN THE SERIES IN NATURAL PHILOSOPHY

Vol. 1. Davydov – Quantum Mechanics
Vol. 2. Fokker – Time and Space, Weight and Inertia
Vol. 3. Kaplan – Interstellar Gas Dynamics
Vol. 4. Abrikosov, Gor'kov and Dzyaloshinskii – Quantum Field Theoretical Methods in Statistical Physics
Vol. 5. Okun' – Weak Interaction of Elementary Particles
Vol. 6. Shklovskii – Physics of the Solar Corona
Vol. 7. Akhiezer *et al.* – Collective Oscillations in a Plasma
Vol. 8. Kirzhnits – Field Theoretical Methods in Many-Body Systems
Vol. 9. Klimontovich – The Statistical Theory of Nonequilibrium Processes in a Plasma
Vol. 10. Kurth – Introduction to Stellar Statistics
Vol. 11. Chalmers – Atmospheric Electricity, 2nd edition
Vol. 12. Renner – Current Algebras and their Applications
Vol. 13. Fain and Khanin – Quantum Electronics Vol. 1. – Basic Theory
Vol. 14. Fain and Khanin – Quantum Electronics Vol. 2. – Maser Amplifiers and Oscillators
Vol. 15. March – Liquid Metals
Vol. 16. Hori – Spectral Properties of Disordered Chains and Lattices
Vol. 17. Saint James, Thomas and Sarma – Type II Superconductivity
Vol. 18. Margenau and Kestner – Theory of Intermolecular Forces
Vol. 19. Jancel – Foundations of Classical and Quantum Statistical Mechanics
Vol. 20. Takahashi – An Introduction to Field Quantization
Vol. 21. Yvon – Correlations and Entropy in Classical Statistical Mechanics
Vol. 22. Penrose – Foundations of Statistical Mechanics
Vol. 23. Visconti – Quantum Field Theory, Volume I
Vol. 24. Furth – Fundamental Principles of Modern Theoretical Physics
Vol. 25. Zheleznyakov – Radioemission of the Sun and Planets
Vol. 26. Grindlay – An Introduction to the Phenomenological Theory of Ferroelectricity
Vol. 27. Unger – Introduction to Quantum Electronics
Vol. 28. Koga – Introduction to the Kinetic Theory – Stochastical Processes in Gaseous Systems
Vol. 29. Galasiewicz – Superconductivity and Quantum Fluids
Vol. 30. Constantinescu and Magyari – Problems in Quantum Mechanics
Vol. 31. Kotkin and Serbo – Collection of Problems in Classical Mechanics
Vol. 32. Panchev – Random Functions and Turbulence
Vol. 33. Talpe – Theory of Experiments in Paramagnetic Resonance
Vol. 34. Ter Haar – Elements of Hamiltonian Mechanics, 2nd Edition
Vol. 35. Clarke and Grainger – Polarized Light and Optical Measurement
Vol. 36. Haug – Theoretical Solid State Physics, Volume 1
Vol. 37. Jordan and Beer – The Expanding Earth
Vol. 38. Todorov – Analytical Properties of Feynman Diagrams in Quantum Field Theory
Vol. 39. Sitenko – Lectures in Scattering Theory
Vol. 40. Sobel'man – Introduction to the Theory of Atomic Spectra
Vol. 41. Armstrong and Nicholls – Emission, Absorption and Transfer of Radiation in Heated Atmospheres

EMISSION, ABSORPTION AND TRANSFER OF RADIATION IN HEATED ATMOSPHERES

BY

BAXTER H. ARMSTRONG
IBM Scientific Center
Palo Alto, California

AND

RALPH W. NICHOLLS
Centre for Research in Experimental Space Science
York University
Toronto, Ontario

PERGAMON PRESS
OXFORD · NEW YORK · TORONTO
SYDNEY · BRAUNSCHWEIG

Pergamon Press Ltd., Headington Hill Hall, Oxford
Pergamon Press Inc., Maxwell House, Fairview Park, Elmsford, New York 10523
Pergamon of Canada Ltd., 207 Queen's Quay West, Toronto 1
Pergamon Press (Aust.) Pty. Ltd., 19a Boundary Street, Rushcutters Bay, N.S.W. 2011, Australia
Vieweg & Sohn GmbH, Burgplatz 1, Braunschweig

First edition 1972

Library of Congress Catalog Card No. 71-176390

Printed in Germany

08 016774 8

Contents

Preface

THIS book is an outgrowth of a research project on air opacities carried out at Lockheed Research Laboratories, Palo Alto, California, from 1957 to 1964. Although at the outset of this project all the formal theory existed that was required for detailed calculations, few such calculations had been made. Of necessity, most of the work in the literature to that date involved analytic approximations, usually on hydrogenic atomic models. The rapid development of the digital computer during the period spanned by this project removed many of the traditional calculational constraints and permitted a dramatic rise in the complexity of the calculations that could be undertaken. This, of course, has been true in virtually all fields of science. But the abrupt transition of interest in the scientific community from atomic physics to nuclear physics in the early 1930s had also resulted in gaps in the reduction to practice of some of the basic formulas of the quantum theory of radiation. Additionally, the defense needs of the United States in World War II and thereafter had led to the growth of an extensive unpublished literature related to opacities, much of which was fragmented and uncorrelated. This situation led to the writing in 1964–6 by us and a number of our colleagues of a series of Defense Atomic Support Agency reports entitled "Thermal Radiation Phenomena", which comprised a pedagogical and practical handbook of the theory and methods that were used by the group involved in the aforementioned research project for calculating opacities. Although these reports brought together and reviewed much previously uncorrelated material, they were written in haste and in a form which was still unavailable to the general public. In addition, the first author conducted research in atmospheric radiation under the sponsorship of the IBM Corporation from 1965 to 1969 which complements the aforementioned opacity theory and fills in some missing gaps, and is therefore included in this volume.

The growth in applications of these methods to fields beyond the realm of weapons, such as satellite technology, aeronomy, astrophysics, etc., has led to the need to publish an organized account of the pedagogy and practice followed in the defense opacity calculations. Two volumes of the *Thermal Radiation Phenomena* reports principally authored and edited by our colleagues were published elsewhere in 1969 (and are referenced in the text that follows). We have undertaken here to revise, develop and bring up to date part of Volume 2 of the *Thermal Radiation Phenomena* reports originally entitled "The Equilibrium Radiative Properties of Air—Theory". This volume is devoted to a review of the basic theory of radiative transfer pertinent to opacity calculations, along with requisite quantum theory of radiation as applied to individual atomic and molecular species. This first volume is therefore not limited to any particular gas or elements, but is of general applicability. In a subsequent volume of this series, the theory and methods will be applied specifically to air. The ultimate application to air shows through occasionally by its subjective influence on our language or by its choice as an example, but this should not be taken to delimit the methods.

We would like to acknowledge the support of the IBM Corporation and the contributions made by Drs. Robert R. Johnston, Paul S. Kelly and S. A. Hagstrom to the original report which are also reflected in these volumes. Drs. John L. Magee, Henry Aroeste, Roland E. Meyerott and Rolf K. M. Landshoff offered considerable assistance and encouragement in that project, and Dr. Meyerott was the originator of the opacity research project at Lockheed in 1957 from which these volumes ultimately devolved. This work was influenced substantially by the work of Dr. John C. Stewart (now deceased) and his colleagues at General Atomic in La Jolla, California, during the aforementioned period 1957 to 1964, as well as by his suggestions and assistance. Special thanks are due to Professor J. Cooper of the University of Colorado for reviewing the manuscript and providing helpful criticism, and to Robert R. Johnston and Dewey R. Churchill of Lockheed Research Laboratories, Palo Alto, for many helpful discussions and much assistance.

IBM Scientific Center, Palo Alto, California BAXTER H. ARMSTRONG
York University, Toronto, Ontario RALPH W. NICHOLLS
December 1970

Table of Symbols

English Letters

Symbol	*Definition*	*Page Introduced*
a	(1) radius of sphere and of cylinder	72
	(2) specific value of linear variable x	91
	(3) normalization constant	172
	(4) coefficient in polynomial expansion	242
a_0	Bohr radius	151
$\boldsymbol{a}$	vector specifying orientation of atom	148
a.u.	atomic units	198
b	(1) parameter in optical depth expression	71
	(2) specific value of linear variable x	91
	(3) coefficient in oscillator equation of motion	104
$b(\nu)$	line shape, or profile, factor as a function of frequency. Variations: $b_x(\nu - \nu_0)$, b_1	26
$b(\omega)$	line shape, or profile, factor as a function of angular frequency	132
c	(1) velocity of light	21
	(2) subscript indicating "continuum" or "continuous"	47
	(3) subscript indicating "column"	64
	(4) subscript indicating "classical"	112
$c(\alpha)$	expansion coefficient	185
d	(1) average frequency spacing between lines	78
	(2) spectroscopic notation for $l = 2$ state	165

Table of Symbols

Symbol	*Definition*	*Page Introduced*
d_U, d_L	degeneracies of upper and lower electronic molecular states, respectively	244
e	(1) base of natural logarithms	15
	(2) electron charge	51
	(3) subscript referring to "electron"	138
eV	electron-volt	52
f	(1) f-number or oscillator strength for a transition. Variations: $f_{\alpha\beta}$, f_{LU}, $f_{v'v''}$, $f_{n'l',n''l''}$, etc.	49
	(2) unspecified mathematical function	52
	(3) state-numbering index or quantum number for final state	186
$f(E, l; r)$	reduced radial wave function	159
$f_c^{(\pm)}(\hat{\boldsymbol{k}} \cdot \hat{\boldsymbol{r}})$	Coulomb scattering amplitude	201
$f(abc; def)$	line-strength factor of Condon and Shortley	167
$f(\eta_i, \eta_f)$	Coulomb matrix element function	201
g	square-integrable function	52
g_i	statistical weight. Variations: g_j, g_L, g_U, g_J	111
$g_{FF}(k_i; k_f)$	Gaunt factor. Variations g_{FF}^B, g_{FF}^{MP}, etc.	202
h	(1) Planck's constant	21
	(2) Sampling interval of independent variable	236
$\hbar$	reduced Planck constant ($h/2\pi$)	112
i	(1) numbering index (often subscripted)	51
	(2) $\sqrt{-1}$	104
	(3) symbolic set of quantum numbers designating a state (frequently an *initial* state)	113
j	(1) numbering index (often subscripted)	51
	(2) symbolic set of quantum numbers	113
$\boldsymbol{j}$	electric current vector	116
j_l	spherical Bessel function	151

Symbol	*Definition*	*Page Introduced*
j_ν	spectral emission coefficient; variations: $j'_\nu, j_\nu^{(s)}$	17
k	(1) Boltzmann constant	21
	(2) normalization constant	80
	(3) force constant	104
	(4) numbering index	113
$\hat{\boldsymbol{k}}$	(1) unit vector along z-direction	120
	(2) wave number unit vector	190
$\boldsymbol{k}_\gamma$	photon propagation or wave-number vector of magnitude k_γ	117
$\boldsymbol{k}_e$	electron propagation or wave-number vector of magnitude k_e	144
k_n	wave-number eigenvalue	155
l	(1) mean free path between collisions. Variation: l_ν	43
	(2) subscript/superscript denoting "line"	47
	(3) summation index or integer variable	56
	(4) orbital angular momentum quantum number, often used as subscript or superscript. Variations: $l_>$, l_i, etc.	116
m	(1) quantum-state index	25
	(2) electron mass	49
	(3) magnetic or azimuthal quantum number	123
δm	mass increment	17
n	(1) index of refraction. Variation: n_r	21
	(2) quantum state index	25
	(3) particle number density. Variation: n_ν	41
	(4) integer variable	155
	(5) principal quantum number	164
n_ω	photon occupation number of cell in phase space. Variations: $\bar{n}_\omega$, $\bar{n}_{\hat{\Omega}}$	117
$n_0(\nu)$	photon flux per unit frequency interval	135

Table of Symbols

Symbol	*Definition*	*Page Introduced*
dn_R	infinitesimal number of photon (radiation) states per unit volume in an energy interval dE	117
n_i	number of ions per cm^3	210
p	(1) spectroscopic notation for orbital angular momentum $l = 1$ state	164
	(2) magnitude of linear momentum	138
$\boldsymbol{p}$	linear momentum vector. Variation: $\boldsymbol{p}_\alpha$	163
q	(1) optical parameter of spectrum line. Variations: $\bar{q}$, $\bar{q}_0$, q_D	89
	(2) index	182
$q(\boldsymbol{r}, \widehat{\Omega}, t)$	time-dependent angular source density	73
$q(\boldsymbol{r}, t)$	angle-averaged source density	73
$q_0(\boldsymbol{r})$	isotropic, steady-state source density	73
$q_{v'v''}$	Franck–Condon factor	242
r	(1) ratio of absorption coefficient contributions	47
	(2) diffusivity factor	82
	(3) magnitude of radius vector	115
$\boldsymbol{r}$	radius vector to a given point	73
$\boldsymbol{r}_{\alpha\beta}$	matrix element of radius vector. Variation: $\boldsymbol{r}_{nlm}^{E'l'm'}$	120
$\bar{r}_{v'v''}$	r-centroid	242
r_0	classical electron radius	104
r_e	equilibrium internuclear distance	253
s	(1) distance measured along a ray	16
	(2) optical parameter of spectrum line in band model	99
	(3) spectroscopic notation for zero orbital angular momentum state	164
	(4) spin parameter	200
$s(nl', n'l')$	multiplet strength factor of Levinson and Nikitin	175

Symbol	*Definition*	*Page Introduced*
t	(1) time	26
	(2) independent or integration variable	62
u	(1) the ratio of photon to thermal energy $h\nu/kT$. Variations: u_M, u_0, u_i, $u^{(1)}$, $u^{(2)}$	45
	(2) radiation energy density	41
	(3) gas mass per cm^2	84
$u(r, \theta, \varphi)$	wave function	151
v	(1) velocity. Variations: $\bar{v}$, $\mathbf{v}$	43
	(2) vibrational quantum number. Variations: v', v''	239
	(3) subscript denoting "volume"	17
v', v''	vibrational quantum number of upper and lower states, respectively	239
w	(1) half half-width of Lorentz-shaped spectrum line. Variations: w_U, w_L, w_n, w_c	49
	(2) transition probability per unit time. Variations: $w_{\alpha\beta}$, w_{ij}, $\bar{w}_{\alpha\beta}$, $w_{\beta\alpha}^{(\varepsilon)}$, $w_{l,l+1}$	113
$w(z)$	complex error function	229
$w_l(k, r)$	reduced wave function	194
w_ω	transition probability per unit time per unit angular frequency interval	131
x	(1) linear coordinate or independent variable	14
	(2) optical parameter of spectrum line (also see q)	79
	(3) parameter in Coulomb matrix element formula. Variation: x_0	202
	(4) frequency scale in units of Doppler width	218
$x_{\alpha\beta}$	matrix element of the linear coordinate x	128
y	(1) ratio of line width to line spacing	81

Table of Symbols

Symbol	*Definition*	*Page Introduced*
	(2) linear coordinate or independent variable	128
	(3) ratio of Lorentz line width to Doppler width	218
$y_{\alpha\beta}$	matrix element of linear coordinate y	128
z	(1) linear coordinate, including height variable for stratified atmosphere	84
	(2) complex variable $x + iy$	229
$z_{\alpha\beta}$	matrix element of linear coordinate z	128
A_{nm}	Einstein coefficients. Variations: A_{UL}, $A_{Lv''J''}^{Uv'J'}$, $A_{\alpha\beta}$	25
A	(1) area	34
	(2) atomic weight	52
	(3) spectroscopic notation for excited molecular electronic state	257
	(4) normalization constant	190
	(5) scattering length	214
$\mathbf{A}$	vector potential (magnitude A_0)	116
$A(x)$	absorption function. Variations: $\bar{A}_\alpha$, $A_\alpha(x)$ etc.	65
A_n	weight coefficients for Gauss–Hermite integration	234
B_{nm}	Einstein coefficient. Variations: B_{UL}, $B_{Lv''J''}^{Uv'J'}$, $B_{\alpha\beta}$, $\mathscr{B}_{\alpha\beta}$	25
B_ν, $B_\nu(T)$	Planck black-body function	21
$B(T)$	frequency-integrated Planck function	36
$B_i(T)$	Planck function integrated over frequency interval i	65
B	Spectroscopic notation for excited molecular electronic state	257
C	numerical constant	58
$C(j_1 j_2 j_3; m_1 m_2 m_3)$	Clebsch–Gordan coefficient	124

Symbol	*Definition*	*Page Introduced*
D	(1) diffusion coefficient	41
	(2) derivative, d/dr	160
$\boldsymbol{D}^{(e)}_{ml;m'l'}$	matrix element	162
D_e	dissociation energy	253
D_1, D_2, etc.	atomic structure parameters	214
$D(\theta, \chi, \varphi)$	molecular dyadic	240
E	(1) electric field strength	24
	(2) neutron energy	231
	(3) abbreviation for $(-1)^n \exp(i\pi z/h)$	236
	(4) molecular energy	240
E_n	energy associated with quantum state n. Variations: E_m, E_α, E_s, E_{vib}, E_{rot}, E_I, E_{cl}, E_{QM}, $E_s^{(UL)}$, $E_A^{(UL)}$, E_i, E_f	25
E_ω	photon energy	198
$E_n(y)$	exponential integral of order n	35
$E(y, x)$	Elsasser band model function	81
F	spectrally integrated flux. Variations: $\bar{F}$, F_x, F_0, etc.	34
F_ν	spectral flux. Variations: $F_{\nu 0}$, $F_{\nu +}$, $\bar{F}_\nu$, F_1	14
$F(\nu)$	folded line profile function	218
$F(lSL; S_pL_p)$	Fractional parentage coefficient. Variation: $F_p(nSL; SL)$	176
$F(a, b; e; x)$	hypergeometric function	202
${}_1F_1(a; b; x)$	confluent hypergeometric function	201
$F_l(kr)$	Coulomb function (regular)	203
$F_e(\boldsymbol{k})$	electron distribution function	192
$G_l(kr)$	irregular Coulomb function	207
$G(\varepsilon_i, l_i, \varepsilon_f, l_f)$	function in Peach formula for Gaunt factor	208
H	volume segment of slab (Fig. 2.5)	37
H_0	unperturbed Hamiltonian	197
H'	perturbation term in Hamiltonian	113
$\boldsymbol{H}$	magnetic field vector	143

Symbol	*Definition*	*Page Introduced*
$H_n(x)$	Hermite polynomial	233
I	(1) frequency-integrated specific intensity. Variations: $I_{\Delta\nu}$, $I^c_{\Delta\nu}$, $I(\theta, x)$, etc.	66
	(2) general notation for definite integral	53
I_0	intensity of collimated beam (energy per cm² per sec)	104
$I_0(\omega)$	spectral intensity of collimated beam (energy per cm² per sec per unit angular frequency interval)	132
$I_\nu(\hat{\Omega})$	specific intensity function. Variations: I_ν, $I_\nu(s, \hat{\Omega})$, $I_\nu(0)$, $I_\nu(s)$, $I^c_{\Delta\nu}$, $I^l_{\Delta\nu}$, $\bar{I}_\nu$, $I_{v'v''}$, I_ω, $I^{(\varepsilon)}_\omega$, $I_\omega(\hat{\Omega})$, etc.	16
$I_l(E, E')$	normalization integral	159
I_n	modified Bessel function	76
J	(1) total angular momentum quantum number	163
	(2) maximum value of integer index	188
	(3) rotational angular momentum quantum number	239
$\boldsymbol{J}$	current vector	41
J_ν	source function	17
$K(x, y)$	Voigt function	218
K_n	modified Bessel function	76
L	(1) slab thickness	16
	(2) subcript or superscript denoting "lower" state	110
	(3) side of normalization box	137
	(4) total orbital angular momentum. Variations: $L^{(i)}$, $L_>$, L_p, Ⱡ	164
$L(\theta)$, L'	chord length	72
$L(x, y)$	imaginary part of complex error function	229
$\mathscr{L}_i$	partial set of quantum numbers belonging to state i (for line transition)	166

Symbol	*Definition*	*Page Introduced*
M	(1) total magnetic quantum number. Variations: M_J, M_L, M_S	164
	(2) mass of atom or molecule	92
$\mathbf{M}$	electric dipole moment vector	238
$M_{\alpha\beta}$	matrix element	125
$M(a, b, x)$	confluent hypergeometric function	233
$\mathcal{M}_i$	partial set of quantum numbers belonging to state i (for multiplet transition)	166
N	number of spectrum lines in a certain interval	80
N_n	occupation number for quantum state n	25
N_v	particle number density. Variations: N_e, $N_v(U)$, $N_v(L)$, etc.	17
$N_0^{(\varepsilon)}(\omega)$	photon flux of given polarization per unit angular frequency interval	135
O	arbitrary operator	189
$O(x)$	order of x	36
P	electric dipole moment of Condon and Shortley	167
$P(\mathbf{r})$	escape probability. Variation: $P^{(v)}(\mathbf{r})$	73
$P(S)$	probability function for line strength S	80
P_0	spatial average escape probability. Variation: P_0^v	73
P_l	Legendre polynomial	123
Q_d	diffusion cross section	198
R	(1) radial distance	37
	(2) line width-to-spacing ratio	94
$\mathbf{R}$	ion coordinate vector	190
$R_i(r)$	radial wave function. Variations: $R_{nlm}(r)$, R_α, $R_l(r)$, $R_{El}(r)$	115
R_n	remainder term	232
$R_e(r)$	electronic transition moment. Variations: $R_{Lv''J''\Lambda''M''}^{Uv'J'\Lambda'M'}$, $\bar{R}_e$, etc.	242

Symbol	*Definition*	*Page Introduced*
$\mathscr{R}(\beta, \alpha)$	radial integral. Variations: $\mathscr{R}_{\beta\alpha}$, $\mathscr{R}(l, l')$	115
Ryd	Rydberg energy unit 13.6 eV	176
S	(1) partial slab volume (Fig. 2.5)	37
	(2) surface area	43
	(3) integrated spectrum line strength	80
	(4) total spin angular momentum	163
	(5) spectroscopic notation for total orbital angular momentum $L = 0$ state. S, P, and D are defined analogously to s, p and d. For details of spectroscopic notation see Allen (1963)	165
$d\mathbf{S}$	surface element of convex body	75
$\mathbf{S}$	electron flux	145
$\bar{S}$	average integrated line strength	80
$S_l(\alpha)$	sum-rule sum	56
S_α	integrated line strength	77
S_{ij}	absolute line strength	163, 164
$S_{v'v''}$	molecular band strength	241
$S_{J''\Lambda''}^{J'\Lambda'}$	Hönl–London factor. Variations: S_J^R, etc.	243
S_{UL}	transition strength matrix element	238
$\mathscr{S}$	relative line strength	166
T	absolute (Kelvin) temperature	21
T_0	complex conjugation operator	191
$\mathrm{Tr}(x)$	transmission function. Variations: $\bar{T}_r$, $\mathrm{Tr}_\alpha(x)$, $\bar{T}$, T_I, T_f, etc.	63
U	subscript or superscript designating upper, or excited state	110
$U(r)$	potential function	252
$U(abcd; ef)$	Jahn coefficient	179
V	(1) volume	27
	(2) potential	189
$\tilde{W}$	wingspread of line	49

Symbol	*Definition*	*Page Introduced*
$W(s)$	equivalent width of line. Variations: $W_\alpha(s)$, $\bar{W}_\alpha(s)$, $\bar{W}$, W	78
$W(x)$	Ladenburg and Reiche function	79
$W(z)$	complex error function	229
W_i	energy of ion	190
$W'(u)$	weighting function in factorization formula	56
$W_P(u)$	weighting function in Planck mean absorption coefficient formula	46
$W_R(u)$	weighting function in Rosseland mean absorption coefficient formula	46
$\bar{W}$	average equivalent width	80
$W(abcd; ef)$	Racah coefficient	167
$W_{a,b}(x)$	Whittaker function	233
X	(1) frequency-to-width ratio	222
	(2) spectroscopic notation for the molecular electronic ground state	257
$Y_{lm}(\theta, \varphi)$	spherical harmonic	123
Z	(1) atomic number	52
	(2) ionic or nuclear charge. Variation: Z_I	151

Greek Letters

α	(1) numbering index	77
	(2) fine-structure constant	199
	(3) screening constant (inverse Debye length)	209
	(4) strength parameter in potential formula	213
	(5) symbolic quantum number or state	112
α_D	half-width of Doppler-broadened line	129
α_ν	atomic absorption coefficient. Variation: $\alpha_{\bar{\nu}}$	17

Symbol	*Definition*	*Page Introduced*
β	(1) symbolic quantum number or state	112
	(2) Doppler line shape parameter	216
	(3) Morse potential parameter	253
γ	(1) symbolic quantum number	167
	(2) Euler's constant 0.577216...	36
	(3) damping constant	104
δ	(1) differential increment	14
	(2) frequency spacing between lines	81
$\delta(x)$	Dirac delta function	16
δ_l	phase shift for general potential function	148
$\delta_{jj'}$	Kronecker delta	125
ε	emissivity. Variations: ε_h, ε_ν, $\varepsilon_{\nu c}$, ε_c, $\bar{\varepsilon}_{ci}$, ε_s, $\varepsilon_{\Delta\nu}$, $\varepsilon^l_{\Delta\nu}$, $\varepsilon^c_{\Delta\nu}$, $\varepsilon_{\nu c\alpha}$	36
ε_ω	radiation energy at angular frequency ω extracted from beam	104
ε_i	(1) ionization potential	211
	(2) energy of electron. Variations: ε_f, $\bar{\varepsilon}$, ε_k, etc.	190
$\hat{\varepsilon}_1$, $\hat{\varepsilon}_2$, $\hat{\varepsilon}$	unit vectors	117
η	(1) photon viscosity	44
	(2) dimensionless Coulomb parameter	201
	(3) ratio of line widths	222
	(4) small parameter in Lippman–Schwinger form of the Schroedinger equation	197
θ	(1) zenith angle or angular coordinate	15
	(2) temperature in eV	211
	(3) electron deflection angle	198
$\varkappa$	mean mass absorption coefficient of Sampson	71
$\varkappa_\nu$	spectral mass absorption coefficient	17
$\varkappa_{\nu'}$	reduced spectral mass absorption coefficient	28

Symbol	*Definition*	*Page Introduced*
$\bar{\varkappa}_R$	Rosseland mean opacity	41
$\bar{\varkappa}_P$	Planck mean opacity	54
$\varkappa_0$	(1) Rosseland opacity upper bound	52
	(2) dimensional constant of reduced absorption coefficient	218
$\bar{\varkappa}_2$	second moment mass absorption coefficient	58
$\bar{\varkappa}$	arbitrary mean absorption coefficient (Figs. 2.11 and 2.12)	59
λ	wavelength. Variations: λ_0, $\lambda_{v'v''}$, $\lambda_{\alpha\beta}$	104
μ	(1) dimensionless screening constant	210
	(2) reduced mass	252
μ_ν	spectral absorption coefficient in cm^{-1}. Variations: $\mu(\nu)$, $\mu\,(u, T)$, μ, $\mu(u)$, μ_c, μ_l, μ_0^{wl}, μ_c^{sl}, $\mu^{(1)}$, $\mu^{(2)}$	14
μ'_ν	reduced spectral absorption coefficient. Variations: μ'_c, μ'_l, etc.	28
$\bar{\mu}_P$	Planck mean absorption coefficient	36
$\bar{\mu}_R$	Rosseland mean absorption coefficient	40
$\mu^{FF}(\omega)$	free–free absorption coefficient	192
$\bar{\mu}_H$, $\bar{\mu}_D$	harmonic and direct mean absorption coefficients	55
$\bar{\mu}_2$	second moment of absorption coefficient	56
$\bar{\mu}_T$	transmission mean absorption coefficient. Other mean absorption coefficients: $\bar{\mu}'$, $\bar{\mu}(s)$, $\bar{\mu}'\,(x, \theta)_i$, etc.	66
μ_0	integrated line absorption coefficient	49
μ_A	reduced mass in atomic units	260
ν	frequency (cycles per sec). Variations: ν_{nm}, ν_0, $\bar{\nu}$, $\nu_{\alpha\beta}$	14
$\bar{\nu}$	wave number λ^{-1} in cm^{-1}	108
$\Delta\nu$	frequency interval	65

Symbol	*Definition*	*Page Introduced*
ξ	(1) mean value of integration variable	91
	(2) Coulomb force parameter	202
π	3.14159...	16
ϱ	mass density	17
$\varrho\,(v'v'')$	correction function for r-centroid approximation	261
ϱ_ν	radiation energy density	122
σ	(1) cross section. Variations: σ_{sc}, $\sigma_{\beta\alpha}(\omega)$, $\sigma(\omega)$, σ (P.E.), $\sigma(R)$, σ_{fi}, σ^{FF}, σ^{el}, etc.	33
	(2) Stefan's constant	36
σ_l	Coulomb phase shift δ_l (Coul.)	151
σ_{ij}	radial dipole integral	116
$d\sigma/d\Omega$	differential cross section	32
τ	(1) time constant reciprocal to damping constant	105
	(2) lifetime of state	216
τ_ν	spectral optical depth. Variations: τ, τ_ν', $\tau_{\nu 1}$, τ_ν^n, etc.	18
τ_R, τ_P	optical depth computed from Rosseland and Planck mean opacities, respectively	71
$\tau_{v'}$	lifetime of vibrational state	247
τ_0	average collision period	106
$d\tau$	element of volume	41
φ	angle coordinate	72
$\varphi(L)$	chord probability function	75
$\varphi_i(R)$	wave function	190
χ	(1) wave function	185
	(2) Euler angle	240
	(3) phase shift function in Peach formula	208
ψ	(1) angular coordinate	120
	(2) wave function. Variations: $\psi_{k_e lm}$, ψ_v, etc.	124
ω	angular frequency (radians per sec). Variations: ω_0, $\Delta\omega$, $\omega_{\alpha\beta}$, etc.	104

Symbol	*Definition*	*Page Introduced*
ω_s	spin weighting function	200
ω_e	molecular vibrational constant	253
$\omega_e x_e$, $\omega_e y_e$, $\omega_e z_e$	molecular vibrational constants	253–6
Δ	finite element or increment of a quantity	34
$\mathbf{\nabla}$	gradient operator	40
Γ	particle flux. Variations: Γ_+, Γ_-, Γ_ν	43
Γ_n	width of neutron level (also Γ)	231
$\Gamma(x)$	gamma function	201
Θ	Zenith angular coordinate	124
Λ	molecular angular momentum quantum number	239
Λ_R	Rosseland mean free path (Λ_l, line contribution; Λ_c, continuous contribution)	41
Σ	spectroscopic notation for molecular state (see Herzberg, 1950, for details of molecular spectroscopic notation)	257
Φ	azimuthal angular coordinate	16
Ψ	wave function	117
$\Psi(X, \eta)$	"Nuclear" form of Voigt profile	222
$\hat{\Omega}$	unit vector for cone of solid angle	15
$d\Omega$	element of solid angle. Variations: $d\Omega_\gamma$, $d\Omega_R$, $d\Omega_e$, etc.	16
$\Delta\Omega$	finite element of solid angle	34

Chapter 1

Introduction

THIS volume is concerned with thermal radiation phenomena in heated gases over a wide range of temperature and density. The principal emphasis is on temperatures above a few thousand degrees, well above, for example, the ambient temperatures of the terrestrial troposphere. There is some unavoidable overlap with lower-temperature studies because some processes and methods are common to both areas. However, this emphasis is our reason for dealing mainly with atomic transitions and with electronic transitions in molecules. For an excellent treatise on the processes of primary importance at lower temperatures such as purely rotational and purely vibrational transitions, the reader is referred to Goody (1964). Thermal radiation phenomena arise from, or are related to, the passage of electromagnetic energy through an atmosphere of some type when significant interaction occurs between the radiation stream and the atmosphere. It is usually implied that a partial or complete thermal equilibrium is maintained in the interaction. The case of complete thermal equilibrium is not too important for it implies a homogeneous system with no net transport of radiation at all. Radiation is a significant mode of energy transfer in all gases at sufficiently high temperatures, and in many situations at low temperatures as well. Since radiative energy transfer is controlled by the absorption coefficient which is, in turn, determined by the microscopic atomic and molecular, and the statistical/thermodynamic properties of the medium, much of this book is concerned with a discussion of these underlying properties. After an elementary introduction to the theory of radiative transfer limited to conditions of local thermodynamic equilibrium, the present volume is concerned with the detailed application of the basic quantum theory of radiation to real atomic and molecular systems. The discussion is limited to elementary particulate interactions; absorption arising from collective modes of excitation of a plasma will not be considered. The radiative transfer problem only reappears occasionally to guide the application into the practical channels which constitute the *raison d'être* of the book.

Although radiation transport is now of wide interdisciplinary interest and application, most of its basic developments were made in an astrophysical milieu (see, for example, Rosseland, 1935). Some of its current applications are in stellar, solar and planetary atmospheres, aeronomy, in meteor, missile, and satellite re-entry phenomena, in combustion physics and chemistry, and in plasma and weapons physics.

A large fraction of the work dating from World War II on specific problems in the above fields of application has been motivated by defense needs and financed by government contracts. As a result, much of the literature on the subject is comprised of unpublished and therefore unrefereed contract reports which are not universally available to the scientific community. Much of the work described in this "grey" literature (Goody, 1964) is important, but some obscurities and errors in an already complex field have propagated through these reports. Other problems due to the particular history of this field have also occurred. For example, a perusal of the reports concerned with opacity calculations shows a considerable repetition of some of the formal arguments involved in justifying the calculations (although not in the calculations themselves), and a lack of assignments of priority, or acknowledgements. There has also been a substantial lack of cross referencing. This has all been due in part, of course, to the classified nature of some of the projects, particularly the earlier ones, and the fact that even the unclassified reports were often not readily available to some of the authors, who then found it necessary to repeat some of the derivations. However, once such a situation has been created it is generally self-stimulating even in the absence of the original causes, due to the large amount of effort which must be expended to ameliorate it. We hope the present series will help to exorcise some of these ghosts; however, all the relevant material has still not been declassified so that this goal cannot yet be completely reached.

The initiation, in 1961, by Prof. S. S. Penner, of the *Journal of Quantita-*

tive Spectroscopy and Radiative Transfer (JQSRT) has done much to provide an appropriate vehicle for the open publication of this work, and the reader is referred to that Journal for recent research papers on topics discussed in this volume.

Formidable experimental difficulties associated with the production and controlled laboratory study of very hot gases ($\sim 10^5$ °K) have limited most experimental work in the field to temperatures below 20,000 °K. Thus, much reliance has had to be placed on theoretical research involving models of increasing realism, complexity and sophistication. As implied above, the basic theory employed has its roots in (a) astrophysical discussions of the transfer of radiation through stellar envelopes, and (b) in the applied quantum mechanics of the radiative properties of atoms and molecules. The bulk transport of radiation through hot gases is usually discussed in terms of the radiation absorption coefficient, which is a phenomenological parameter of the material through which the radiation passes. This absorption coefficient in turn can be specified as a function of wavelength and absorber gas properties by recourse to quantum theory and statistical mechanics. From a detailed knowledge of the "spectral" absorption coefficient of the gas, realistic mean absorption coefficients can be derived in terms of which some aspects of radiation transport may more conveniently be discussed. In addition, if conditions of local thermodynamic equilibrium prevail, absorption coefficients are related by Kirchhoff's law to the emission coefficients of the gas. At low temperatures there are many contributors to the absorption while at very high temperatures there are fewer contributing effects and therefore the situation is conceptually somewhat simpler. Figure 1.1 illustrates some significant subdivisions and interrelations towards the high-temperature limit of our considerations. As one progresses downwards in temperature, the effects and interactions proliferate and become, at our present state of knowledge, more fragmented and diverse.

The macroscopic absorption coefficient is comprised of two factors. These are (a) the populations of the absorbing species and (b) the cross section per particle, or the microscopic "absorption coefficient". The first of these factors is obtainable from statistical mechanics, and a full discussion as well as tables of properties for air are given, in *The Equilibrium*

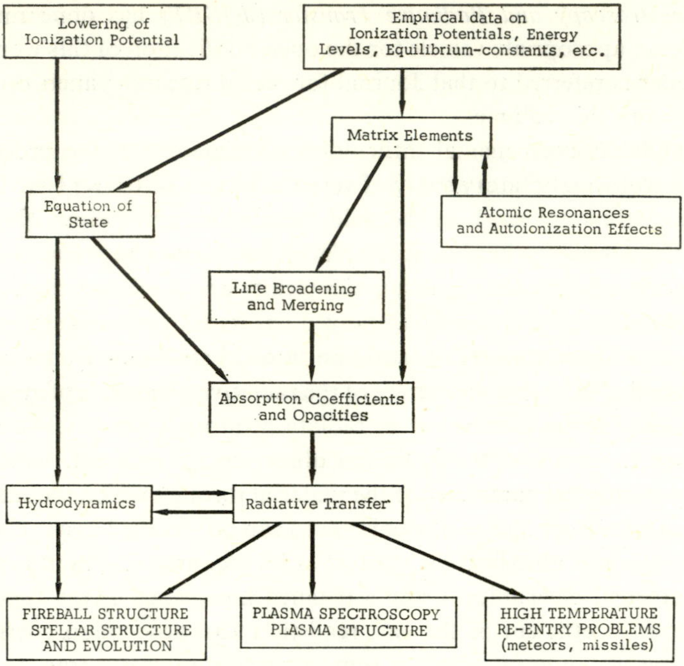

FIG. 1.1. Some significant subdivisions and interconnections of high-temperature radiation physics.

Thermodynamic Properties of High Temperature Air by Gilmore (1967). In order that the statistical mechanics treatment be realistic, accurate information must also be available on the thermochemical properties of the absorbers, and on equilibrium constants, when chemical processes (dissociation, etc.) that can occur are also temperature-dependent. The second of these factors requires a quantum mechanical description of atomic/molecular structure to determine radiative transition probabilities and is discussed in detail in subsequent chapters of the present volume. Realistic atomic and molecular models are needed for adequate calculations of this second factor. Careful experimental measurements of atomic and molecular properties are also needed in order to assess or verify the quality of the models selected. Much progress has been made during the past two decades in the approximate, yet realistic quantum mechanical analyses of complex atomic and molecular systems.

It should be emphasized here that both the above-mentioned aspects of the problem of the theoretical calculation of absorption coefficients require the strong support of experimental programs that can provide the basic atomic and molecular data on which the calculations often depend. Thus, experimental work in this field (among others, of course) occupies a dual role, since it must provide some of the basic input information as well as serve its traditional role in verifying the outcome of calculations.

It has come as a surprise to many that all such basic measurements were not made long ago. The growth of research institutes of "Laboratory Astrophysics" is indicative of the large amount of atomic and molecular physics left to be done.

The study of absorption coefficients may be further subdivided quite naturally into two parts as a function of temperature. At low temperatures (below about 12,000 °K for air, for example) most of the absorbing species are molecular, and both the statistical mechanics of absorber populations and the quantum theory of absorbing transitions must be couched in molecular terms. At higher temperatures one must usually deal with a multispecies ionic gas. Theory (both statistical and quantum mechanics) is far more developed for atomic than for molecular species, and calculations can be more readily carried out on both aspects of the absorption coefficient, as is shown in later chapters.

The realistic and comprehensive theoretical studies that can now be carried out or have been performed in the past decade have only become possible due to a number of circumstances. The availability of large-scale, high-speed digital computers, the experimental provision of relatively good critical experimental data, and the development of detailed and cogent theoretical models and methods, have all been necessary conditions which had to be established before such work as is discussed in this series could be seriously undertaken. It is interesting to note that the accurate calculation of opacities covers a remarkably wide range of physical phenomena and theories. While crude opacity values are relatively easy to obtain, accurate values are very difficult to calculate. Even though, for most aspects of the calculations, the basic physical processes are reasonably well understood, the carrying out of the necessary calculations is a tedious and complicated task, and long-term efforts are required.

As a commentary on the recent development of a relatively modest scientific field, it is worth emphasizing how much the need for opacity values (in view of the character of these quantities) particularly for air has stimulated research on such a wide and otherwise diverse variety of topics.

The plan of this volume is as follows: Chapter 2 reviews the basic theory of radiative transfer to establish definitions and to present the overall scope of the problem. Mean absorption coefficients are introduced and the inequalities and bounds which they satisfy are discussed. Emissivities for several geometries are discussed, and a limiting relation is given between the Planck mean absorption coefficient and the emissivity for an arbitrary geometry. This relation, due to Johnston (Johnston and Platas, 1969) is obtained from a formulation taken over from neutron transport theory. Transmission functions and other quantities such as "diffusivity factors" which are useful in cases of intermediate optical depths are also discussed. Chapter 3 reviews and describes the elementary quantum theory of radiation by atoms. The formal results of the theory are reduced to the formulas for specific radiative processes and some clarifying and comparative comments are made on the equivalent formulae derived by a number of authors. Detailed calculations for discrete, photoelectric, and free–free processes are presented, with considerable attention paid to such things as the wave-function normalization and the angular factors in the matrix elements. The phenomenology of line profiles is reviewed, including a comprehensive discussion of the Voigt profile. Chapter 4 extends the theory of Chapter 3 to take account of radiative transitions by molecular species, with emphasis on electronic transitions.

Historical Sketch

Before plunging into the detailed theory it will be valuable for purposes of orientation to sketch briefly in this chapter the development of theoretical research on absorption coefficients of hot gases made during this century.

The importance of radiation absorption coefficients of heated gases in the quantitative description of stellar atmospheres was increasingly realized during the decade 1915–26. During this period it became clear that the purely convective models of stellar atmospheres were not adequate to account for the observed facts, and that in some way the radiation flux must be included in any realistic account of stellar structure. Schwarzschild (1906) had in fact already begun to lay the foundation of radiative stellar models. Between 1915 and 1926 Eddington (1926) made a systematic application of radiative transfer theory to stellar atmospheres, and with E. A. Milne (1924) was primarily responsible for establishing the direct physical significance of radiative stellar models. The work of this period, including the early calculations of absorption coefficients, is described in Eddington's classic book (1926) which stands as a monument to and closes the initial period of development of radiative stellar models. The historical significance of Eddington's book is heightened by its position relative to the formulation of quantum mechanics. It appeared just at the onset of the major period of development of the subject and before the results of this development were available. Also, it appeared just after Kramers' (1923) remarkable derivation of semiclassical absorption coefficients, and Rosseland's (1924) discovery of the correct mean absorption coefficient to use for stellar interiors. These two developments enabled Eddington and Milne to provide a solid physical basis for the radiative aspects of stellar structure. It is interesting to note that Milne independently obtained the same results as Kramers (1923) for the semiclassical photoelectric cross section. His paper (Milne, 1924) appeared in the same journal, the *Philosophical Magazine*, just two months after Kramers'. Because of Kramers' prior publication, the semiclassical cross sections are ordinarily attributed to him, and Milne's paper is now remembered primarily for the connection between the photoelectric and recombination cross sections first given in it.

The next period of development involved the application of quantum mechanics to the study of absorption coefficients. Among the principal papers in which complete quantum mechanical discussion of continuous radiative processes were made are those of Oppenheimer (1928, 1929), Gaunt (1930), and Stobbe (1930). The classic paper of Born and Oppenheimer (1927) is of fundamental importance in the discussion of molecular contributions.

Following the formal quantal descriptions of elementary processes which contribute to the spectral absorption coefficients, Strömgren (1932, 1933) carried out the first systematic and detailed calculations of the mean absorption coefficient defined by Rosseland as the most appropriate for discussion of radiative transfer in stellar interiors.

The next important development was the paper of Menzel and Pekeris (1935) which provided a very extensive analysis of hydrogenic matrix element formulae with tables of numerical values to correct the Kramers semiclassical approximation. In addition, they discussed, analyzed, and made improvements to previous approximate formulae for the spectral and mean absorption coefficients. Papers by Marshak and Bethe (1940) and Morse (1940) for the first time considered interparticle interactions in the calculation of occupation numbers for the study of opacity and thermodynamic properties of gases at high temperatures and densities. These authors adopted the suggestion of Slater and Krutter (1935) to apply the Fermi–Thomas method to the calculation of occupation numbers and/or other thermodynamic properties under conditions of temperature and pressure of astrophysical interest. This was the origin of the use of the ion-sphere model in opacity calculations. Its original use, by Wigner, Seitz, Slater and others, was in the theory of metals. But, as pointed out by Slater and Krutter (1935), it is also advantageous for thermodynamic calculations under conditions of high temperature and pressure. It permits more detailed consideration to be made of interparticle interactions than the previous "excluded volume" type corrections† which are more or less intuitive in nature. The Thomas–Fermi calculations of Slater and Krutter (1935), using the ion–sphere model, provided the necessary ground work for the Marshak–Bethe (1940) and Morse (1940) calculations. This original ion–sphere theory was derived heuristically without any statistical justification. Mayer (1947) was the first to work out a detailed statistical mechanics foundation for this theory.

† Discussions of studies of truncation with early cutoffs and corrections for the partition function may be found in Fowler (1936). See also Fermi (1924) and Urey (1924).

With the development of nuclear weapons during World War II, it was recognized that conditions in and around the fireball of a nuclear explosion would be similar to those in a star. Thus, in order to predict these conditions, it was necessary to consider the opacity both of the materials used in the construction of a nuclear weapon, and of the air surrounding such a weapon, after heating caused by the nuclear explosion has taken place. The first calculations of the opacity of air were therefore carried out under the Manhattan project by Hirschfelder and Magee (1945). Later, a much more extensive general development of opacity calculation theory was provided by Mayer (1947), although he did not carry out any explicit calculations for air in that (1947) report. The Hirschfelder–Magee calculations covered a very extensive temperature–density range, but with a limited number of contributing components. This work was later extended to include a few more components and to include a consideration of Planck mean absorption coefficients as well as Rosseland means (Magee and Hirschfelder, 1958/1947). Their calculations were based on Morse's method with some simplifications (i.e. neglect of excited states and pressure ionization).

As stated above, Mayer (1947) gave the first statistical discussion, based on the canonical ensemble, of the ion–sphere theory for obtaining occupation numbers. He also included explicit consideration of free–bound electron–electron interactions in his treatment, thus giving more accurate and detailed formulae for the occupation number calculations than had previously been available. In addition, he considered the effects of line transitions on the opacity for the first time.

Because of the complexity of this problem, previous investigators had limited themselves to the so-called "continuous" opacity, that is, absorption due to the photoelectric and free–free processes, and scattering. At the suggestion of E. Teller, Mayer performed the first realistic estimates of the line effect, both by means of a statistical theory which he developed (see also Goody, 1952, who developed it independently) and by direct calculation (Kivel and Mayer, 1965/1954). Specific lines or groups of idealized lines had been used for some time in low-temperature radiation transport studies (Elsasser, 1938) but these were traditionally confined to relatively narrow frequency regions in contrast to the extremely broad regions involved in a high-temperature opacity calculation. Further history of opacity calculations has been given by Mayer (1964). After 1945 opacity calculations were continued under government sponsorship, but much of this work, at least through the early 1960s, remains in unpublished reports, as was mentioned above when we referred to it as grey literature. That which did get published in the open literature shows a continued development towards a more detailed inclusion of lines, a more accurate treatment of interparticle interactions, and a more accurate treatment of matrix elements away from the scaled hydrogenic values (in the case of atoms) almost universally used up to 1960 (Keller and Meyerott, 1955; Meyerott, 1956; Armstrong *et al.*, 1961; Breene and Nardone, 1962; Cox, 1964; Stewart, 1964; Stewart and Rotenberg, 1965; Johnston *et al.*, 1965; Armstrong, Johnston and Kelly, 1965; Churchill and Meyerott, 1965; Stewart and Pyatt, 1966; Churchill *et al.*, 1966; Armstrong *et al.*, 1967; Landshoff and Magee, 1969). In the case of the molecular absorption calculations included in the above references, it is still necessary to use empirical data to evaluate the matrix elements. The improvements in these calculations have been in the direction of more comprehensive inclusion of absorbing systems, and in the inclusion of finer spectral detail. Work analogous to the government-sponsored, defense-motivated studies cited above have also been carried on in the Soviet Union, e.g. see Biberman and Norman (1963), Vorobyov and Norman (1964), and Zel'dovich and Raizer (1966). Astrophysical studies of opacity and absorption coefficients were slowed by World War II, and did not really begin to flourish until well into the 1960s. This work has been extensively reviewed by Cox (1965) who also describes the large-scale opacity calculations carried out at Los Alamos. Cox's results still make use of hydrogenic matrix elements, but cover an extremely broad range of density, temperature, composition and contributing effects, including lines. Improvements on the hydrogenic approximation for large-scale astrophysical opacity calculations were begun by Peach (1962), who applied the "quantum defect" methods of Burgess and Seaton (1960) to a calculation of the continuous absorption coefficients of nitrogen and oxygen at 10,000–13,000 °K. By generalizing the Coulomb approximation of Bates and Damgaard (1949), Burgess and Seaton obtained a general and very widely applicable method of calculating photoionization cross sections from empirical energy-level data. The accuracy and simplicity of this method lends itself well to large-scale calculations of the types required for multicomponent gases at high temperatures. Thus, to a large extent it frees the investigator from the limitations of the scaled hydrogenic matrix elements without imposing the burden of massive Hartree–Fock type calculations (e.g. Hartree, 1957).

Later, Peach (1965) developed a generalization of the quantum-defect method to free–free transitions and applied the method to a large number of atomic systems (Peach, 1967a, b, c; 1970). This type of calculation, made possible by the development of modern high-speed digital computers, has culminated in the availability of very accurate continuous absorption coefficients and opacity values for the study of the structure of stellar atmospheres, as typified by the papers of Travis and Matsushima (1968) and Peach (1970). Accurate, nonhydrogenic treatment of lines in astrophysics was initiated by Mihalas and Morton (1965), and this has been followed by calculations of increasing complexity up to the inclusion of autoionization lines by Watson (1969). A comprehensive astrophysical calculation of opacities was carried out subsequent to Cox's by Carson, Mayers and Stibbs (1968). These authors give an historical review as does Cox and an extensive description of the many theoretical components involved in the calculation. They use the Thomas–Fermi atomic model for both the statistical mechanics and for the wave functions and matrix elements required. They were apparently unaware of opacity applications of this model that preceded them (Stewart and Rotenberg, 1965; Stewart and Pyatt, 1966), as well as certain other features such as the configuration splitting of absorption edges due to the electron coupling in many electron atoms (Armstrong *et al.*, 1961; Armstrong *et al.*, 1967). More recent astrophysical calculations have been made by Watson (1970), by Cox and Stewart (1970), and by Norris and Baschek (1970).

As stated at the beginning of this introduction, our interests in this book overlap, at the low end of temperatures of interest to astrophysics, the study of radiative processes in the ambient terrestrial atmosphere. There is, of course, a large body of literature devoted to this study. In addition to the book by Goody (1964) more recent books by Kondratyev (1965, 1969) have appeared which primarily address the experimental aspects of the field in contrast to Goody's book which is primarily theoretical. There is also, of course, a strong overlap between our interests and the field of plasma spectroscopy. The reader should consult the book by Griem (1964) and the excellent review article by Cooper (1966) for the details of this discipline. Applications of radiative transfer and opacity notions to engineering problems have been discussed by Bond, Watson, and Welch (1965) and in a more recent book by Penner and Olfe (1968). These books also contain extensive discussions of otherwise unpublished opacity calculations such as those of Mayer, Bernstein and Dyson, Stewart and Pyatt, and of Freeman.

Chapter 2

Elementary Radiative Transfer

IN THIS chapter, a non-rigorous review of the theory of radiative transfer indicates how the subject is dominated by the concept of spectral absorption coefficient. An attempt is made to provide some physical insight into the requirements which the applications of transfer theory impose on the calculation of absorption coefficients, and to demonstrate the physical significance of the Planck and Rosseland mean absorption coefficients. For this reason, certain elementary solutions to the transfer equation have been emphasized. An excellent and comprehensive discussion of radiative transfer, which also emphasizes the physical aspects, has recently been given by Goody (1964) (see also Morse and Feshbach, 1953, sect. 2.2). The formal theory particularly as it applies to scattering is more extensively developed by Chandrasekhar (1939, 1950) and Kourganoff (1952). A good reference for the present purpose, although unavailable in the open literature, is the report by H. Mayer (1947). Aller (1963) given an excellent account of approximate solutions.

We first undertake in Section 2.1 a brief discussion of the formulation of the transfer equation and the definitions involved. In this initial discussion *we neglect scattering*; however, a few remarks on this subject are given later, in Section 2.3. In Section 2.2 we consider the definition of local thermodynamic equilibrium in some detail, in order to clarify the roles of induced and spontaneous emission in radiative transfer.

In many problems the spectral distribution of the radiation is not of primary concern. For these cases an appropriately defined mean value of the absorption coefficient may be useful. The manner in which the mean value is calculated depends, of course, on the characteristics of the problem under consideration. A limiting case of physical interest is that of an optically thin sample of gas, i.e. a sample whose dimensions are small compared with the mean free path of radiation in the gas. Consideration of this case leads to the definition of the Planck mean absorption coefficient. Conversely, the equally important optically thick case is conveniently described in terms of the Rosseland mean absorption coefficient. These two limiting cases are discussed in detail in Sections 2.4 and 2.5 and the respective mean absorption coefficients are derived.

Section 2.6 undertakes a discussion of the general features of these mean absorption coefficients and the bounds that can be derived for them by means of the Schwartz inequality. This is followed in Section 2.7 by a discussion of the problem of radiation transfer for intermediate optical depths. In view of the complications involved, we limit these considerations to isothermal, uniform-density conditions. Emissivities for various geometries are discussed in some detail, including those for plane slabs, spheres, and cylinders. A relation is derived, in the limit of small optical depth, which connects the Planck mean absorption coefficient and the emissivity for an arbitrary shape through the mean chord length for that shape. We include a brief mention of band models in this section to lay the groundwork for the definition and applications of the concept of diffusivity factor presented in the last section. This factor can be employed to more simply account for the angular integration of the radiation intensity in a plane-parallel atmosphere than by direct integration.

2.1. The Equation of Radiative Transfer (Without Scattering)

Consider a collimated beam of radiation of frequency ν incident on an extended volume of gas which absorbs the radiation, but does not emit. If F_ν is the flux of radiant energy per unit frequency interval, that is, energy per (cm^2 $\times$ sec $\times$ unit frequency interval) this flux is assumed to be attenuated by each thin slab of gas of thickness δx in proportion to the product of the magnitude F_ν and the thickness δx. This is illustrated in Fig. 2.1. Thus, we can write

$$\delta F_\nu = -\mu_\nu F_\nu \delta x. \tag{2.1.1a}$$

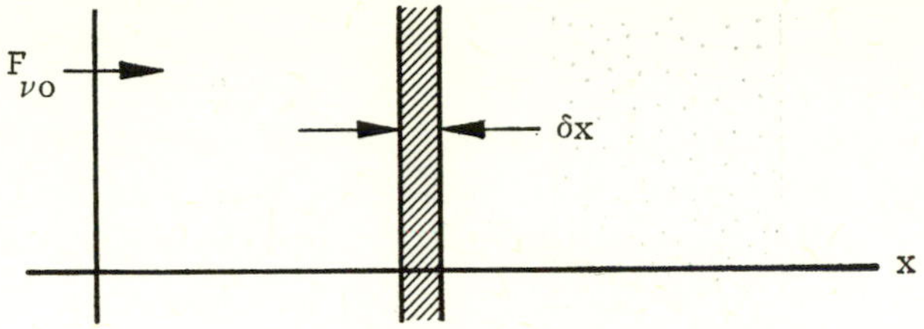

FIG. 2.1. Illustration of a collimated beam of radiation $F_{\nu 0}$ incident on a plane-parallel slab of gas (shaded section).

This assumption defines the constant of proportionality μ_ν, the linear or volume *absorption coefficient*, as

$$\mu_\nu = \lim_{\delta x \to 0} \left\{ - \frac{\delta F_\nu}{F_\nu \delta x} \right\}. \tag{2.1.1b}$$

This phenomenological rule, known as Lambert's, Bouguer's, or Beer's law, leads, when integrated, to the exponential decay law:

$$F_\nu = F_{\nu 0}\, e^{-\mu_\nu x} \tag{2.1.2a}$$

which is borne out in many experimental circumstances. An excellent discussion, including some of the early history, of this law is given by Nielsen, Thornton, and Dale (1944). From the usual interpretation of transport parameters, $l_\nu \equiv \mu_\nu^{-1}$ is thought of as a mean free path for absorption. Thus, eqn. (2.1.2a) can be rewritten

$$F_\nu = F_{\nu 0}\, e^{-x/l_\nu} \tag{2.1.2b}$$

Experimentally, re-emission follows absorption, so that δF_ν as given in eqn. (2.1.1) must be augmented by the radiation emitted in the slab δx and the result is the equation of transfer, eqn. (2.1.4), below. Furthermore, the emitted flux will in general not be so well collimated as our ideal parallel incident beam, so the definition of F_ν must be generalized to energy per unit solid angle across unit surface normal to a specified direction θ, Φ with a unit vector $\hat{\boldsymbol{\Omega}}$.† This is illustrated in Fig. 2.2 which shows an infinitesimal pencil of beams within an element of solid angle

† We will generally use the caret to indicate unit vectors.

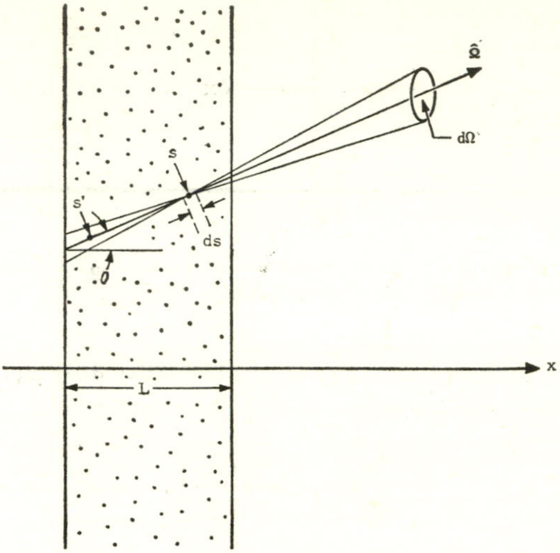

FIG. 2.2. Illustration of an infinitesimal cone of radiation in a slab of gas of thickness L.

$d\Omega$ about the direction $\widehat{\boldsymbol{\Omega}}$ traversing a slab of gas of thickness L which extends to infinity in directions perpendicular to x. The relationship between the flux $F_{\nu 0}$ of the collimated (or parallel) beam of radiation and the general intensity function $I_\nu(\theta, \Phi)$ which allows for an angular distribution is established through the Dirac δ-function (Chandrasekhar, 1950, p. 22):

$$I_\nu(\theta, \Phi) = F_{\nu 0}\, \delta(\cos\theta_0 - \cos\theta)\, \delta(\Phi - \Phi_0) \tag{2.1.3a}$$

where θ_0, Φ_0 is the direction of the parallel beam, and the δ-functions satisfy ($d\Omega \equiv \sin\theta\, d\theta\, d\Phi$)

$$\int_{4\pi} \delta(\cos\theta_0 - \cos\theta)\, \delta(\Phi - \Phi_0)\, d\Omega = 1. \tag{2.1.3b}$$

$I_\nu(\theta, \Phi) \equiv I_\nu(\widehat{\boldsymbol{\Omega}})$ is called the *specific intensity* and has dimensions of energy per unit time per unit frequency interval per unit solid angle per unit area normal to $\widehat{\boldsymbol{\Omega}}$. The resulting equation of radiative transfer which takes account of absorption and emission in the slab can then be written as

$$\frac{dI_\nu(\widehat{\Omega})}{ds} = j_\nu(\widehat{\Omega})\,\varrho - \mu_\nu I_\nu(\widehat{\Omega}) \tag{2.1.4}$$

where s denotes length measured in the direction $\widehat{\Omega}$. The energy per unit time radiated in the direction $\widehat{\Omega}$ by a unit mass of gas per unit frequency per unit solid angle has been denoted by $j_\nu(\widehat{\Omega})$, the *emission coefficient*, and ϱ is the mass density. For convenience, the emission coefficient is often replaced by the *source function* defined as

$$J_\nu(\widehat{\Omega}) = \varrho j_\nu(\widehat{\Omega})/\mu_\nu. \tag{2.1.5a}$$

With this definition the equation of radiative transfer becomes

$$\frac{1}{\mu_\nu}\frac{dI_\nu(\widehat{\Omega})}{ds} = J_\nu(\widehat{\Omega}) - I_\nu(\widehat{\Omega}). \tag{2.1.5b}$$

The decrease in beam intensity $(-\delta I_\nu)$ can be expressed in a number of equivalent ways, each of which defines a different absorption coefficient. All of the resulting equations are derived from eqn. (2.1.1) which invokes simple proportionality to beam intensity I_ν (by virtue of eqn. (2.1.3a)) and the increment of path length δx. Equation (2.1.1) thereby defines the linear (or volume for unit cross section) absorption coefficient μ_ν whose dimensions are $(\text{length})^{-1}$. The effect of mass density ϱ or number density N_v of absorbing particles is seen in eqns. (2.1.6a) and (2.1.6b) below which define, respectively, the *mass absorption coefficient* $\varkappa_\nu$ and *atomic absorption coefficient* $\alpha_\nu \cdot \varkappa_\nu$ has dimensions of $(\text{length})^2\,(\text{mass})^{-1}$. α_ν has dimensions $(\text{length})^2$. It is thus often treated as an absorption cross section which interpretation is pursued further in the next chapter.

$$\delta I_\nu = -\varkappa_\nu I_\nu\,(\varrho\delta x) \tag{2.1.6a}$$

$$\delta I_\nu = -\alpha_\nu I_\nu\,(N_v\delta x) \tag{2.1.6b}$$

We note that $\varrho\delta x$ is the mass increment δm per unit area, and similarly $N_v\delta x$ is the number increment δN_v per unit area. Comparison between eqns. (2.1.1), (2.1.6a), and (2.1.6b) leads to

$$\mu_\nu = \varkappa_\nu\varrho = \alpha_\nu N_v = \frac{1}{l_\nu} \tag{2.1.6c}$$

which relates most of the absorption coefficient parameters in common use.

One absorption coefficient parameter not previously discussed is a dimensionless quantity, the optical depth τ_ν. Its infinitesimal increment is defined by

$$d\tau_\nu = \mu_\nu \, ds \tag{2.1.7}$$

and, therefore, the total optical depth between points s' and s is

$$\tau_\nu \, (s', s) \equiv \int_{s'}^{s} \mu_\nu \, ds. \tag{2.1.8a}$$

(One should note that it is customary in astrophysics to measure optical depth backwards along the line ss'. This would require a minus sign in the above definitions.) The foregoing definitions of radiative transfer quantities along with others which we do not consider here, have been conveniently summarized by Aller (1963).

The equation of transfer (2.1.5b) may now be written

$$\frac{dI_\nu}{d\tau_\nu} = J_\nu(\widehat{\Omega}) - I_\nu(\widehat{\Omega}). \tag{2.1.8b}$$

Equation (2.1.8b), which governs the transport of electromagnetic energy through an atmosphere, is often difficult to solve. An extensive literature pertaining to it has accrued over the years, principally in the fields of astrophysics and meteorology. Some classical references are Schuster 1905), Schwarzschild (1906), Eddington (1926), Milne (1930), Chandrasekhar (1939; 1950), Elsasser (1942), Mayer (1947), Kourganoff (1952), Unsöld (1955), Aller (1963) and Goody (1964). For engineering applications two good references are Bond, Watson and Welch (1965) and Penner and Olfe (1968). Exact solutions can be obtained for only a very limited number of model problems so that approximate methods of solution are of paramount importance. As stated by Goody (1964), the physical content of the equation of transfer is very meager. Under the conditions of local thermodynamic equilibrium, most of the physics of the situation is in fact contained in the absorption coefficient μ_ν, the discussion of which constitutes the primary subject of this book.

A general solution to eqn. (2.1.8b) can be obtained formally as follows (Goody, 1964). Consider a beam along the direction $\hat{\Omega}$ as indicated in Fig. 2.2. Multiply eqn. (2.1.8b) by $e^{\tau_\nu(s',s'')}$ to obtain

$$\frac{d}{d\tau_\nu}\left[e^{\tau_\nu(s',s'')} I_\nu(s'',\hat{\Omega})\right] = e^{\tau_\nu(s',s'')} J_\nu(s'',\hat{\Omega}). \qquad (2.1.9)$$

Integrating this equation along the direction $\hat{\Omega}$ from $s'' = s'$ to $s'' = s$, we obtain

$$e^{\tau_\nu(s',s)} I_\nu(s,\hat{\Omega}) = I_\nu(s',\hat{\Omega}) + \int_{s'}^{s} e^{\tau_\nu(s',s'')} J_\nu(s'',\hat{\Omega})\, d\tau_\nu.$$

Now, using eqn. (2.1.8a), the above equation is readily converted to

$$I_\nu(s,\hat{\Omega}) = e^{-\tau_\nu(s',s)} I_\nu(s',\hat{\Omega}) + \int_{s'}^{s} e^{-\tau_\nu(s'',s)} J_\nu(s'',\hat{\Omega})\, d\tau_\nu. \qquad (2.1.10)$$

Thus the intensity $I_\nu(s,\hat{\Omega})$ at s is equal to the intensity $I_\nu(s',\hat{\Omega})$ incident at some previous point (s') exponentially diminished by absorption, plus the integrated intensity emitted between s'' and s', each element again diminished by the absorption which takes place ahead of s''. In the limiting case of an optically deep medium, we can take s' at $-\infty$ and set the incident intensity term equal to zero. Equation (2.1.10) then becomes

$$I_\nu(s,\hat{\Omega}) = \int_{-\infty}^{s} e^{-\tau_\nu(s'',s)} J_\nu(s'',\hat{\Omega})\, d\tau_\nu \qquad (2.1.11)$$

which shows how the radiation emerging from an emitting gas is limited by absorption in the gas.

The foregoing very brief discussion of the equation of radiative transfer is a standard one, dealing with concepts and quantities usually met with and defined in the literature and which date back to Schuster (1905) and Schwarzschild (1906). It is interesting but difficult to attempt to place this equation in a broader perspective of physical laws and formulations, since its relationship to other formulations as well as its fundamental justification (since it is a phenomenological approach) is not so clear cut. Its relationship to the Boltzmann transport equation is, however, straightforward: the transfer equation is the time-independent photon transport equation (Sampson, 1965b). From this we can make some further inferences regarding its relation to the geometrical optics approximation. We also append a few references to discussions of the relationship of the transfer equation as an intensity formulation to Maxwell's equations as the amplitude formulation. We will not consider the time-dependent case at all.

Relationship to Geometrical Optics

Elementary geometrical optics, viz. the eikonal equation (Born and Wolf, 1959), deals with the phase of a wave only and neglects absorption; the equation of radiative transfer neglects phases completely, dealing only in intensities, their absorption, emission and scattering. In wedding absorption to geometrical optics, one must be careful because, in the presence of absorption, surfaces of equal phase (the basis of ordinary geometrical optics) do not necessarily coincide with the surfaces of equal amplitude, or intensity (Epstein, 1930; Stone, 1963, sect. 15.4). Bekefi (1966) has an excellent discussion bearing on the relationship between geometrical optics and the transfer equation. He includes a careful discrimination between the direction of the wave propagation vector and the direction of the energy flow. Inadequate attention to this point has caused some confusion. For example, Zheleznyakov (1967) has derived an equation for energy transfer "in the geometrical optics approximation" by replacing the force term in the Boltzmann transport equation (which causes the distribution function to depend on changes in velocity) by a term from geometrical optics that gives the change in the distribution function due to the change in *direction* of the (group) velocity. But the equation from geometrical optics that he used was actually that for the wave propagation vector (see Zheleznyakov, 1967, p. 851, eqn. (13)), which differs from the group velocity vector when the medium is not isotropic. This was later clarified by Enomé (1969) who reaffirmed the conclusions of Bekefi (1966). Bekefi derives the transfer equation "in the geometrical optics approximation" to be

$$n_r^2 \frac{d}{ds}\left(\frac{I_\nu}{n_r^2}\right) = j_\nu - \mu_\nu I_\nu \qquad (2.1.12)$$

where n_r, the "ray refractive index" is equal to n, the ordinary refractive index, in an isotropic medium (i.e. one in which the propagation vector and group velocity lie in the same direction) and is given by a complicated formula which we will not quote, in the anisotropic case. This latter case is of great importance, for example, for a plasma in an external magnetic field. There is an important restriction on the above equation, namely, that the *emission and absorption on the right hand side above, be weak*. For the definition of weakness, the reader is referred to Bekefi (1966). Bekefi also gives a detailed derivation of the effect of an index of refraction n on the intensity of black body radiation. This effect is to multiply the customary (vacuum) Planck function by n_r^2, so that the black body intensity becomes

$$B_\nu(T) = \frac{2h\nu^3 n_r^2}{c^2}(e^{h\nu/kT} - 1)^{-1}. \tag{2.1.13}$$

where h and k are Planck's and Boltzmann's constants and T is the absolute temperature. The reason for the appearance of n_r in this formula is that the change in the velocity of light c produces a change in the number density of states in phase space (see also Condon, 1968). Since this number density also multiplies the matrix elements for the emission and absorption coefficients that appear in eqn. (2.1.12), they also depend on n_r. The formulas are given by Bekefi. However, considerations such as those given by Bekefi (1966) are still of a somewhat *ad hoc* nature because they combine a wave or interference property (the index of refraction) with an intensity formulation which neglects wave properties.

The Boltzmann transport equation is a classical, not a quantum equation. The wave nature of the particles for which it is derived is completely neglected. Its use for photons implies that light is being treated as a gas of independent photons. Since there is no wave aspect in the formulation, all interference properties of the photons are neglected and they are treated like classical material particles except that, in between acts of emission, absorption or scattering they always travel with the vacuum velocity of light. On the other hand, geometrical optics, although it does not include diffraction *per se*, does take into account some microscopic aspects of interference phenomena—namely, those which enter via the index of refraction. Lax, for example, has explicitly pointed out (Lax, 1951, p. 297) that the inclusion of refraction in geometrical optics distinguishes it as a higher approximation than the Boltzman photon transport equation.

This inclusion of the refraction, or bending, of light rays is, in fact, the primary purpose, or *raison d'être* of geometrical optics. Refraction arises from the superposition of the amplitudes of all the scattered secondary wavelets arising from the atoms or molecules comprising the medium. These scattered waves interfere destructively in all but the forward (or near forward) direction for randomly distributed scatterers. The scattering which occurs at large angles such as, for example, in the earth's atmosphere, is due to the incoherence which is produced by recoil of the molecules (or by random thermal motion). Because of this incoherence, the intensities rather than amplitudes must be added for such large angle scattering and destructive interference no longer cancels the large-angle contribution. For an excellent discussion of this problem, see Goldberger and Watson (1964, p. 766). The coherence required by the presence of n_r in eqn. (2.1.12) is not clearly evident in the derivations, nor is the reason clear why the restriction of weak absorption is required here and not in the straight photon Boltzmann equation. This requirement must stem, however, from the fact that the presence of n_r in the equation requires that the field amplitudes and velocity of energy propagation must, in principle, be identifiable. That is to say, there must be a well-defined coherent wave path. In the presence of large absorption and emission, this is difficult to do because of the rapid fluctuations in the field amplitudes. In the straight photon Boltzmann equation, the observability of the field amplitudes and energy flow velocities is given up completely, and only the gross rate of energy transfer is considered. (When absorption is strong, even in the absence of interference, energy is not transported at the photon velocity c, because of continual absorption and re-emission.)

In principle, these questions can be answered by a direct reduction of the basic amplitude (Maxwell) equations of the field in the presence of absorption, emission, and scattering to an intensity transfer equation. However, this is a difficult and complicated venture (Foldy, 1945; Lax, 1951; Goldberger and Watson, 1964; Dolginov, Gnedin and Silant'ev, 1970) for which adequate elementary physical interpretation is not yet available. Perhaps it is worth noting that the index of refraction is the natural material property parameter for an amplitude formulation. In the elementary quantum theory of radiation, scattering and absorption are proportional to the intensity of the radiation rather than the amplitude. Hence, they cannot be inserted directly into the wave equation without making it nonlinear. The occurrence of the index of refraction in the wave equation allows one to parametrize these effects in a way that is quite satisfactory and particularly suited to coherent situations involving more-or-less fixed scattering positions and small attenuations. The cross sections are the natural property parameters for the intensity formulation, so that the fundamental terms expressing the scattering, absorption or emission of radiation appear directly in the radiative transfer equation and the result is a simple linear† differential equation for the intensity. Thus, although the radiative transfer equation is a lower approximation than an amplitude equation, it is much simpler and is ideally suited to the elementary quantum theory of radiation problems which satisfy the requirements for the existence of a constant transition probability per unit time (Schiff, 1968, p. 401; Heitler, 1954). Its eminent success as a phenomenological equation can undoubtedly be attributed to the fact that the physical conditions that lend validity to the intensity formulation are much the same as those required for the existence of a constant transition probability per unit time. It is also highly compatible with the uncertainty relations between field quantities. Absorption and emission require that we specify to some extent the number of quanta in the field (the intensity function I_ν specifies it precisely). When one does this, the field strengths and phases become uncertain and must be permitted to fluctuate (Heitler, 1954). Conversely, if we fixed on the field strengths (amplitudes) as our observables, the number of quanta in the field becomes uncertain, so that an attenuation measurement would, in principle, become impaired. In many cases of macroscopic observations these uncertainty limitations will be insignificant due to the satisfaction of classical conditions (e.g. a large number of photons in the field). However, the extraordinary compatibility of the radiative transfer equation with the simple fundamental features of the quantum theory of radiation means that at the very least it will have a much broader degree of validity than most phenomenological equations. Perhaps the best way of emphasizing the fundamental simplicity and

† If we make the simplifying assumption that the optical properties of the medium do not depend on the intensity.

significance of the radiative transfer equation is to reiterate that it is an "independent particle" equation. Both the photon and the material particles are described individually and the absorption and emission by each atom or molecule is independent of the others. The elementary theories of gases and of radiation are also independent particle theories. This provides the basis for the aforementioned compatibility. An amplitude or wave formulation of electromagnetic theory becomes basically a many-body or collective theory in the presence of an assembly of particles. (One cannot distinguish which part of the electric field strength E belongs to a given photon!) Therefore, it is not surprising to find it more complicated in many respects. In transparent solid media the transmission of light is basically a collective effect and the amplitude formulation is naturally suited to this problem. But refractive interference is possibly the only *simple* collective effect and absorption and emission are difficult to accommodate (in a fundamental way) in a collective theory.

2.2 Local Thermodynamic Equilibrium, Einstein Coefficients, and Induced Emission

One of the most common situations in which a solution to the equation of radiative transfer is sought is that in which local thermodynamic equilibrium (LTE) is assumed. This assumption is often introduced through the assertion that Kirchoff's law

$$j_\nu = \varkappa_\nu B_\nu(T) \tag{2.2.1}$$

holds point-wise in a medium. This implies that at each point in the system a local temperature T is defined but that it may vary from point to point over the system (Kourganoff, 1952; Chandrasekhar, 1939, 1950; Aller, 1963). In this formula j_ν is the radiant energy emitted per unit mass per unit time per unit solid angle per unit frequency interval and $\varkappa_\nu$ is the similarly defined mass absorption coefficient. We recall from Section 2.1 that $\varkappa_\nu \varrho$ is defined such that

$$dI_\nu = -\varkappa_\nu \varrho I_\nu \, ds \tag{2.2.2}$$

is the change in the emerging radiation intensity after traversal of a distance ds in the medium to which $\varkappa_\nu$ pertains, if there is no emission. $B_\nu(T)$

is the usual Planck function $2h\nu^3 e^{-h\nu/kT}/c^2\,(1 - e^{-h\nu/kT})$, ν the frequency, and T is the absolute temperature.

Complications arise in these equations through the neglect of induced emission effects which cause j_ν to depend somewhat on incident intensity. Chandrasekhar (1939, chap. 5, sect. 3). points out that incorrect results are obtained if j_ν of eqn. (2.2.1) is substituted directly into the transfer eqn. (2.1.4). *Ad hoc* corrections are then necessary to compensate for the neglect of induced emission effects. These problems may be avoided and natural allowance be made for the effect of induced emission if a local temperature (and thus LTE) is defined through a Boltzmann equation relating the occupation numbers N_n and N_m of two quantum states involved† in absorption or emission (Chandrasekhar, 1939, eqn. (108ff.) and discussion following eqn. (118), Mayer, 1947)

$$\frac{N_m}{N_n} = \exp\,(h\nu_{nm}/kT) \tag{2.2.3}$$

where $h\nu_{nm} = E_n - E_m$ is the energy separation between the states. For simplicity statistical weights have been assumed to be unity in this discussion. LTE so defined makes no assumption about the radiation field and therefore it need not be in equilibrium with the matter. In practice it is often not in equilibrium except at the center of strong spectral features. Equation (2.2.1) implicitly includes the radiation field but eqn. (2.2.3) does not.

To follow the consequences of this definition it is necessary to introduce the Einstein coefficients A_{nm} and B_{nm}‡ for spontaneous and induced emission, respectively. We define these coefficients as follows. $A_{nm}b(\nu)/4\pi$ is the probability per unit time, per unit frequency interval, per unit solid angle that an atom or molecule in excited state n emits a photon with

† In any real physical situation there will usually be many absorbing and emitting states, and often many different atomic and molecular species. Since the transition and species can be treated one at a time and then added, we limit ourselves to single events and single species in this chapter. This is conceptually much simpler, and involves no loss of generality if the reader bears in mind that the appropriate sums must ultimately be carried out.

‡ The Einstein B-coefficient B_{nm} and the Planck function $B\,(T)$ should not be confused.

energy centered about $h\nu_{nm} = E_n - E_m$. The line-shape factor $b(\nu)$ has been included so that our results can be expressed *per unit frequency interval.* The line shape is the probability distribution in frequency for the photon, and represents the probability per unit frequency interval that the photon produced in the transition $n \to m$ has the frequency ν. The usual definitions of the Einstein coefficients (see, for example, Chandrasekhar, 1939; also sect. 3.1 of the sequel) which assume infinitesimally sharp lines do not incorporate this factor and consequently require cumbersome double definitions of emission and absorption coefficients and of intensities. The normalization condition

$$\int_0^\infty b(\nu)\, d\nu = 1$$

is satisfied by $b(\nu)$, and it will be further discussed later in this chapter. By using this factor in our definitions at this point, all quantities can be expressed as functions of frequency, rather than some appearing integrated over the line profile.

From the foregoing definitions we see that the energy dE_S emitted spontaneously in a time dt into the solid angle $d\Omega$ and the frequency interval $d\nu$ by the excited atom or molecule is

$$dE_S(\nu) = \frac{h\nu A_{nm} b(\nu)}{4\pi}\, dt\, d\Omega\, d\nu. \tag{2.2.4a}$$

Since the spontaneous emission is isotropic, the total number of photons spontaneously emitted per second per unit frequency interval per atom into all angles is $A_{nm}b(\nu)$. Now it turns out that the probability of the emission of a photon $h\nu_{nm}$ is augmented by stimulated emission if the atom is immersed in a field of radiation containing a distribution of frequencies about ν_{nm}. B_{nm} is defined so that $c^{-1}B_{nm}I_\nu(\widehat{\Omega})\, d\Omega\, d\nu\, dt$ is the probability that an excited atom or molecule in state n is stimulated by an external intensity of radiation $I_\nu(\widehat{\Omega})$ to emit a photon with energy centered about $h\nu_{nm}$ in the direction $\widehat{\Omega}$, the frequency interval $d\nu$ and dt. Thus, the stimulated or induced energy emitted by the single system into the various differential intervals is dE_I which is given by

$$dE_I(\nu) = (h\nu)\, c^{-1} B_{nm}\, b(\nu)\, I_\nu(\widehat{\Omega})\, d\Omega\, d\nu\, dt. \tag{2.2.4b}$$

The coefficients A_{nm} and B_{nm} are properties of the individual atomic or molecular systems, and their evaluation is a problem in the quantum theory of radiation which will be discussed in Chapter 3. We assume, while discussing radiation transport, that they are known quantities. Equations (2.2.4a) and (2.2.4b) are, respectively, the spontaneous and stimulated radiant energy contributions emitted per atom or molecule. The total radiant energy emitted by an element of volume dV (with mass $dm = \varrho\, dV$) will then be obtained by adding these two terms and multiplying by the number of particles within the element dV that are in the quantum state n. Thus, if N_n is the number density of particles in the state n,

$$b(\nu)\, N_n\, dV\, [A_{nm}/4\pi + c^{-1} B_{nm} I_\nu(\widehat{\Omega})]\, h\nu\, d\Omega\, d\nu\, dt \qquad (2.2.5)$$

is the energy emitted by the volume element dV in time dt into the element of solid angle $d\Omega$ and frequency interval $d\nu$ about ν_{nm}. The energy emitted per unit mass of material will be this quantity divided by the mass $\varrho\, dV$ of the element. The energy per unit mass per unit solid angle, frequency interval and time is called the emission coefficient j_ν. Thus, the emission coefficient can be written as

$$j_\nu = \frac{N_n}{\varrho}\left[\frac{A_{nm}}{4\pi} + c^{-1} B_{nm} I_\nu(\widehat{\Omega})\right] h\nu\, b(\nu). \qquad (2.2.6)$$

It is shown in Chapter 3 that if A_{nm}, B_{nm} and B_{mn} are the Einstein coefficients for spontaneous emission, stimulated emission and (stimulated) absorption for transitions between levels n(upper) and m(lower), each assumed for simplicity to be of unit statistical weight, then

$$\frac{A_{nm}}{B_{nm}} = \frac{8\pi h\nu^3}{c^3}, \qquad (2.2.7a)$$

$$B_{nm} = B_{mn}. \qquad (2.2.7b)$$

The B-coefficients are defined here relative to radiation energy density rather than to radiation intensity (Aller, 1963; Allen, 1963). We may further relate the absorption coefficients μ_ν and $\varkappa_\nu$ to B_{mn} by appeal to eqns. (2.1.1), (2.1.3a) and (2.1.6a). Thus, from these equations and the discussion above, we find

$$-\delta I_\nu = \mu_\nu I_\nu\, \delta x = \varrho \varkappa_\nu I_\nu\, \delta x = N_m I_\nu\, \delta x B_{mn} h\nu\, b(\nu)\, c^{-1}$$

or

$$\mu_\nu = \varrho\varkappa_\nu = N_m B_{mn} h\nu\, b(\nu)\, c^{-1}. \qquad (2.2.7c)$$

Note at this point that the assumption has crept in that the particles in the medium are emitting and absorbing independently of each other except for the effects of interparticle interactions present in the line-shape function $b(\nu)$. The assumption occurs when we write the emission and absorption coefficients as products of number densities and transition probabilities.

From eqns. (2.2.6) and (2.2.7a, b, c) we obtain

$$A_{nm} = \frac{8\pi h\nu^3}{c^3} \frac{\mu_\nu}{N_m(h\nu)\, b(\nu)}$$

and

$$j_\nu = \frac{n_n}{\varrho}\left[\frac{2\nu^2}{c^2}\frac{\mu_\nu}{N_m} + \frac{\mu_\nu}{N_m h\nu} I_\nu(\widehat{\Omega})\right] h\nu. \qquad (2.2.8)$$

This may be rewritten in a more compact form using the definitions of LTE (eqn. (2.2.3)) and the Planck function:

$$B_\nu(T) \equiv \frac{2h\nu^3}{c^2} \frac{1}{e^{h\nu/kT} - 1}. \qquad (2.2.9)$$

Thus, the emission coefficient becomes

$$j_\nu = \frac{e^{-h\nu/kT}}{\varrho}\left[\mu_\nu B_\nu(T)(e^{h\nu/kT} - 1) + \mu_\nu I_\nu(\widehat{\Omega})\right]$$

which reduces to

$$\varrho j_\nu = \mu'_\nu B_\nu(T) + e^{-h\nu/kT} \mu_\nu I_\nu(\widehat{\Omega}) \qquad (2.2.10a)$$

where we have defined

$$\mu'_\nu = \mu_\nu(1 - e^{-h\nu/kT}). \qquad (2.2.11)$$

$\varkappa'_\nu$ may similarly be defined so that eqn. (2.2.10a) becomes

$$j_\nu = \varkappa'_\nu B_\nu(T) + e^{-h\nu/kT} \varkappa_\nu I_\nu(\widehat{\Omega}). \qquad (2.2.10b)$$

Equation (2.2.10b) explicitly shows how the augmented emission coefficient j_ν depends on I_ν as a result of stimulated emission. In this expression for j_ν, which should be used when a change of quantum state occurs, the first term is the contribution to the emission coefficient from spontaneous emission and the second is the contribution from stimulated emission. That is,

$$j_\nu^{\text{(spontaneous)}} = \varkappa'_\nu B_\nu(T) \tag{2.2.12a}$$

and

$$j_\nu^{\text{(stimulated)}} = \varkappa_\nu \, e^{-h\nu/kT} \, I_\nu(\widehat{\Omega}). \tag{2.2.12b}$$

There are therefore two equivalent† ways of formulating the principle of local thermodynamic equilibrium, viz.

$$\frac{N_m}{N_n} = \exp\left(\frac{h\nu}{kT}\right), \tag{2.2.13a}$$

$$j_\nu^{\text{(spontaneous)}} = \varkappa'_\nu B_\nu(T). \tag{2.2.13b}$$

Equation (2.2.13b) is more useful here since it permits the transfer equation without scattering to be written in similar form to eqn. (2.1.4):

$$\frac{1}{\varrho}\frac{dI_\nu}{ds} = j_\nu^{\text{(spontaneous)}} + j_\nu^{\text{(induced)}} - \varkappa_\nu I_\nu. \tag{2.2.14}$$

The foregoing equations lead to the following forms for the transfer equation which also imply existence of LTE:

$$\frac{1}{\varrho}\frac{dI_\nu}{ds} = \varkappa'_\nu (B_\nu - I_\nu) \tag{2.2.15a}$$

or

$$\frac{dI_\nu}{ds} = \mu'_\nu (B_\nu - I_\nu). \tag{2.2.15b}$$

Equation (2.2.15b) is very similar to eqn. (2.1.5b) and its formal solution may thus be written down from inspection of eqn. (2.1.10) replacing J_ν by B_ν and μ_ν by μ'_ν.

† Note that we have not included scattering in this definition; this is sometimes done in astrophysics by means of a generalization of eqn. (2.2.13b).

2.3. Scattering

The type of scattering of primary interest in the study of high-temperature plasmas is coherent scattering from the ground and low-lying states of atoms and/or molecules and their ions. By *coherent scattering* is meant scattering that occurs without a change in frequency of the photon involved, and without a change of state of the atomic system involved.† (For an excellent discussion of this, see Lax, 1951.) Therefore, we will confine ourselves here to a few remarks concerning coherent, non-relativistic scattering which are intended to point out the difference between transfer by (coherent) scattering and transfer by absorption/emission. For more extensive discussions of the general subject of scattering, see, for example, Chandrasekhar (1950), Van de Hulst (1957) and Kerker (1969). Explicit forms of the cross sections involved for coherent scattering from hydrogen, and for Compton scattering, will be considered in a subsequent volume, along with mention of the relativistic corrections required for Compton scattering at sufficiently high temperatures.

Once the assumption of LTE is made, and the emission and absorption terms in the transfer equation appropriately simplified, these terms will differ from the scattering terms which appear in the transfer equation. The reason is that emission into a beam by scattering is not accompanied by a transition of the atom or molecule and is thus not directly affected by excited states of the system other than the one which the system is in at the moment of scattering. That is to say, absorption and emission occur with a change of internal state of the system, whereas scattering does not. The two types of processes, i.e. scattering and absorption/emission will be related and pass one into the other for short-lived excited states. For such states, scattering cannot be distinguished from a real absorption followed almost simultaneously by a real emission with the emitting system reverting to its initial state (Heitler, 1954). This happens near a reso-

† The terminology here is somewhat imprecise due to the complications of the subject. In problems of interest to us, the coherence typified here is usually destroyed by recoil and thermal motion which cause small frequency shifts. Thus, no interference occurs and intensities may be added. This was discussed briefly in Section 2.1; see also Sekera (1968).

nance line and is further complicated by the fact that a photon with energy near enough to a real excitation energy can produce the real excitation, or transition, with the help of a transfer of energy from or to a nearby electron or ion (cf. Baranger, 1962). Thus, if we compute all real absorption and emission transitions, including those which need an outside interaction of energy 0 to ΔE to occur, we will have included the scattering produced by virtual states which fail to conserve energy by an amount between 0 and ΔE.

Mayer (1947) and Rudkjøbing (1947) have shown that there is no correction of the form $1 - e^{-h\nu/kT}$ to be applied to non-relativistic scattering. One should not expect it since the atom does not change state and, hence,

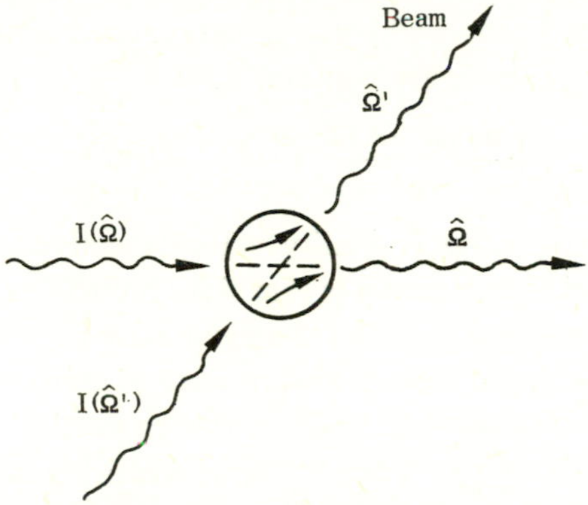

FIG. 2.3. Illustration of induced scattering. $I(\hat{\Omega}')$ represents the beam intensity in the particular direction $\hat{\Omega}'$; scattering induced by the intensity $I(\hat{\Omega})$ diverts part of this beam into the direction $\hat{\Omega}$. The effect is symmetric between the directions $\hat{\Omega}$ and $\hat{\Omega}'$.

there is no upper state to re-emit back into the beam, but it can be demonstrated rigorously. Mayer does this by showing that the induced scattering *into* the beam is exactly cancelled by induced scattering out of the beam. Pictorially what he demonstrates is as shown in Fig. 2.3. Scattering *into* the direction $\widehat{\Omega}'$ from the direction $\widehat{\Omega}$ depends on an integral over the differential scattering cross section†

$$\frac{d\sigma(\hat{\boldsymbol{\Omega}} \to \hat{\boldsymbol{\Omega}}')}{d\Omega}$$

of the form

$$\int_{\hat{\Omega}} \frac{d\sigma(\hat{\boldsymbol{\Omega}} \to \hat{\boldsymbol{\Omega}}')}{d\Omega} I_\nu(\hat{\boldsymbol{\Omega}})\, d\Omega \tag{2.3.1}$$

over all other directions $\hat{\boldsymbol{\Omega}}$, which is the ordinary scattering term, and an induced scattering term proportional to

$$\int_{\hat{\Omega}} \frac{d\sigma(\hat{\boldsymbol{\Omega}} \to \hat{\boldsymbol{\Omega}}')}{d\Omega} I_\nu(\hat{\boldsymbol{\Omega}})\, I_\nu(\hat{\boldsymbol{\Omega}}')\, d\Omega\,. \tag{2.3.2}$$

That is to say, this second term depends on the beam intensity in the scattered direction, which is why it is called induced scattering. However, the scattering out of the beam contains a term

$$\int \frac{d\sigma(\hat{\boldsymbol{\Omega}}' \to \hat{\boldsymbol{\Omega}})}{d\Omega'} I_\nu(\hat{\boldsymbol{\Omega}}')\, I_\nu(\hat{\boldsymbol{\Omega}})\, d\Omega' \tag{2.3.3}$$

which, with the proper proportionality factors, exactly cancels the previous term. The implication apparently is that since induced scattering does occur, one might expect a correction to appear analogous to the $1 - e^{-h\nu/kT}$ correction obtained in the case of absorption and emission. However, it is not just induced emission that leads to this correction. If this were the case we would need only to correct the absorption term in the transfer equation and no distinct emission term correction would appear. Since both absorption and induced emission are proportional to the intensity I_ν, they can always be lumped together. The existence of spontaneous emission is also required—an emission term independent of the incident intensity. This is what is lacking in the case of scattering. There is no spontaneous scattering. As Mayer (1947) points out, the scattering terms in the transfer equation are proportional to $n_{\hat{\Omega}}(1 + n_{\hat{\Omega}})$ ($n_{\hat{\Omega}}$ = number of "radiation oscillators" in the direction $\hat{\boldsymbol{\Omega}}$). Thus, there is no term independent of $n_{\hat{\Omega}}$ as there is in the case of emission (which is simply proportional to $1 + n_{\hat{\Omega}}$).

† For a definition of the differential cross section see, for example, Schiff (1968, p. 110) or Massey and Burhop (1969, p. 5).

The diminution of the beam $I_\nu(\widehat{\Omega})$ due to scattering within the element of distance ds can thus be written

$$\delta I_\nu^{(sc)}(\widehat{\Omega}) = -N_v \sigma_{sc} I_\nu(\widehat{\Omega})\, ds \tag{2.3.4}$$

where σ_{sc} is the *total* cross section for scattering by the particles with number density N_ν. In addition to this depletion, the beam will be augmented by the amount

$$\delta I_\nu^{(sc)}(\widehat{\Omega}) = N_v \int_{\hat{\Omega}'} I_\nu(\widehat{\Omega}') \frac{d\sigma(\widehat{\Omega}' \to \widehat{\Omega})}{d\Omega'}\, d\Omega'. \tag{2.3.5}$$

If we add the above terms to the transfer equation, eqn. (2.2.15b), we obtain

$$\frac{dI_\nu}{ds} = \mu_\nu'(B_\nu - I_\nu) + N_v \int \frac{d\sigma(\widehat{\Omega}' \to \widehat{\Omega})}{d\Omega'} I_\nu(\widehat{\Omega}')\, d\Omega' - \mu_\nu^{(sc)} I_\nu \tag{2.3.6}$$

where we have defined $\mu_\nu^{(sc)}$ by:

$$\mu_\nu^{(sc)} \equiv N_v \sigma_{sc}. \tag{2.3.7}$$

2.4. Emission from a Gas Sample in the Optically Thin Limit: the Planck, or "Emission Mean" Absorption Coefficient

Before defining the Planck mean absorption coefficient, let us investigate the flux emitted by a plane-parallel slab of gas in order to define the optically thin limit in several more or less equivalent ways. Consider the radiation from an isolated slab of gas at temperature T, as in Fig. 2.2. The total monochromatic radiation flux per unit area leaving one face of the slab within the solid angle $\Delta\Omega$ is

$$F_{\nu+} = \int_{\Delta\Omega} I_\nu \cos\theta\, d\Omega \tag{2.4.1}$$

where θ is measured from the x-direction normal to the slab. Multiplying the transfer equation, eqn. (2.2.15), on both sides by $d\Omega$ using $dx = ds \cos\theta$, and integrating over the hemisphere of outward directions, we obtain

$$\int_{\Delta\Omega} \frac{dI_\nu}{dx} \cos\theta \, d\Omega = \Delta\Omega \, \mu'_\nu B_\nu - \mu'_\nu \int_{\Delta\Omega} I_\nu \, d\Omega . \tag{2.4.2}$$

From eqn. (2.4.1) and the definition of the mean outward (partial) intensity $\bar{I}_{\nu+}$, viz.

$$\bar{I}_{\nu+} \equiv \frac{1}{\Delta\Omega} \int_{\Delta\Omega} I_\nu \, d\Omega , \tag{2.4.3}$$

we can rewrite eqn. (2.4.2) as

$$\frac{dF_{\nu+}}{dx} = \Delta\Omega \, \mu'_\nu (B_\nu - \bar{I}_{\nu+}) . \tag{2.4.4}$$

The optically thin approximation now consists in neglecting† $\bar{I}_{\nu+}$ relative to B_ν, and taking the derivative $dF_{\nu+}/dx$ as the ratio of finite increments of emitted flux δF_ν and slab thickness δx. With these approximations we obtain from eqn. (2.4.4) the integrated flux or total radiant energy emitted within the solid angle $\Delta\Omega$,

$$\delta F_+ \cong \Delta\Omega \, \delta x \int \mu'_\nu B_\nu (T) \, d\nu . \tag{2.4.5}$$

The same result could be obtained by integrating eqn. (2.2.10) for the emission coefficient upon neglect of (a) self-absorption, and (b) induced emission. The flux from a volume element $\delta V = \delta A \delta s$ cut out by the walls of an infinitesimal slab is (see Fig. 2.4)

$$\delta F_+ \cong \int_\nu \int_\Omega j_\nu \, \varrho \delta V \cos\theta \, d\nu \, d\Omega / \delta A . \tag{2.4.6}$$

But $\delta s = \delta x / \cos\theta$ and $\varrho j'_\nu \cong \mu'_\nu B_\nu (T)$. Thus

$$\delta F_+ \cong \delta x \int_\nu \int_\Omega \mu'_\nu B_\nu (T) \, d\nu \, d\Omega = \Delta\Omega \, \delta x \int \mu'_\nu B_\nu (T) \, d\nu .$$

If $\Delta\Omega$ is the entire hemisphere, then eqn. (2.4.6) becomes

$$\delta F_+ = [2\pi \int \mu'_\nu B_\nu (T) \, d\nu] \, \delta x .$$

† Recall that we are considering an isolated slab; hence, there is no incident intensity at the boundary.

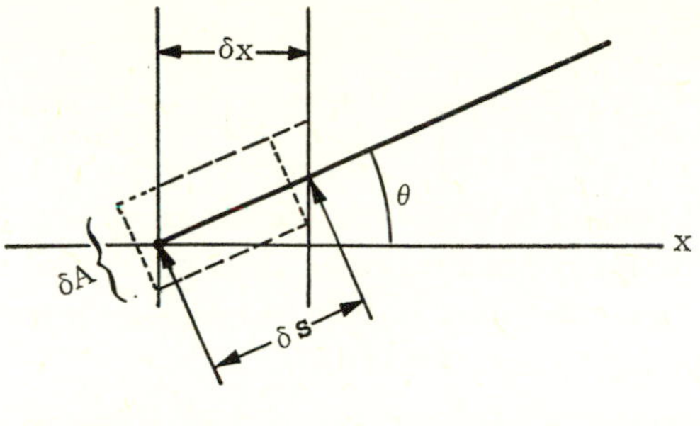

FIG. 2.4. Flux emission from an element of volume in a slab of thickness δx.

The result can also be obtained as follows from the general solution to the transfer equation, eqn. (2.1.10), using the approximation $\mu_\nu'\delta x \ll 1$. With the additional assumption that the incident intensity vanishes, eqn. (2.1.10) yields the total outward flux from one side of an isothermal, isobaric slab as†

$$F = 2\pi \int d\nu\, B_\nu(T) \int_\theta (1 - e^{-\mu_\nu' \delta x/\cos\theta}) \cos\theta \sin\theta\, d\theta. \quad (2.4.7)$$

We wish to show that the correct result to first order in $\mu_\nu'\delta x$ is obtained by expanding the exponential, in spite of the fact that $\cos\theta \to 0$ within the range of integration. With a change of variables eqn. (2.4.7) becomes

$$F = 2\pi \int d\nu B_\nu (T) [\tfrac{1}{2} - E_3 (\mu_\nu' \delta x)] \quad (2.4.8)$$

where

$$E_n(y) = \int_1^\infty \frac{dx}{x^n} e^{-xy} \quad (2.4.9)$$

are the usual exponential integrals (Chandrasekhar, 1950; Armstrong, 1969a). By use of the relations

$$\left.\begin{aligned} nE_{n+1}(y) &= e^{-y} - yE_n(y) \\ E_1(y) &= -\gamma - \log y + \sum_{n=1}^{\infty} (-1)^{n-1} \frac{y^n}{n \cdot n!} \end{aligned}\right\} \quad (2.4.10)$$

† We recall from the discussion following eqn. (2.2.15) that J_ν is to be replaced by B_ν and μ_ν by μ_ν'.

where γ = Euler's constant, we find

$$F = 2\pi \int d\nu\, B_\nu(T) \{\mu'_\nu \delta x + O[(\mu'_\nu \delta x)^2 \log \mu'_\nu \delta x] + O[(\mu'_\nu \delta x)^2]\} \quad (2.4.11)$$

which agrees with eqn. (2.4.5) to first order in $\mu'\delta x$ (with $\Delta\Omega = 2\pi$). This result shows that radiation within the angular range near $\theta = \pi/2$, which is not involved in many practical cases, does not contribute to F in first order. There is a contribution in first order to $\bar{I}_\nu$, however, where the result corresponding to eqn. (2.4.11) is

$$\bar{I}_\nu = (1 + \gamma)\, \mu'_\nu \delta x + (\mu'_\nu \delta x) \log(\mu'_\nu \delta x) + O[(\mu'_\nu \delta x)^2].$$

The radiative energy emitted by one face of a slab of perfect radiator in thermal equilibrium at temperature T can be obtained from eqn. (2.4.7) by letting $\delta x \to \infty$. The result is, for the energy emitted per unit area,

$$\int_\nu \int_{2\pi} B_\nu(T) \cos\theta\, d\Omega\, d\nu = \int_{2\pi} B(T) \cos\theta\, d\Omega = \sigma T^4 \quad (2.4.12)$$

where

$$B(T) \equiv \int B_\nu(T)\, d\nu = \sigma T^4/\pi \quad (2.4.13)$$

and σ is Stefan's constant, $2\pi^5 k^4/15c^2h^3$.

We can define the (flux) emissivity ε of a slab relative to this standard as the ratio of the total radiant energy one face of the slab emits to that emitted by one face of a blackbody slab at the same temperature. Thus for a thin slab, ε is given by the ratio

$$\varepsilon \equiv \frac{\delta F_+}{\sigma T^4} = \frac{2\pi\, \delta x \int \mu'_\nu B_\nu(T)\, d\nu}{\sigma T^4}. \quad (2.4.14)$$

Since this quantity depends on the slab thickness, it is somewhat more satisfactory to define $\varepsilon/\delta x$ as the *emissivity per unit length.* If we define the Planck mean absorption coefficient (or emission mean) $\bar{\mu}_P(T)$ as

$$\bar{\mu}_P(T) = \frac{\int \mu'_\nu B_\nu(T)\, d\nu}{\int B_\nu(T)\, d\nu} = \frac{\pi \int \mu'_\nu B_\nu(T)\, d\nu}{\sigma T^4}, \quad (2.4.15)$$

the emissivity per unit length of an optically thin slab becomes

$$\varepsilon/\delta x = 2\bar{\mu}_P(T). \quad (2.4.16)$$

A similar quantity, the "hemispherical emissivity" ε_h is sometimes used (Penner, 1959). It is defined as the ratio of the radiant flux emitted by a hemispherical volume (with radius δR) of gas into a "point" collector at the center of its base. We will show that $\varepsilon_h/\delta R$ is one-half the thin-slab emissivity per unit length;

$$\varepsilon_h/\delta R = \bar{\mu}_P(T). \tag{2.4.17}$$

In general, the relation between the thin emissivity and the mean absorption coefficient depends on the geometry of the emitting sample through a numerical factor; we will discuss this in more detail in Section 2.7. We can calculate the hemispherical emissivity with reference to Fig. 2.5. The radiation arriving at 0 from the shaded volume H within the hemispherical

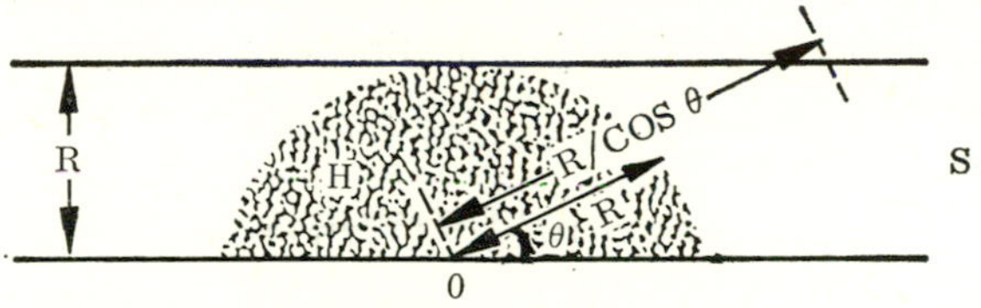

FIG. 2.5. Thin-slab and hemispherical geometries.

shell contributes to the hemispherical emissivity. The radiation arriving at 0 from the remainder of the slab, S, is the additional contribution which arises in the thin-slab definition. We shall show that these two contributions are equal in the limit of small optical depth.

Consider the isothermal solution to the transfer equation (eqn. (2.1.11)) divided into the contributions from regions H and S:

$$\begin{aligned} I_\nu &= \int_0^{\tau_\nu} B_\nu(T)\, e^{-\tau_\nu'}\, d\tau_\nu' = \int_0^{\tau_{\nu 1}} B_\nu(T)\, e^{-\tau_\nu'}\, d\tau_\nu' + \int_{\tau_{\nu 1}}^{\tau_{\nu 2}} B_\nu(T)\, e^{-\tau_\nu'}\, d\tau_\nu' \\ &= B_\nu(T)\,[1 - e^{-\tau_{\nu 1}}] + B_\nu(T)\,[e^{-\tau_{\nu 1}} - e^{-\tau_{\nu 2}}] \end{aligned} \tag{2.4.18}$$

where

$$\begin{aligned} \tau_{\nu 1} &= \mu_\nu' R, \\ \tau_{\nu 2} &= \mu_\nu' R/\cos\theta. \end{aligned} \tag{2.4.19}$$

We obtain the flux crossing the surface at 0 by multiplying I_ν of eqn. (2.4.18) by $\cos\theta\, d\Omega\, d\nu$, and integrating over $d\Omega$ and $d\nu$. The result is

$$F = F_1 + F_2 \tag{2.4.20}$$

where

$$F_1 = \pi \int B_\nu(T)\,(1 - e^{-\mu_\nu' R})\, d\nu \tag{2.4.21}$$

is the contribution from the hemispherical region H and

$$F_2 = \int_\nu \int_{2\pi} B_\nu(T)\,(e^{-\tau_{\nu 1}} - e^{-\tau_{\nu 2}}) \cos\theta\, d\Omega\, d\nu \tag{2.4.22}$$

is the contribution from the remainder of the slab, region S.

We make the approximation of small optical depth, viz. $\tau_\nu \ll 1$, and expand the exponentials. The justification of this expansion has been given in eqn. (2.4.11).

This leads to

$$F_1 = \pi R \int B_\nu(T)\, \mu_\nu'\, d\nu \tag{2.4.23}$$

and

$$F_2 = \int_\nu \int_{2\pi} B_\nu(T)\,[-\mu_\nu' R + \mu_\nu' R/\cos\theta] \cos\theta\, d\Omega\, d\nu = \pi R \int B_\nu(T)\, \mu_\nu'\, d\nu. \tag{2.4.24}$$

From eqns. (2.4.23) and (2.4.24) we see immediately that $F_1 = F_2$. From the definition of the Planck mean absorption coefficient (eqn. (2.4.15)), the slab emissivity is

$$\frac{F}{\sigma T^4} = 2\bar{\mu}_P(T)\, R \tag{2.4.25}$$

while the hemispherical emissivity $F_1/\sigma T^4$ by the above reasoning is just one-half this quantity, or $\bar{\mu}_P R$. It is worth emphasizing that the condition for the validity of the Planck mean as a measure of the emissivity of a thin slab in LTE (see eqn. (2.4.16)) is $\mu_\nu' \delta x \ll 1$ and not $\bar{\mu}_P \delta x \ll 1$, as is sometimes stated. This latter condition is not as stringent. The existence of very strong lines which contribute dramatically to the magnitude of $\bar{\mu}_P$ can produce a significant difference between these two criteria. For such lines a medium can be optically thin in the wings of the lines and strongly self absorbing in the centers. The use of the second criterion alone would permit the overestimation of the energy loss from a slab in such cases because photons in the line centers would be counted as lost to the medium when actually they are reabsorbed. Another point worthy of emphasis is that in an optically thick medium the photons help maintain the equilibrium over the energy levels of the particles. In an optically thin medium, the radiation is lost and does not interact with the medium and help maintain equilibrium. Thus, the LTE conditions necessary for the usefulness of the Planck mean are harder to maintain than in the optically thick case due to the very conditions that establish its validity. If particle collisions are sufficiently frequent, LTE can be maintained without the help of the radiation field; however, one must be careful in any particular situation to verify that this is the case. In addition to yielding the emissivity of a thin slab, the Planck mean is useful in establishing what is known as the condition of *radiative equilibrium.* For a discussion of this application, see Kourganoff (1952, p. 243).

We will return to a discussion of the optically thin limit in Section 2.7 after we consider the case of intermediate optical depths. By use of results obtained from neutron transport theory, we will obtain there a formula for the emissivity which provides, in the optically thin limit, the result for a general emitting geometry.

2.5. Emission from a Gas Sample in the Optically Thick Limit: the Rosseland or "Diffusion Mean" Absorption Coefficient

Whereas the Planck mean is of interest when the intensity $I_\nu \ll B_\nu$, another mean becomes of importance in the "opposite" limit, namely,

$$I_\nu \cong B_\nu. \tag{2.5.1}$$

We have previously noted that when $I_\nu \ll B_\nu$ and LTE prevails, the radiation is *far* from being in equilibrium with the matter. On the other hand, eqn. (2.5.1) will be satisfied when the radiation is nearly in equilibrium with the matter.

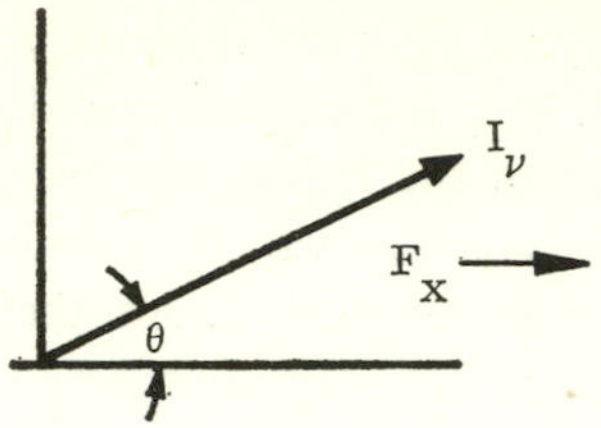

FIG. 2.6. Intensity I_ν and x-component of flux F_x.

Inserting eqn. (2.5.1) as a zeroth-order approximation into the transfer equation, eqn. (2.2.15), we obtain

$$I_\nu \cong B_\nu - \frac{1}{\mu'_\nu} \frac{dB_\nu}{ds} \tag{2.5.2}$$

for a first-order approximation. With eqn. (2.4.1) the integrated net flux† in the x-direction can be given, as illustrated in Fig. 2.6, by

$$F_x = -\int \frac{1}{\mu'_\nu} \frac{dB_\nu}{ds} \cos\theta \, d\Omega \, d\nu, \tag{2.5.3}$$

since the first term of eqn. (2.5.2) is isotropic and thus does not contribute to the net flux. Substituting $s = x/\cos\theta$, the above equation yields

$$F_x = -\int \frac{1}{\mu'_\nu} \frac{dB_\nu}{dx} \cos^2\theta \, d\Omega \, d\nu = -\frac{4\pi}{3} \int \frac{1}{\mu'_\nu} \frac{dB_\nu}{dx} \, d\nu \tag{2.5.4}$$

with similar expressions for F_y and F_z. Thus, we obtain

$$\boldsymbol{F} = -\frac{4\pi}{3} \int \frac{1}{\mu'_\nu} \nabla B_\nu(T) \, d\nu = -\left(\frac{4\pi}{3} \int \frac{1}{\mu'_\nu} \frac{dB_\nu}{dT} \, d\nu \right) \nabla T. \tag{2.5.5}$$

If the Rosseland mean absorption coefficient is now defined as

$$\frac{1}{\bar{\mu}_R} \equiv \int \frac{1}{\mu'_\nu} \frac{dB_\nu(T)}{dT} \, d\nu \left[\int \frac{dB_\nu(T)}{dT} \, d\nu \right]^{-1} \tag{2.5.6}$$

† We are here considering angular integrations over the complete sphere.

eqn. (2.5.5) becomes

$$\boldsymbol{F} = -\frac{4\pi}{3}\frac{1}{\bar{\mu}_R}\left(\int \frac{dB_\nu}{dT}\,d\nu\right)\nabla T = -\frac{4\pi}{3}\frac{1}{\bar{\mu}_R}\nabla B(T). \qquad (2.5.7)$$

In terms of the Rosseland mean free path $\Lambda_R = \bar{\mu}_R^{-1}$, and the radiation density

$$u = \frac{4\pi}{c}\int B_\nu(T)\,d\nu,$$

eqn. (2.5.7) can be written

$$\boldsymbol{F} = -\frac{c\Lambda_R}{3}\nabla u. \qquad (2.5.8)$$

This equation is the basis of the definition of the Rosseland mean free path Λ_R, the Rosseland mean absorption coefficient $\bar{\mu}_R$, and the Rosseland mean opacity $\bar{\varkappa}_R$ where the latter is defined as

$$\bar{\varkappa}_R = \bar{\mu}_R/\varrho. \qquad (2.5.9)$$

Note that in contrast to the Planck mean, the Rosseland mean is an *inverse* mean and thus emphasizes small values of the absorption coefficient. The Planck mean emphasizes large values of absorption coefficient.

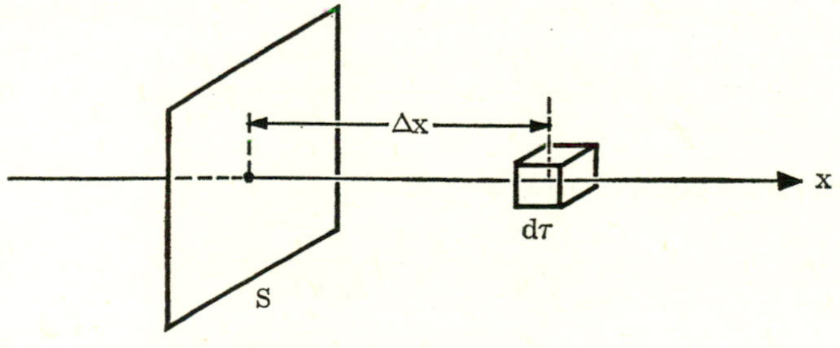

FIG. 2.7. Transport across a surface area, S.

Equation (2.5.8) is analogous to that of the diffusion equation (Kennard, 1938)

$$\boldsymbol{J} = D\nabla n \qquad (2.5.10)$$

where J is a current, n is a particle density and D is the diffusion coefficient. Thus Λ_R is an effective diffusion length and $c\Lambda_R/3$ is the diffusion coefficient. Equation (2.5.8) describes the diffusion of photons through the gas.

The limits of validity of eqn. (2.5.7) can be obtained by examining the approximations implicit in eqns. (2.5.1) and (2.5.2) as follows. The transfer equation (eqn. (2.2.15)) may be written as

$$I_\nu = B_\nu - \frac{1}{\mu'_\nu}\frac{dI_\nu}{ds}. \tag{2.5.11}$$

Also we may set

$$\frac{dI_\nu}{ds} = \frac{dI_\nu}{dT}\frac{dT}{ds}.$$

If $I_\nu \cong B_\nu$ (eqn. (2.5.1)) then $(1/\mu'_\nu)\, dI_\nu/ds$ must be small compared to B_ν; thus, from eqns. (2.5.1) and (2.5.11), we have

$$\frac{1}{\mu'_\nu}\frac{dI_\nu}{dT}\frac{dT}{ds} \ll B_\nu(T), \tag{2.5.12}$$

which is the basic requirement of the approximation. But as $I_\nu \cong B_\nu$, we can also write

$$\frac{1}{\mu'_\nu}\frac{dT}{ds} \ll B_\nu(T)\left[\frac{dB_\nu(T)}{dT}\right]^{-1}. \tag{2.5.13}$$

In view of the formula for the Planck function, eqn. (2.2.9), we have that

$$B_\nu(T)\Bigg/\frac{dB_\nu(T)}{dT} = \frac{(1 - e^{-h\nu/kT})\, kT^2}{h\nu}. \tag{2.5.14}$$

Therefore, from eqns. (2.5.13) and (2.5.14),

$$\frac{1}{\mu'_\nu}\frac{|dT/ds|}{T} \ll \frac{1 - e^{-h\nu/kT}}{h\nu/kT} \tag{2.5.15}$$

which expresses the limit of validity of the diffusion approximation in terms of of the characteristic length $T/(dT/ds)$ associated with the temperature variation. For $h\nu \ll kT$, this requirement simplifies to

$$\mu'_\nu\left(T\Bigg/\frac{dT}{ds}\right) \gg 1.$$

A derivation of the Rosseland diffusion theory can also be given by analogy to simple kinetic diffusion theory (Born, 1946, pp. 274; see also Kennard, 1938, pp. 188ff.) if one notes that the specific intensity I_ν has the nature of an energy distribution function (cf. 2.4 of Morse and Feshbach, 1953), or that $I_\nu/h\nu$ has the nature of a particle distribution function. With reference to Fig. 2.7, let n be the density of particles (in this case photons) at the surface S. We assume that the density varies only with x so that the density at $d\tau$ is $n + \Delta x(dn/dx)$. It is well known from elementary kinetic theory that the particle flux Γ across a given surface within a gas is given approximately by

$$\Gamma = \tfrac{1}{4} n\bar{v} \tag{2.5.16}$$

where $\bar{v}$ is the average velocity of the particles (Kennard, 1938, p. 62). If one retraces this derivation giving all the particles the same velocity c the same result is obtained. In view of the change in density along x, the rate at which particles collide within $d\tau$ and thereafter cross S (i.e. photons are absorbed and re-emitted to cross S) will be increased by the amount $\frac{1}{4}\Delta x\,(dn/dx)\,c$. Thus, the flow of photons towards $-x$ will be

$$\Gamma_- = \frac{1}{4} nc + \frac{1}{4} (\Delta \bar{x}_-)\, c\, \frac{dn}{dx}$$

where $\Delta\bar{x}_-$ is the average value of Δx at the last point of collision for each particle. With a similar expression for Γ_+, and by use of $\Delta\bar{x}_- = \frac{2}{3}l$, $\Delta\bar{x}_+ = -\frac{2}{3}l$, where l is the mean free path between collisions (Kennard, 1938, p. 140), we obtain the net particle flow across the surface S:

$$\Gamma = \Gamma_+ + \Gamma_- = \frac{1}{3}\, cl\, \frac{dn}{dx}. \tag{2.5.17}$$

Generalizing this expression to three dimensions and setting the mean free path l for photons of frequency ν equal to $1/\mu'_\nu$, we obtain the photon number flux per unit frequency interval:

$$\boldsymbol{\Gamma}_\nu = \frac{c}{3\mu'_\nu} \nabla n_\nu. \tag{2.5.18}$$

This result can be compared directly with the standard form of the diffusion equation, eqn. (2.5.10), to yield $c/3\mu_\nu'$ as the effective photon diffusion coefficient. For photons in equilibrium [or approximate equilibrium according to eqn. (2.5.1)] the particle density per unit frequency interval n_ν is given by

$$n_\nu = \left(\frac{1}{h\nu}\right)\frac{4\pi}{c}B_\nu(T) \tag{2.5.19}$$

with the result that the photon energy flux $h\nu\boldsymbol{\Gamma}_\nu = \boldsymbol{F}_\nu$ becomes

$$\boldsymbol{F}_\nu = \frac{4\pi}{3\mu_\nu'}\nabla B_\nu(T) \tag{2.5.20}$$

in agreement with eqn. (2.5.5). Performing the frequency integration leads again to the definition of the Rosseland mean free path of eqn. (2.5.6).

When one is concerned with forces or the transfer of momentum in a gas, the diffusion of particles leads to a viscosity of the gas. By analogy, one would expect the photon diffusion approximation to lead to a viscosity for photons. This is indeed the case as was first shown by Jeans (1926a, 1926b; see also Milne, 1930). Hazlehurst and Sargent (1959) give an improved value of this photon viscosity η:

$$\eta = \frac{6}{15}\frac{\Lambda_R}{c^2}\sigma T^4 \tag{2.5.21}$$

where σ is Stefan's constant. Although we derived the Rosseland mean free path for considerations of the flow of energy, we find here an additional application of this parameter when we are concerned with the flow of momentum.

The foregoing derivations of the Rosseland diffusion theory have been given for the physical insight which they convey. The more rigorous standard derivation (Goody, 1964; Chandrasekhar, 1939) is to set the source function

$$J_\nu\,(s'', \hat{\boldsymbol{\Omega}})\ \{\equiv J_\nu\,[\tau_\nu\,(s'', s)]\}$$

in eqn. (2.1.11) equal to $B_\nu\,[\tau_\nu\,(s'', s)]$ and expand $B_\nu(\tau_\nu)$ in the Taylor series

$$B_\nu\,[\tau_\nu\,(s'', s)] = B_\nu(s) + \left.\frac{dB_\nu}{d\tau_\nu}\right|_s \tau_\nu\,(s'', s) + \left.\frac{d^2B_\nu}{d\tau_\nu^2}\right|_s \frac{\tau_\nu^2\,(s'', s)}{2} + \cdots. \quad (2.5.22)$$

The integration indicated in eqn. (2.1.11) can now be carried out and the flux computed to obtain the result given in eqn. (2.5.7).

2.6. General Features, Inequalities and Bounds for the Planck and Rosseland Mean Absorption Coefficients

The Planck and Rosseland mean absorption coefficients defined in Sections 2.4 and 2.5 can be written as

$$\bar{\mu}_P = \frac{15}{\pi^4}\int_0^\infty u^3 e^{-u}\mu\,(u, T)\,du \quad (2.6.1)$$

and

$$(\bar{\mu}_R)^{-1} = \frac{15}{4\pi^4}\int_0^\infty \frac{u^4 e^{2u}}{\mu\,(u, T)\,(e^u - 1)^3}\,du, \quad (2.6.2)$$

respectively, where we have written $\mu(\nu) \equiv \mu\,(u, T)$ to avoid the suggestion that $\mu(\nu)$ depends only on u, and where $u \equiv h\nu/kT$. These formulae can be obtained from eqns. (2.4.15) and (2.5.6) by use of the expression for the integral of the Planck function

$$B = \int_0^\infty B(\nu)\,d\nu = \frac{2\pi^4k^4T^4}{15c^2h^3} \quad (2.6.3)$$

and by interchanging the order of differentiation and integration in the integral

$$\int_0^\infty \frac{dB_\nu}{dT}\,d\nu \quad (2.6.4)$$

to write

$$\int_0^\infty \frac{dB_\nu}{dT}\,d\nu = \frac{d}{dT}B = \frac{8\pi^4k^4T^3}{15c^2h^3}. \quad (2.6.5)$$

In eqs. (2.6.1) and (2.6.2), the factor $1 - e^{-u}$ that differentiates between μ and μ' has been absorbed into the weighting functions. We will label these weighting functions $W_P(u)$ and $W_R(u)$. Equations (2.6.1) and (2.6.2) can thus be written

$$\bar{\mu}_P = \int_0^\infty W_P(u)\, \mu\,(u, T)\, du, \tag{2.6.6}$$

$$(\bar{\mu}_R)^{-1} \equiv \Lambda_R = \int_0^\infty \frac{W_R(u)}{\mu\,(u, T)}\, du. \tag{2.6.7}$$

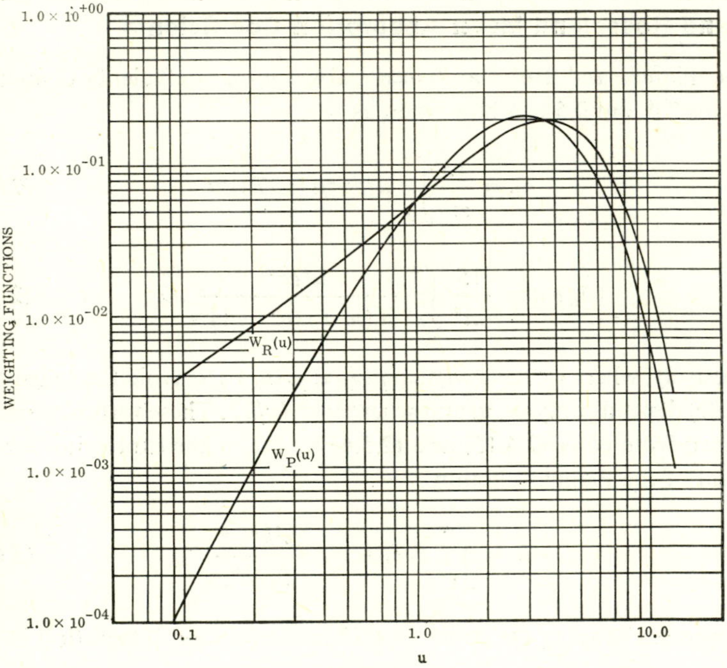

FIG. 2.8. Planck and Rosseland weighting functions as given by eqns. (2.6.8) of the text.

Graphs of the weighting functions

$$W_P(u) \equiv \frac{15}{\pi^4}\, u^3 e^{-u} \tag{2.6.8a}$$

and

$$W_R(u) = \frac{15}{4\pi^4} \frac{u^4 e^{2u}}{(e^u - 1)^3} \tag{2.6.8b}$$

are shown in Fig. 2.8. By inspection or differentiation it can be easily verified that $W_P(u)$ has its maximum at $u = 3$. We can similarly show that $W_R(u)$ has a maximum where

$$u_M = 4\left(\frac{e^{u_M} - 1}{e^{u_M} + 2}\right); \tag{2.6.9}$$

u_M is thus just a little less than 4 (since $(e^u - 1)/(e^u + 2) < 1$). More accurately, eqn. (2.6.9) can be solved graphically to obtain $u_M \cong 3.74$.

The weighting function curves shown in Fig. 2.8 clearly show the frequency regions, at a given temperature, that dominate the calculation of each mean. Since $\bar{\mu}_P$ is a direct mean and the contributions to it are additive, further comment is unnecessary. The Rosseland mean coefficient $(\bar{\mu}_R)^{-1}$, is however, a harmonic mean, or physically, a mean free path rather than strictly a mean absorption coefficient, and as such merits further discussion.

Effect of Lines on the Rosseland Mean

Following Mayer (1947) we divide the absorption coefficient into two parts:

$$\mu = \mu_c + \mu_l, \tag{2.6.10}$$

where μ_c is the continuous, or slowly varying part and μ_l is the rapidly varying line contribution. If we further introduce the notation

$$r(u) \equiv \mu_l/\mu_c, \tag{2.6.11}$$

such that $\mu = \mu_c(1 + r)$, eqn. (2.6.7) for Λ_R can now be rewritten as

$$\Lambda_R = \int_0^\infty \frac{W_R(u)\,du}{\mu_c(1 + r)}. \tag{2.6.12}$$

Thus, we can write

$$\Lambda_R = \int_0^\infty \frac{W_R(u)}{\mu_c}\,du - \int_0^\infty \left(\frac{r(u)}{1 + r(u)}\right) \frac{W_R(u)}{\mu_c}\,du \tag{2.6.13a}$$

or

$$\Lambda_R \equiv \Lambda_c - \Lambda_l, \tag{2.6.13b}$$

where eqn. (2.6.13b) defines a notation for the continuous and line contributions to the Rosseland mean free path respectively, in terms of the two integrals in eqn. (2.6.13a). This expression shows explicitly how the lines reduce the mean free path from the value Λ_c that it would have in the presence of continuous processes alone.

The effect of a strong line can be shown in the following way. We write $r(u)/(1 + r(u))$ as $1/\{1 + 1/r(u)\}$ and note that a strong line implies

$$\frac{1}{r(u_0)} \ll 1, \tag{2.6.14}$$

where u_0 is the position of the line.

Thus, we can expand this factor as follows:

$$\frac{r(u_0)}{1 + r(u_0)} = 1 - \frac{1}{r(u_0)} + \left(\frac{1}{r(u_0)}\right)^2 + \cdots. \tag{2.6.15}$$

The line contribution to Λ_R from the small frequency region Δu_0 about u_0 where μ is large is, therefore,

$$\Delta\Lambda_l = \frac{W_R(u_0)}{\mu_c\,(u_0 T)}\,\Delta u_0 \left[1 - \frac{1}{r(u_0)}\right] \tag{2.6.16}$$

to first order in $1/\{r(u_0)\}$. The zero-order term precisely cancels the contribution of Λ_c, leaving only a first-order contribution to Λ_R from this interval. Thus to lowest order the line eliminates the entire transmission over the interval. Its zero-order effect on Λ_R depends only on the width of the spectral region which it blacks out, and is independent of the strength of the line. The first-order correction term is

$$\Delta\Lambda_R = \frac{W_R(u_0)}{r(u_0)\,\mu_c}\quad \Delta u_0 = \frac{W_R(u_0)}{\mu_l}\,\Delta u_0 \tag{2.6.17}$$

which, as one would expect, does not depend upon μ_c, since $\mu_l \gg \mu_c$ in this region. For a simple model of rectangular, non-overlapping lines of widths Δu_i and heights μ_i

$$\mu_l(u_i, T) = \mu_i, \quad |u - u_i| \le \frac{\Delta u_i}{2} \qquad (2.6.18)$$
$$= 0, \quad |u - u_i| > \frac{\Delta u_i}{2},$$

these results imply that to first order in $[r(u_0)]^{-1}$, the Rosseland mean free path may be written as

$$\Lambda_R = \int_0^\infty \frac{W_R(u)\,du}{\mu_c} - \sum_i \frac{W_R(u_i)}{\mu_c}\Delta u_i + \sum_i \frac{W_R(u_i)}{\mu_l}\Delta u_i. \qquad (2.6.19)$$

This equation follows from eqn. (2.6.13) if the first two terms of eqn. (2.6.15) are used for $r(u)/[1 + r(u)]$. The last term of eqn. (2.6.19) is negligible if the lines are sufficiently strong.

For Lorentz and Doppler line shapes, the integrations over the line profiles can still be performed (Mayer, 1947) leading to an approximate result with the same form as eqn. (2.6.19) but with $\Delta u_0/2$ being replaced by $\pi/2$ or 1.0 times a quantity $\tilde{W}$, called the wingspread of the line. Mayer defines this quantity as the distance between the line position $u_0 = h\nu_0/kT$ and the frequency $u = h\nu/kT$ at which $r = 1$, i.e. where the line absorption coefficient is equal to the continuum absorption coefficient.

For the Lorentz dispersion shape

$$b(\nu) = \frac{w/\pi}{(\nu - \nu_0)^2 + w^2}, \qquad (2.6.20a)$$

where

$$\int b(\nu)\,d\nu = 1, \qquad (2.6.20b)$$

with conventional (half) half-width w, the wingspread is given approximately by

$$\tilde{W} \cong \frac{hw}{kT}\left(\frac{\mu_0}{\pi w \mu_c}\right)^{1/2} \qquad (2.6.21)$$

where

$$\mu_0 = \int \mu_l(\nu)\,d\nu = \frac{\pi e^2 N_v f}{mc} \qquad (2.6.22)$$

is the absorption coefficient integrated over the line or, loosely speaking, the integrated line strength, and e, m and c are the electron charge and mass, and the velocity of light, respectively. The last expression above, which evaluates the absorption coefficient integrated over the line profile in terms of fundamental atomic parameters, except for the factor f arises from the classical theory of radiation. This theory will be discussed in Section 3.1. The factor f, called the "f-number" of the line is a quantum-theory correction to the classical result, and will be treated in Section 3.3 [eqn. (3.3.58 ff)]. Thus, for the Lorentz shape, the first approximation to Λ_l for a single line is, from eqn. (2.6.19),

$$\Lambda_l = \left(\frac{\pi}{2}\right)(2\tilde{W})\frac{W_R(u_0)}{\mu_c} \tag{2.6.23}$$

so that a line of this type approximately blacks out a frequency interval $\Delta\nu = \pi\tilde{W}kT/h$, or

$$\Delta\nu\,(\text{blacked out}) = w\left(\frac{\pi\mu_0}{w\mu_c}\right)^{1/2} = \left(\frac{\pi w\mu_0}{\mu_c}\right)^{1/2}.$$

(For further discussion, including the problem of *overlapping* strong lines, and the formulas for the Doppler-broadened case, see Mayer, 1947.)

We will conclude this section with a brief discussion of the weak line case. From eqn. (2.6.11) we define a weak line by

$$r(u) \ll 1. \tag{2.6.24}$$

In this approximation, we can expand $r/(1 + r)$ as

$$\frac{r(u)}{1 + r(u)} = r(u)\,[1 - r(u) + \cdots]. \tag{2.6.25}$$

Thus, to the same order of approximation that strong lines black out a frequency interval completely (neglect of $1/\{r(u_0)\}$), weak lines may be neglected completely. The first-order contribution of a weak line follows from eqn. (2.6.25) and eqns. (2.6.13a, b)

$$\Lambda_l \cong \int_0^\infty r(u_0)\frac{W_R(u)}{\mu_c}\,du = \int_0^\infty \frac{\mu_l(u_0)}{\mu_c^2} W_R(u)\,du \tag{2.6.26}$$

If, as is usually the case, the variation of the line profile is quite rapid compared to variations in $W_R(u)$ and μ_c, the integration in eqn. (2.6.26) can be performed immediately. The result is

$$\Lambda_l = \frac{\mu_0}{\mu_c^2} W_R(u)\Big|_{u=u_0} \tag{2.6.27}$$

where again μ_0 is the integrated line absorption coefficient. From this result it is evident that the contributions of weak lines to Λ_R are *additive and independent of the line width.* The contribution of a given line depends only on its strength as defined in eqn. (2.6.22), in distinction to the strong-line case where just the opposite situation prevails. We can now state a consistent first-order approximate formula for Λ_R in the presence of

weak $(\mu_l/\mu_c \ll 1)$ and of strong, $(\mu_l/\mu_c \gg 1)$

non-overlapping rectangular lines. The formula is

$$\Lambda_R \cong \int_0^\infty \frac{W_R(u)\,du}{\mu_c(u,T)} - \sum_{i(s)} \frac{W_R(u_i)}{\mu_c(u_i,T)} \Delta u_i + \sum_{i(s)} \frac{W_R(u_i)\,\Delta u_i}{\mu_l^{sl}(u_i,T)} - \sum_{j(w)} \frac{\mu_0^{wl}(j)}{\mu_c^2(u_j,T)} W_R(u_j). \tag{2.6.28}$$

In this formula, the Δu_i are the (reduced) widths of the strong lines and the $\mu_0^{wl}(j) = (\pi e^2/mc)\, N_v f_j$ are the integrated weak-line absorption coefficients. The letters *wl* and *sl* have been used in appropriate places in the formula to distinguish weak lines and strong lines, respectively.

Although these formulas and their equivalents for more realistic line profiles are descriptive and qualitatively very useful in providing physical intuition, it has been found both practical and convenient in recent calculations of mean absorption coefficients (Stewart and Pyatt, 1961; Freeman, 1963; Armstrong *et al.*, 1967) to perform analytic or detailed numerical integrations over the line profiles with the aid of high-speed digital computers. If this is done, lines of intermediate strength can be accurately accounted for, together with overlapping lines.

Huebner (1964) has presented a valuable review of the subdivisions of the temperature, density and atomic number domain which itemizes the principal contributing effects to the opacity within these various regimes, and includes comments on some methods of calculation.

INEQUALITIES AND BOUNDS ON MEAN ABSORPTION COEFFICIENTS

In calculations of the Rosseland and Planck mean absorption coefficients, it is advantageous to have limits for the values to be expected for these quantities. This is particularly true when one includes considerations of the line contribution, since it has such a wildly fluctuating, complicated frequency dependence. Bernstein and Dyson (1959) have presented a theorem which places an upper limit on the Rosseland mean opacity. This theorem is based on the Schwartz inequality for two functions f and g, say, whose product and squares are integrable.

$$[\int fg\,dx]^2 \leq \int f^2\,dx \int g^2\,dx. \tag{2.6.29}$$

If f^2 is chosen to be the Rosseland mean integrand, and g^2 is chosen proportional to the absorption coefficient $\mu(\nu)$ then $\int fg\,dx$ is a definite integral independent of $\mu(\nu)$, and the integral† $\int \mu(\nu)\,d\nu$, which was introduced previously in eqn. (2.6.22), can be evaluated by means of the Thomas–Reiche–Kuhn f-number sum rule (Bethe and Salpeter, 1957). Their result may be given as (see Bond, Watson and Welch, 1965)

$$\bar{\varkappa}_R \leq \varkappa_0 \tag{2.6.30}$$

where $\varkappa_0$ is equal to

$$\frac{\pi e^2 h N_v Z}{10.9 mckT\varrho}$$

and has the numerical value

$$\varkappa_0 = \frac{Z}{A}\left(\frac{\text{Ryd}}{kT}\right) 4.43 \times 10^5\ \text{cm}^2/\text{g}. \tag{2.6.31}$$

In this relation Z is the atomic number, A is the atomic weight, and Ryd is 13.6 eV when kT is also expressed in eV. At low density, the upper limit of eqn. (2.6.30) is many orders of magnitude higher than the continuous opacity, which should constitute the dominant contribution (at sufficiently low density). Hence it would be convenient if an upper limit could be obtained closer to the true value even if, perhaps, more effort

† As used here, this integral must be taken over all frequencies, not just over a single-line profile as in eqn. (2.6.22).

might be needed in its calculation. An advance in this direction was made by Armstrong (1962) by applying the Schwartz inequality in a different manner, as follows. We recall the definition of the Planck mean absorption coefficient:

$$\bar{\mu}_P = \frac{15}{\pi^4} \int_0^\infty e^{-u} u^3 \mu(u) \, du \tag{2.6.32}$$

where, as above,

$$u = h\nu/kT.$$

We can apply Schwartz's inequality with

$$f^2 = \frac{15}{4\pi^4} \frac{u^4 e^{2u}}{\mu(u) (e^u - 1)^3} \tag{2.6.33}$$

and

$$g^2 = \frac{15}{\pi^4} u^3 e^{-u} \mu(u) \tag{2.6.34}$$

to derive

$$\frac{\bar{\mu}_P}{\bar{\mu}_R} \geq I^2, \tag{2.6.35}$$

where the definite integral I is given by

$$I = \frac{15}{2\pi^4} \int_0^\infty \frac{u^{7/2} e^{u/2} \, du}{(e^u - 1)^{3/2}}, \tag{2.6.36}$$

and has the value 0.9743 as obtained by numerical integration. With this value for I, eqn. (2.6.35) yields

$$\bar{\mu}_R \leq 1.053 \bar{\mu}_P. \tag{2.6.37}$$

The Planck mean $\bar{\mu}_P$ is much more tedious to compute than the Bernstein–Dyson $\varkappa_0$, but it generally gives a lower value and is a quantity of some interest in its own right. It does not depend on the detailed shape of the lines, in distinction to the Rosseland mean, so it it is still much easier to compute than this latter quantity. An empirical test of the relative sizes of these bounds is afforded by the absorption coefficients calculated by

Armstrong *et al.* (1967). Using their results for nitrogen, we have calculated the ratio of these two bounds, $\varkappa_0/1.05\bar{\varkappa}_P$, and the results are shown in Table 2.1 as a function of temperature and density. From this comparison, the inequality given by eqn. (2.6.37) is seen to be nearer an equality than the one given by eqn. (2.6.30), particularly towards low density and high temperature where the atoms tend to be stripped of most of their electrons. Physically, it is easy to understand the relationship between $\bar{\mu}_P$ and $\bar{\mu}_R$.

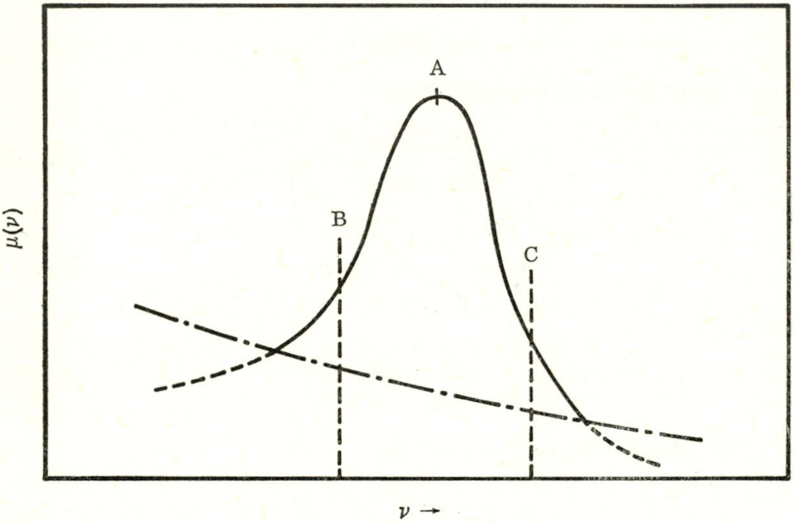

FIG. 2.9. Schematic illustration of spectrum line superimposed on continuum background. The region between *B* and *C* is the line core, and the regions to the left and right of *B* and *C*, respectively, are the line wings.

Figure 2.9 shows the general case of a line superimposed on the continuum. The Planck mean obtains its dominant contributions when μ_ν is large as at *A*. On the other hand, $\bar{\mu}_R$ obtains its dominant contributions from regions of *small* values of μ_ν as at *B* and *C*. We can realistically make the contribution of this line to $\bar{\mu}_R$ as small as we please by passing to physical conditions such that the line becomes progressively more narrow and peaked (low density, low temperature). The Planck mean is invariant to this distortion, depending only on the area under the curve—

TABLE 2.1. *A comparison of the Dyson upper bound, $\varkappa_0$, and the Planck mean upper bound, $1.053\,\bar{\varkappa}_P$, for the Rosseland mean absorption coefficients of nitrogen.* The number in parenthesis is the density in g/cm³ corresponding to the ratio above it. The superscript on the density is the power of 10 to which it should be raised.

kT	$\varkappa_0/1.053\bar{\varkappa}_P$					
2	8.11	15.1	18.1	19.0	21.1	24.5
	(5.44^{-3})	(7.34^{-5})	(7.71^{-6})	(8.96^{-7})	(9.07^{-8})	(9.22^{-9})
5	2.38	3.54	4.59	5.83	7.95	14.3
	(5.94^{-3})	(1.28^{-4})	(1.22^{-5})	(1.31^{-6})	(1.51^{-7})	(1.74^{-8})
10	2.28	5.50	13.0	55.0	3.56×10^2	2.58×10^3
	(9.55^{-3})	(2.05^{-4})	(2.03^{-5})	(2.28^{-6})	(2.28^{-7})	(3.77^{-8})
15	2.91	13.3	73.1	5.11×10^2	3.70×10^3	2.73×10^4
	(1.29^{-2})	(2.96^{-4})	(3.23^{-5})	(4.00^{-6})	(5.19^{-7})	(6.89^{-8})
20	4.46	38.2	2.58×10^2	1.87×10^3	1.37×10^4	1.01×10^5
	(1.64^{-2})	(4.17^{-4})	(4.82^{-5})	(6.08^{-6})	(7.95^{-7})	(1.06^{-7})

which is preserved under all but the most severe physical conditions.† At the opposite, but physically unrealistic, extreme, we could distort the line shape to be flat (i.e. constant in frequency) over a width w much greater than the span of the two weighting functions but still such that μw = constant. Then the weighting integrations could be performed, giving unity in both cases, and the two means would be equal. Perhaps a clearer explanation, however, resides in the fact that a harmonic mean of any function, say

$$\bar{\mu}_H \equiv 1\Big/\left(\frac{1}{\Delta x}\int_{\Delta x}\frac{dx}{\mu(x)}\right), \tag{2.6.38}$$

is always less than the corresponding direct mean

$$\bar{\mu}_D \equiv \frac{1}{\Delta x}\int_{\Delta x}\mu(x)\,dx. \tag{2.6.39}$$

This is easily seen by setting

$$f = \frac{1}{\sqrt{\{\mu(x)\}}} \quad \text{and} \quad g = \sqrt{\{\mu(x)\}}$$

† As long as the line width is small compared to kT, which is usually the case.

in the Schwartz inequality, eqn. (2.6.29), and integrating over the range Δx. This immediately yields

$$\bar{\mu}_H \leq \bar{\mu}_D. \tag{2.6.40}$$

Except for the factor 1.053 this is the same as eqn. (2.6.37). Hence, it appears that the weighting functions are somewhat superfluous. The equality occurs when both functions are constant over Δx.

Another approach to the problem of bounds has been taken by Liberman (1962) who wrote a general factorization formula

$$\int_0^\infty \frac{W'(u)\, du}{\mu'(u)} \int_0^\infty u^l \mu'(u)\, du \geq \left\{\int_0^\infty [W'(u)\, u^l]^{1/2}\, du\right\}^2 \tag{2.6.41}$$

where $\mu'(u) = \mu(u)\,(1 - e^{-u})$ as usual and

$$W'(u) = \frac{15}{4\pi^4} \frac{u^4 e^u}{(1 - e^u)^2}. \tag{2.6.42}$$

The integral on the right-hand side of eqn. (2.6.41) has the value for $l = 0$ of 3.302, and it is the square of this quantity that introduces the factor 10.9 into the definition of $\varkappa_0$ used in eqn. (2.6.30). For $l = 2$, its value is 19.112 which will be needed in eqn. (2.6.43). On the left-hand side of eqn. (2.6.41) the first integral is the reciprocal of $\bar{\mu}_R$, and the second integral may be evaluated by means of the f-number sum rules

$$\sum_\beta f_{\beta\alpha}\,(E_\beta - E_\alpha)^l = S_l(\alpha)$$

for $l = 2, 1, 0, -1$, and -2. The S_l are relatively simple functions of basic atomic parameters for a given state α. The "f-number" will be defined later in Chapter 3. (See also Bethe and Salpeter, 1957, p. 358.) The Bernstein–Dyson theorem corresponds to $l = 0$. It appears that the bound eqn. (2.6.41) yields for $l = 2$ is also superior to the Bernstein–Dyson result as was shown by Liberman by a computation for hydrogen compared with the Bernstein–Dyson continuum opacity values. Liberman's result for $l = 2$ may be witten as

$$\bar{\mu}_R \leq \bar{\mu}_2 \equiv \frac{1}{(19.112)^2} \int_0^\infty u^2 \mu'(u)\, du. \tag{2.6.43}$$

Armstrong (1965c) has noted that since $e^{-u}(1+u) \leq 1$, one can write

$$\int u^2 (1+u) e^{-u} \mu(u)\, du \leq \int u^2 \mu(u)\, du \tag{2.6.44}$$

from which it easily follows that

$$\int u^3 e^{-u} \mu(u)\, du \leq \int u^2 (1-e^{-u}) \mu(u)\, du. \tag{2.6.45}$$

Equations (2.6.32) and (2.6.43) may then be combined to give

$$\bar{\mu}_P \leq 56.3 \bar{\mu}_2, \tag{2.6.46}$$

which offers an opportunity to examine the relative utility of $\bar{\mu}_P$ and $\bar{\mu}_2$ as upper bounds for $\bar{\mu}_R$.

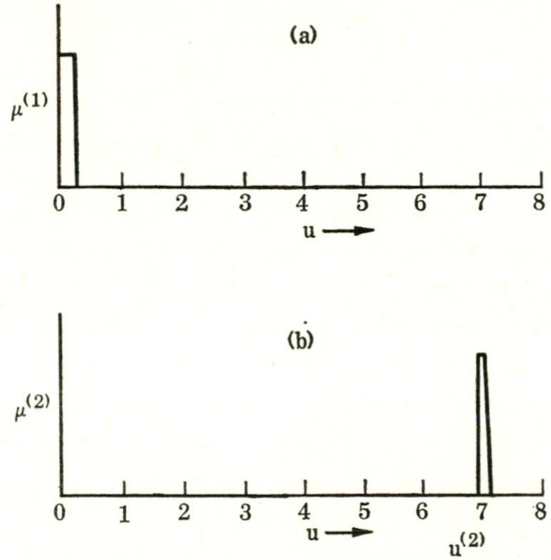

FIG. 2.10. Simplified model absorption coefficients which illustrate the dominant trend in (a) high-temperature absorption and (b) low-temperature absorption.

Let us consider the two idealized model absorption coefficients shown in Fig. 2.10. For the first model portrayed in Fig. 2.10 we understand that $\mu = 0$ unless $u \ll 1$. In this case we note that

$$u^2 (1 - e^{-u}) \cong u^3 \cong u^3 e^{-u},$$

and therefore, calling this first model $\mu^{(1)}(u)$, we can write

$$\int u^3 e^{-u} \mu^{(1)}(u)\, du \cong \int u^2 \mu'(u)\, du,$$

with the result that $\bar{\mu}_P^{(1)} \cong 56.3\bar{\mu}_2$.

For the second model, we understand that the absorption coefficient peaks at a value of $u^{(2)} \gg 1$. (In Fig. 2.10 this value is arbitrarily represented as $u^{(2)} = 7$.) We can now write that

$$\int u^3 e^{-u} \mu^{(2)}\, du \cong C \int u^2 \mu'\, du,$$

with

$$C \approx e^{-u^{(2)}} [1 + u^{(2)}] \ll 1.$$

This yields

$$\bar{\mu}_P^{(2)} \approx 56.3 C \bar{\mu}_2 .$$

For $u^{(2)} \sim 6$, $56.3C$ is equal to unity, and for $u^{(2)} \gg 6$, $56.3C \ll 1$.

Since $u = h\nu/kT$, these models are idealized representations of high-temperature $[\mu^{(1)}]$ and low-temperature $[\mu^{(2)}]$ absorption, with "high" and "low" defined by the relative size of kT and the range of typical ionization edges and absorption frequencies taken for $h\nu$. We can thus conclude qualitatively that for temperatures of the order of typical atomic ionization potentials and larger, $\bar{\mu}_2 < \bar{\mu}_P$ and thus $\bar{\mu}_2$ constitutes a better bound for $\bar{\mu}_R$. On the other hand, for temperatures below these ionization potentials, the inequality in eqn. (2.6.37) is expected to be more useful.

These conclusions are also indicated from the behavior of the functions that weight $\mu(\nu)$ in the defining integrals for $\bar{\mu}_P$ and $\bar{\mu}_2$. The function $u^3 e^{-u}$ cuts off the high-frequency absorption (compared with kT), whereas $u^2(1 - e^{-u})$ does not, but rather increases monotonically. Thus, at low temperatures where most of the absorption potential lies far ahead of kT in the form of lines and edges, $\bar{\mu}_P < \bar{\mu}_2$. However, at high temperatures kT has already passed most of the absorption potential (which lies toward $u = 0$ in the free–free continuum), and $\mu(\nu)$ is decreasing monotonically, with the result that $\bar{\mu}_2 < \bar{\mu}_P$.

Some confirmation of these predictions can be obtained by a comparison of results. Liberman has computed $\bar{\varkappa}_2 = \bar{\mu}_2/\varrho$ for hydrogen for a range of densities at $kT = 5$, 10 and 20 eV. Figures 2.11 (5 and 10 eV) and 2.12

(20 eV) show his results in comparison with the Stewart–Pyatt (1961) results for $\bar{\varkappa}_P$ and $\bar{\varkappa}_R$, which include line contributions. For $kT = 5$ eV, $\bar{\varkappa}_P$ is about 1.5 times $\bar{\varkappa}_2$ for the densities shown, and this ratio clearly increases with kT, while a rough calculation indicates that $\bar{\varkappa}_P$ becomes approximately equal to $\bar{\varkappa}_2$ at about 3–4 eV, and then presumably $\bar{\varkappa}_P < \bar{\varkappa}_2$

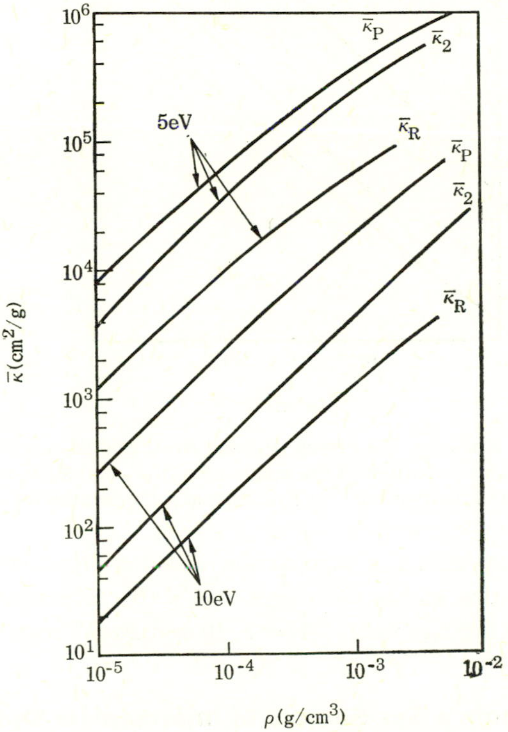

FIG. 2.11. Upper bounds on the Planck ($\bar{\varkappa}_P$) and Rosseland ($\bar{\varkappa}_R$) mean absorption coefficients: A comparison of the magnitudes of $\bar{\varkappa}_P$ and $\bar{\varkappa}_R$ as a function of density with Liberman's $\bar{\varkappa}_2$ for hydrogen at temperatures of 5 and 10 eV.

when $kT \gtrsim 3$ eV. Also, as kT increases, $\bar{\varkappa}_R$ approaches $\bar{\varkappa}_2$. The suspicious behavior of $\bar{\varkappa}_R$ vs. $\bar{\varkappa}_2$ at the low-density end of the 20 eV curve (where $\bar{\varkappa}_R$ appears to be on the verge of crossing $\bar{\varkappa}_2$) is probably due to the Compton scattering contribution included in the Stewart–Pyatt (1961) results, but

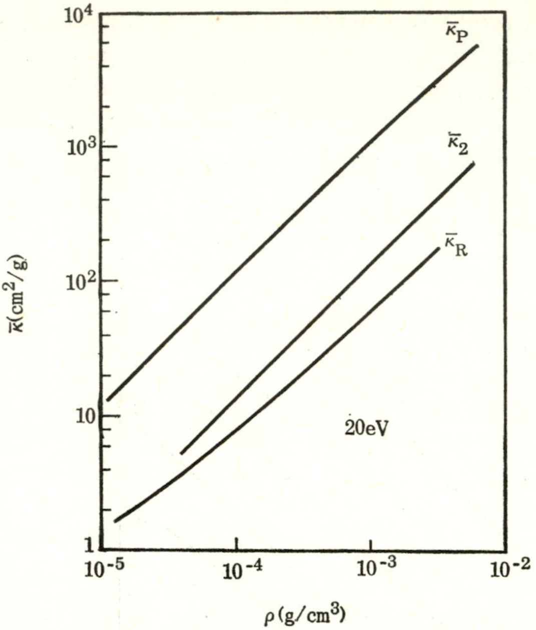

FIG. 2.12. Upper bounds on the Planck ($\bar{\varkappa}_P$) and Rosseland ($\bar{\varkappa}_R$) mean absorption coefficients: A comparison of the magnitudes of $\bar{\varkappa}_P$ and $\bar{\varkappa}_R$ as a function of density with Liberman's $\bar{\varkappa}_2$ for hydrogen at a temperature of 20 eV.

omitted in the sum rule. For comparison with the Dyson bound, eqn. (2.6.30), we note that $\varkappa_0$ has the values 1.21×10^6, 6.03×10^5 and 3.01×10^5, for these examples where $kT = 5$, 10 and 20 eV, respectively.

2.7. Emission from a Gas Sample of Intermediate Optical Depth: the Transmission Mean Absorption Coefficient

Radiative transfer in the realistic general case of intermediate, or finite, optical depth, where one cannot make the approximation that the radiating gas sample is optically thin or optically thick, is quite difficult to handle mathematically. Resort is usually made to numerical integrations of the transfer equation (Hillendahl, 1969) involving elaborate approximation techniques or to alternative methods as the Invariant Imbedding techniques of Ambartsumian and Chandrasekhar (Goody, 1964; Chandrasekhar, 1950). Mean absorption coefficients are no longer so useful in expressions for the net radiative flux, since the coefficients become dependent on path length and on the geometrical configuration of the sample. Intensities and fluxes are thus computed directly.

A formal simplicity can be given to certain of the intensity and flux expressions however, by defining mean absorption coefficients, which are of convenience in calculation. The mathematics for the general case of intermediate optical depth is sufficiently complex that the only situations for which closed analytic solutions exist are those for gas samples with extremely simple geometric configurations at constant density and temperature.† We will, therefore, consider explicitly only these conditions and configurations. Consideration of more general cases can be facilitated by means of "transmission functions" which we will also define and discuss briefly.

Consider an isothermal, plane-parallel slab of gas of uniform density, of thickness L, and of infinite extent perpendicular to the x-axis (Fig. 2.2). Now select an infinitesimal pencil of radiation through the slab at an angle θ to the x-axis. The outward intensity $I_{\nu+}$ along such a pencil can readily be obtained from first principles. (See, for example, Penner, 1959, pp. 13–15). We make use of the general solution, eqn. (2.1.10), with the incident intensity $I_\nu(s;\widehat{\Omega})$ set equal to zero, and the lower limit s' taken as zero. For $x \leq L$, since the density and temperature are assumed to be constant and there is azimuthal symmetry in Φ, we can write eqn. (2.1.10) or (2.1.11) as ($J_\nu = B_\nu(T)$):

$$I_\nu(L, \theta) = B_\nu(T) \int_0^{\mu_\nu' L/\cos\theta} e^{-\mu' s}\, d(\mu' s). \tag{2.7.1}$$

The integration can be performed immediately, leading to the result

$$I_\nu(L, \theta) = B_\nu(T)\,(1 - e^{-\mu_\nu' L/\cos\theta}). \tag{2.7.2}$$

In terms of the normal optical depth

$$\tau_\nu^n \equiv \int_0^x \mu_\nu'\, dx,$$

† Certain simple cases where the density or temperature is not uniform may be effectively reduced to uniform-case expressions by approximation. See, for example, Armstrong (1968a).

where x, as in Fig. 2.2, is measured normal to the slab, or in terms of our oiriginal path-length variable s, this result has the equivalent forms:

$$
\begin{aligned}
I_\nu(\tau_\nu^n, \theta) &= B_\nu(T)\,[1 - e^{-\tau_\nu{}^n/\cos\theta}],\\
I_\nu(s, \widehat{\Omega}) &= B_\nu(T)\,[1 - e^{-\mu_\nu{}'s}].
\end{aligned}
\tag{2.7.3}
$$

Penner (1959, p. 14) calls the last of these expressions the "basic law for uniformly distributed radiators". We note that for $s = R$ the last of these expressions is also appropriate for the intensity observed at the center of the base of a hemisphere of radius R.

If we multiply the intensity as given by eqn. (2.7.3) by $\cos\theta$ and integrate over solid angle in order to compute the flux from one face of the slab we find

$$F_\nu = 2\pi B_\nu(T) \int_0^{\pi/2} (1 - e^{-\mu_\nu{}'x/\cos\theta}) \cos\theta \sin\theta \, d\theta. \tag{2.7.4}$$

This is usually expressed in terms of the exponential integrals

$$E_n(y) \equiv \int_1^\infty \frac{e^{-yt}}{t^n}\, dt$$

previously introduced in Section 2.4, by use of the transformation $t = 1/\cos\theta$. With these substitutions eqn. (2.7.4) becomes [cf. eqn. (2.4.8)]

$$F_\nu = \pi B_\nu(T)\,[1 - 2E_3(\mu_\nu' x)]. \tag{2.7.5}$$

The total flux over all frequencies is obtained from this expression by integration over ν:

$$F = \sigma T^4 \left[1 - \frac{2\pi}{\sigma T^4} \int B_\nu(T)\, E_3(\mu_\nu' x)\, d\nu \right]. \tag{2.7.6}$$

This expression has been used with the assumption of a constant absorption coefficient to obtain a simple estimate of the effects of finite optical thickness (Davis, 1964). If μ_ν' is assumed to be a constant $\bar{\mu}'$ independent of ν, eqn. (2.7.6) becomes

$$\bar{F} = \sigma T^4\,[1 - 2E_3(\bar{\mu}' x)]. \tag{2.7.7a}$$

Computations of $\bar{F}/F$ by Strack (1962) using $\bar{\mu}' = \bar{\mu}_P$, the Planck mean, indicate that this is a useful approximation only under certain rather limited circumstances. For further calculations of $\bar{F}$ see Davis (1964). The correct value of $\bar{\mu}'$ that makes this expression exact, often called the "transmission mean", is

$$\bar{\mu}_T' \equiv -\frac{\cos\theta}{x} \ln\left[\int_0^\infty \frac{B_\nu(T)}{B(T)} e^{-\mu_\nu' x/\cos\theta} \, d\nu\right] \qquad (2.7.7b)$$

as can be readily verified by substitution in eqn. (2.7.6) [where $B(T)$ is defined as in eqn. (2.4.13)]. In the limit as $x \to 0$, this expression reduces to the Planck mean as given by eqn. (2.4.15).

The Transmission Function and Emissivities

Because of the frequent appearance of the exponential function $e^{-\mu_\nu s}$ involving the absorption coefficient and path length, it is convenient to define and name it as an independent quantity. The customary name† applied is intensity *transmission function* (Elsasser, 1942; Chandrasekhar, 1950) since it is the ratio of the diminished intensity to the original intensity under isothermal, homogeneous conditions, i.e.

$$e^{-\mu_\nu s} = \frac{I_\nu(s)}{I_\nu(0)} \equiv \mathrm{Tr}\,(\mu_\nu s) \qquad (2.7.8)$$

(cf. eqn. (2.1.2)) in the presence of pure absorption alone. Although, strictly speaking, one can have pure absorption alone only under non-equilibrium circumstances, under more general circumstances this exponential function constitutes an integrating factor for the transfer equation [cf. eqns. (2.1.9) and (2.1.10)] as long as induced emission can be neglected. From eqns. (2.2.11) and (2.2.12) one sees that this is possible as long as $h\nu/kT \gg 1$. This is generally true for optical or infrared radiation at atmospheric temperatures, so that the transmission function $e^{-\mu_\nu s}$ is customarily used in such low-temperature applications (cf., for instance, Goody, 1964). At higher temperatures one cannot generally neglect induced re-emission into the beam, so that eqn. (2.7.8) no longer constitutes the most

† A flux transmission function can also be defined and used. This will appear in Section 2.8.

useful definition of a function from which radiative intensities and fluxes can be more or less directly obtained. Under conditions of LTE, the modified absorption coefficient $\mu_\nu' = \mu_\nu (1 - e^{-h\nu/kT})$ replaces μ_ν so that $e^{-\mu_\nu s}$ is replaced by $e^{-\mu_\nu' s}$ as the integrating factor for the transfer equation. This latter factor is also called a transmission function even though, strictly speaking, it is not the simple ratio of two intensities.†

For gases in LTE, a direct connection of $e^{-\mu_\nu' s}$ with observation is afforded by the definition of emissivity for a non-thin slab. Let us define the spectral emissivity of an isothermal gas sample as the ratio of the flux it emits per unit frequency interval to the flux per unit frequency interval which a blackbody emits. Then from eqn. (2.7.3) we find for the spectral emissivity $\varepsilon_{\nu c}$ of an infinitesimal pencil of gas of length s the result

$$\varepsilon_{\nu c} \equiv \frac{I_\nu (s, \widehat{\Omega})}{B_\nu(T)} = 1 - e^{-\mu_\nu' s} \equiv 1 - \mathrm{Tr}\,(\mu_\nu' s). \qquad (2.7.9a)$$

For such an infinitesimal pencil the intensity will all be in the direction of the pencil; hence, the flux and intensity are the same. Approximately the same result would be obtained for a long, narrow cylinder or column of gas. Hence, the emissivity given by eqn. (2.7.9a) is also known as the *beam* or *column* (spectral) emissivity and has been given a subscript c. It can also be called the *hemispherical*‡ emissivity, since if we hold s constant, integrate $F_\nu = I_\nu \cos\theta$ and $B_\nu \cos\theta$ over a hemisphere and take the ratio, we obtain the emissivity that would be observed at the center of the base of a hemisphere as at 0 in Fig. 2.5. Of course, this "high-temperature" definition $e^{-\mu_\nu' s}$ can still be used at low temperatures as long at LTE prevails or $h\nu \gg kT$, whereas the reverse is not true, viz. $e^{-\mu_\nu s}$ is no longer appropriate at high temperatures. The definition of emissivity given by eqn. (2.7.9a) is very similar to the definition of *absorption function*

† The Rosseland mean absorption coefficient can be calculated from a set of values of this transmission function by use of the relation

$$\int_0^\infty \exp\,(-\mu_\nu' s)\, ds = \mu_\nu'^{-1}.$$

This has been pointed out by Stewart and Penner (Penner and Olfe, 1968, p. 190).

‡ All these names are somewhat confusing, but we mention them since they are in the literature. In view of the defining equation, eqn. (2.7.9a), it would probably be most straightforward to call this quantity the *intensity emissivity*.

$A(\mu_\nu s)$ often employed in low-temperature radiation studies. If there is negligible re-emission into the beam, the fraction of the original beam intensity $I_\nu(0)$ which is absorbed (under isothermal, homogeneous conditions) is given by

$$A(\mu_\nu s) = [I_\nu(0) - I_\nu(s)]/I_\nu(0) = 1 - \exp(-\mu_\nu s)$$

or

$$A(\mu_\nu s) = 1 - \mathrm{Tr}(\mu_\nu s). \tag{2.7.9b}$$

Although this definition is formally almost identical to that given in eqn. (2.7.9a), the difference in interpretation and validity should be noted. If there is appreciable re-emission into the beam, eqn. (2.7.9b) loses its direct significance.

To define a total emissivity, we need to take the ratio of the total (frequency-integrated) intensity to the integrated Planck function as in Section 2.4. Thus the total (isothermal) column emissivity can be written as

$$\varepsilon_c = \frac{\int d\nu\,(1 - e^{-\mu_\nu' s})\,B_\nu(T)}{\int B_\nu(T)\,d\nu} = \frac{\int d\nu\,[1 - \mathrm{Tr}(\mu_\nu' s)]\,B_\nu(T)}{B(T)}. \tag{2.7.10}$$

Experimentally, one would expect to observe intensities that are effectively averages over small frequency intervals, such as line or band emissivities. The frequency intervals of concern here for defining and computing such emissivities are those which contain from a fraction of a spectrum line to perhaps a few lines. Under most conditions the Planck functions will normally not vary substantially over such an interval and the average of the intensity over $\Delta\nu_i$ can be written as

$$\bar{I}_{\nu_i}^{\Delta\nu} = \bar{\varepsilon}_{ci} B_{\nu_i}(T) = [1 - \overline{\mathrm{Tr}_i(\mu_\nu' s)}]\,B_{\nu_i}(T) \tag{2.7.11}$$

where

$$\bar{\varepsilon}_{ci} = 1 - \overline{\mathrm{Tr}_i(\mu_s' s)} \tag{2.7.12a}$$

and

$$\overline{Tr_i(\mu_\nu' s)} = \frac{1}{\Delta\nu_i}\int_{\Delta\nu_i} e^{-\mu_\nu' s}\,d\nu \tag{2.7.12b}$$

with the result that the total column emissivity becomes (cf. Goody, 1964), if we choose equal intervals $\Delta\nu_i$,

$$\varepsilon_c = \frac{\sum_i \bar{\varepsilon}_{ci} B_i(T)}{\sum B_i(T)} \tag{2.7.13}$$

where i rather than ν_i is used to label the frequency intervals and

$$B_i(T) = \int_{\Delta\nu_i} \frac{B_\nu(T)\, d\nu}{\Delta\nu_i}.$$

The column (or intensity) emissivity can be converted to a slab (or flux) emissivity ε_S by integration of I_ν and B_ν over a hemisphere of outward directions as in the preceding eqn. (2.7.4). Thus, we can write

$$\varepsilon_S = \frac{\int_{2\pi} d\Omega \int d\nu\, (1 - e^{-\mu_\nu' x/\cos\theta})\, B_\nu(T) \cos\theta}{\pi \int B_\nu(T)\, d\nu}$$

$$= 1 - \frac{2\pi}{\sigma T^4} \int d\nu B_\nu(T)\, [E_3(\mu_\nu' x)] \tag{2.7.14}$$

which is, of course, just the ratio $F/\sigma T^4$ as given by eqn. (2.7.6).

Although, as mentioned previously, mean absorption coefficients are not as significant for finite optical depth as in the limiting cases, it is still instructive to define them. This amounts merely to a re-expression of the formulas for flux and intensity that one obtains directly from the transfer equation, but it provides a certain amount of conceptual continuity between the limiting cases. Also, because of the rapid, violently fluctuating frequency dependence of the spectral absorption, it is often essential to define mean absorption coefficients to make a given calculation tractable, as well as experimentally meaningful.

The frequency integrated intensity of eqn. (2.7.3) which is

$$I(\theta, x) = \int (1 - e^{-\mu'_\nu x/\cos\theta})\, B_\nu(T)\, d\nu \tag{2.7.15}$$

can be expressed in the same form as the spectral intensity by the proper definition of a mean absorption coefficient which we will call $\bar{\mu}_T$, the transmission mean. Thus, we can write

$$I(\theta, x) = (1 - e^{-\bar{\mu}_T x/\cos\theta})\, B(T) \tag{2.7.16}$$

if we define $\bar{\mu}_T$ as in eqn. (2.7.7a) (Mayer, 1964)

$$\bar{\mu}_T(x, \theta) = -\frac{\cos\theta}{x} \ln \int_0^\infty \frac{B_\nu(T)}{B(T)}\, e^{-\mu_\nu' x/\cos\theta}\, d\nu \tag{2.7.17}$$

$$= -\frac{\cos\theta}{x} \ln \int \frac{B_\nu(T)}{B(T)}\, \mathrm{Tr}\,(\mu_\nu' x/\cos\theta)\, d\nu. \tag{2.7.18}$$

Analogously, the (frequency) average transmission function $\overline{\mathrm{Tr}\,(\mu_\nu' s)}$ of eqn. (2.7.12b) can be re-expressed as an average absorption coefficient by means of the definition

$$\bar{\mu}(s) = -\frac{1}{s} \ln \{\overline{\mathrm{Tr}\,(\mu_\nu' s)}\}. \tag{2.7.19}$$

The total emissivities can then be expressed simply in terms of $\bar{\mu}_T$, which, in turn, can then be expressed in terms of the $\bar{\mu}$ of eqn. (2.7.19). The hemispherical, or column (intensity) emissivity becomes

$$\varepsilon_c = \frac{I\,(s, \widehat{\Omega})}{B(T)} = 1 - e^{-\bar{\mu}_T s} \tag{2.7.20}$$

while the slab ("flux") emissivity takes the form

$$\varepsilon_s = \frac{\int I\,(\theta, x) \cos\theta \, d\Omega}{\pi B\,(T)} = \frac{1}{\pi} \int (1 - e^{-\bar{\mu}_T x/\cos\theta}) \cos\theta \, d\Omega. \tag{2.7.21}$$

Although this is an exact expression, it is not of great practical value because of the labor involved in obtaining $\bar{\mu}_T\,(x, \theta)$.

The connection between $\bar{\mu}(s)$ and $\bar{\mu}_T\,(x, \theta)$ can be seen by considering averages over limited spectral regions across which B_ν does not vary appreciably [viz. regions over which the transmission function may be averaged as in eqn. (2.7.11)]. For such a region $\Delta\nu$, which we label with the index i, we have:

$$\begin{aligned} \bar{\mu}_T\,(x, \theta) &= -\frac{\cos\theta}{x} \ln \left[\sum_i \frac{B_i}{B(T)} \int_{\Delta\nu_i} e^{-\mu_\nu' x/\cos\theta} \, d\nu \right] \\ &= -\frac{\cos\theta}{x} \ln \left[\sum_i \frac{B_i\,\Delta\nu_i}{B(T)} \, [\overline{\mathrm{Tr}_i\,(\mu_\nu' x/\cos\theta)}] \right] \\ &= -\frac{\cos\theta}{x} \ln \left[\sum_i \frac{B_i\,\Delta\nu_i}{B(T)} \, e^{-\bar{\mu}(x/\cos\theta)_i x/\cos\theta} \right]. \end{aligned} \tag{2.7.22}$$

Rearranging and taking the exponential of both sides yields

$$e^{-\bar{\mu}_T(x,\theta)x/\cos\theta} = \sum_i \frac{B_i\,\Delta\nu_i}{B(T)} \, e^{-\bar{\mu}'(x,\theta)_i (x/\cos\theta)}. \tag{2.7.23}$$

For the hemispherical or column emissivity, one obtains from eqn. (2.7.20) and eqn. (2.7.23)

$$\varepsilon_c = 1 - \frac{1}{B(T)} \sum_i B_i \, \Delta\nu_i \, e^{-\bar{\mu}'(s)_i s}. \tag{2.7.24}$$

Equations (2.7.16)–(2.7.24) are primarily of computational interest. In this respect they are not trivial since they [together with eqns. (2.7.11) and (2.7.12) for the transmission function average] assist in determining the type of spectral average one should use in order to avoid specifying any more spectral detail than is essential for a given calculation. They provide criteria for ascertaining which spectral features of the absorption coefficient are most important under a given set of circumstances. Thus, we can see that the spectral average needed for a geometry which is neither optically thin nor optically thick is that of the absorption coefficient in the *exponential* $e^{-\mu_\nu' s}$. Spectral regions within $\Delta\nu$ for which $\mu_\nu' s \gg 1$ will contribute zeros to the average

$$\frac{1}{\Delta\nu} \int e^{-\mu_\nu' s} \, d\nu$$

and to an emissivity such as given by eqn. (2.7.10) or eqn. (2.7.20). Thus, one only needs to know accurately the width of such regions. Much less accuracy is needed for the strength of μ_ν since one only has to ascertain that $\mu_\nu' s \gg 1$. On the other hand, for very weak regions where $\mu_\nu' s \ll 1$ the integrand $e^{-\mu_\nu' s}$ is near unity. In this case one can expand the exponential and the contribution to an emissivity such as eqn. (2.7.24) is proportional to $\mu_\nu' s$; therefore, the strengths of these regions are of greatest importance and their widths are relatively unimportant. The relative importance of the widths and strengths of small spectral regions, particularly lines, is analogous but less distinct here than in the calculation of Rosseland and Planck mean absorption coefficients as was discussed in Section 2.6. The geometric parameter or length s does not enter the Planck and Rosseland mean so that the relative importance of different spectral features is fixed once and for all by the basic assumptions. For the intermediate optical depths now being considered, the appearance of the parameter s provides another variable which affects the importance of a given spectral feature in the absorption coefficient as a function of optical depth. Similar considerations are applicable to the gross contributions

$$\frac{B_i\, e^{-\mu_i(x)x}}{B}$$

to the emissivity as given by eqn. (2.7.24) for different i. The only difference is that the gross contributions given by eqn. (2.7.24) have the Planck function B_i as an additional weighting factor compared to the fine contributions (for which B_ν is constant) which make up

$$\frac{1}{\Delta\nu}\int e^{-\mu_\nu' s}\, d\nu\, .$$

Thus, if one is calculating an integrated flux or total emissivity away from the Planck function maximum, where B_i is small, much less accuracy is required in the specification of μ_ν than for regions near the Planck maximum.

In considering lines superimposed upon a background continuum, it again is of interest to inquire into the degree of separability that exists between the two types of emission along the lines suggested by Mayer (1947) for the Rosseland mean as discussed in Section 2.6. What we can show is that the total intensity radiated by an atmosphere equals the intensity of the continuum emission, plus the intensity of line emission diminished by the factor $\exp(-\mu_c' s)$, where μ_c is the continuum absorption coefficient which is assumed to be essentially constant over some frequency interval $\Delta\nu$. The intensity $I_\nu(s, \widehat{\boldsymbol{\Omega}})$ in the direction $\widehat{\boldsymbol{\Omega}}$ due to an emitting column of gas of length s is given by

$$I_\nu(s, \widehat{\boldsymbol{\Omega}}) = B_\nu(T)\,(1 - e^{-(\mu_c' + \mu_l')s}) \tag{2.7.25}$$

where μ_c' and μ_l' are the continuum and line contributions to the absorption coefficient, respectively. Integrating eqn. (2.7.25) over the interval $\Delta\nu$ we get

$$I_{\Delta\nu}(s, \widehat{\boldsymbol{\Omega}}) = \int_{\Delta\nu} d\nu\, B_\nu(T) - e^{-\mu_c' s}\int_{\Delta\nu} B_\nu(T)\, e^{-\mu_l' s}\, d\nu \tag{2.7.26}$$

where we have assumed μ_c' is constant over $\Delta\nu$.

The intensity due to lines alone satisfies the equation

$$\int_{\Delta\nu} d\nu\, B_\nu(T)\, e^{-\mu_l' s} = \int_{\Delta\nu} d\nu\, B_\nu(T) - I^l_{\Delta\nu}.$$

Substituting this equation into eqn. (2.7.26), we get

$$I_{\Delta\nu} = I^c_{\Delta\nu} + I^l_{\Delta\nu}\, e^{-\mu_c' s} \tag{2.7.27}$$

where the continuum intensity $I^c_{\Delta\nu}$ is given by

$$I^c_{\Delta\nu}(s, \widehat{\Omega}) = \int_{\Delta\nu} d\nu\, B_\nu(T)\,(1 - e^{-\mu_c' s}).$$

Note that eqn. (2.7.27) holds regardless of whether the lines overlap or not.

The importance of eqn. (2.7.27) is that it gives us a ready idea as to the relative importance of continuum versus line effects. Thus if, for example,

$$\mu_c' s > 0.1$$

the continuum spectrum will cause an apparent reduction in the line contribution, and in regions where

$$\mu_c' s \gg 1$$

the lines may be disregarded entirely for most practical purposes.

Equation (2.7.27) may be put into an equivalent form in terms of partial (intensity) emissivities:

$$\varepsilon_{\Delta\nu} = \frac{I^c_{\Delta\nu}}{\int_{\Delta\nu} d\nu\, B_\nu(T)} + \frac{I^l_{\Delta\nu}\, e^{-\mu_c' s}}{\int_{\Delta\nu} d\nu\, B_\nu(T)} = 1 - (1 - \varepsilon^l_{\Delta\nu})\, e^{-\mu_c' s} \tag{2.7.28}$$

or

$$\varepsilon_{\Delta\nu} = \varepsilon^l_{\Delta\nu} + \varepsilon^c_{\Delta\nu} - \varepsilon^l_{\Delta\nu}\, \varepsilon^c_{\Delta\nu} \tag{2.7.29}$$

where the line and continuum partial emissivities are given, respectively, by

$$\varepsilon^l_{\Delta\nu} = \frac{I^l_{\Delta\nu}}{\int_{\Delta\nu} B_\nu(T)\, d\nu}$$

and

$$\varepsilon_{\Delta\nu}^{c} = \frac{I_{\Delta\nu}^{c}}{\int_{\Delta\nu} B_\nu(T)\,d\nu} = 1 - e^{-\mu_c' s}.$$

Sampson (1965a) has suggested an approximate method of handling the case of intermediate optical depth by use, in the transfer equation, of a mean absorption coefficient that interpolates between the Planck mean absorption coefficient in the limit of small optical depth and the Rosseland mean coefficient in the limit of large optical depth. His mean absorption coefficient is defined as

$$\varkappa(x) = \left(\frac{b}{b + \tau_P(x)}\right)\left(\frac{b + \tau_R(x)}{b + \tau_P(x)}\right)\bar{\varkappa}_P(x) + \frac{\tau_P(x)}{b + \tau_P(x)}\,\bar{\varkappa}_R(x)$$

as a function of the spatial variable x based on the expression

$$\tau(x) = \frac{b + \tau_R(x)}{b + \tau_P(x)}\,\tau_P(x)$$

for the optical depth. In these expressions, b is an *ad hoc* parameter of order unity in value, τ_R and τ_P are computed from eqn. (2.1.8a) by use of $\bar{\varkappa}_R$ and $\bar{\varkappa}_P$, respectively, for the absorption coefficient and these latter quantities are the Rosseland and Planck mean absorption coefficients defined and discussed in the preceding three sections of this chapter. Sampson compared results for flux and flux divergence calculations from these approximations with exact calculations over a range of dimensionally scaled variables for free–free absorption. In his example, the temperature was given a linear dependence on distance, and it is this temperature variation that introduces the spatial dependence upon x in the above formulas. His results for the approximation were generally within a factor 2 of the exact result over the range of his calculations. Although undoubtedly of value for some types of estimates, this method is limited by the lack of a fundamental measure of its accuracy, so we do not pursue it further herein.

Emissivities for General Geometries

We can also compute the emission from some other simple geometries under isothermal, isobaric conditions. First, let us consider a sphere.

From a given point on the surface of the sphere, radiation is emitted along the paths of all chords that can be drawn through the given point to other points on the sphere. Let the length of one of these chords that intersects at an angle θ to the normal to the sphere at the point of emission be $L(\theta)$. Then by our previous results, the intensity emitted along this path is $I_\nu = B_\nu(T)\,(1 - e^{-\mu_\nu L(\theta)})$. Furthermore, the length of the chord for a sphere of radius a is $L(\theta) = 2a \sin \varphi/2$, where φ is the angle subtended by the chord at the center of the sphere. From straightforward geometry, we determine that $\theta = \pi/2 - \varphi/2$. Hence, the flux at a given point on the surface of the sphere,

$$F_\nu = \int I_\nu \cos \theta \, d\Omega,$$

since there is azimuthal symmetry about the normal to the sphere, is given by

$$F_\nu = 2\pi B_\nu(T) \int_0^{\pi/2} [1 - e^{-\mu_\nu a \cos \varphi}] \sin \varphi \cos \varphi \, d\varphi$$

which can readily be evaluated to yield

$$= \frac{2\pi B_\nu(T)}{(2\mu_\nu a)^2} [2\,(\mu_\nu a)^2 - 1 + e^{-2\mu_\nu a}(1 + 2\mu_\nu a)]. \qquad (2.7.30)$$

We can easily pass to the optically thin limit $\mu_\nu a \ll 1$ from this formula, with the result

$$F_\nu \,(\text{optically thin}) = \frac{4\pi}{3} \mu_\nu a B_\nu(T).$$

We can also integrate this over frequency to express it in terms of an optically thin total emissivity $\varepsilon = \int F_\nu \, d\nu/\sigma T^4$, which becomes

$$\varepsilon = \frac{4a}{3} \bar{\mu}_P \qquad (2.7.31)$$

where $\bar{\mu}_P$ is the Planck mean absorption coefficient. For other geometries, the calculation becomes generally much more difficult. Johnston and Platas (1969) have noted that some results derived from neutron diffusion theory can be used in this context. Case, de Hoffman and Placzek (1953) define a neutron *angular source density* $q\,(\mathbf{r}, \hat{\boldsymbol{\Omega}}, t)$ which represents the number of particles of direction $\hat{\boldsymbol{\Omega}}$ produced per unit volume, unit time and unit solid angle. The *source density* $q(\boldsymbol{r}, t)$ is the integral of the angular source density over all directions. These authors then define the *average escape probability* P_0 for the steady state as

$$P_0 = \frac{\int_V q_0(\boldsymbol{r})\, P(\boldsymbol{r})\, dV}{\int_V q_0(\boldsymbol{r})\, dV}.$$

In this expression, the time variable t has been suppressed because of the assumption of a steady state, and the subscript zero has been appended to $q(\boldsymbol{r})$ to indicate the assumption of an isotropic emission of particles. The factor $P(\boldsymbol{r})$ is the probability that a particle emitted at $\boldsymbol{r}$ escapes from the emitting sample under consideration, and V indicates the volume of the sample over which the integration is to be performed. For simplicity, we limit our considerations to convex bodies. Now, under conditions of LTE, the radiation analog to $q_0(\boldsymbol{r})$ is the quantity $4\pi\varrho j_\nu^{(s)}/h\nu$, where $j_\nu^{(s)}$ is the spontaneous emission coefficient defined earlier. In order to make this analogy, we will require that the induced emission be incorporated into the absorption as is customary. All this means is that we treat the first term on the right-hand side of eqn. (2.2.15a) as the total emission and the second term as the total absorption. Therefore, for our radiation problems, we can define the average *photon* escape probability $P_0^{(\nu)}$ by

$$P_0^{(\nu)} = \frac{4\pi \int_V [\varrho j_\nu^{(s)} P^{(\nu)}(\boldsymbol{r})/h\nu]\, dV}{4\pi \int_V [\varrho j_\nu^{(s)}/h\nu]\, dV}$$

where $P^{(\nu)}(\boldsymbol{r})$ is the point-wise probability that a photon emitted within the sample volume at the point $\boldsymbol{r}$ escapes from this volume (i.e. it is not absorbed). We now observe that $h\nu$ times the numerator in the above equation is the energy emitted per unit time and frequency interval by the sample. That is to say, it is $\bar{F}_\nu S$, where S is the surface area of the sample and $\bar{F}_\nu$ is the emitted flux averaged over the surface area. Therefore, we can write the definition of $P_0^{(\nu)}$ as

$$P_0^{(\nu)} = \frac{\bar{F}_\nu S}{4\pi \int_V \mu'_\nu B_\nu\,(T)\, dV}$$

where we ave substituted the LTE value of $j_\nu^{(s)}$ in the denominator. Since we are only concerned here with isothermal, isobaric conditions, the volume integration can be performed immediately so that we can write this result as

$$\bar{F}_\nu = \pi P_0^{(\nu)}\left(\frac{4V}{S}\right)\mu'_\nu B_\nu(T). \tag{2.7.32a}$$

or

$$\bar{F} = \pi\left(\frac{4V}{S}\right)\int P_0^{(\nu)}\mu'_\nu B_\nu(T)\,d\nu. \tag{2.7.32b}$$

Since $\pi B_\nu(T)$ is the energy emitted per unit time and frequency interval by a perfect radiator, we can use this result to define a spectral emissivity ε_ν and a total emissivity ε (averaged over the surface of the emitting sample):

$$\varepsilon_\nu = \frac{\bar{F}_\nu}{\pi B_\nu} = P_0^{(\nu)}\left(\frac{4V}{S}\right)\mu'_\nu$$

$$\varepsilon = \frac{\bar{F}}{\pi B} = \left(\frac{4V}{S}\right)\frac{\int P_0^{(\nu)}\mu'_\nu B_\nu(T)\,d\nu}{\int B_\nu(T)\,d\nu}. \tag{2.7.33}$$

These simple results are quite informative. The quantity $4V/S$ is equal to the mean chord length over the geomtery of the emitting sample (Case, de Hoffman and Placzek, 1953; Weinberg and Wigner, 1958) a relation that dates back to Cauchy in the nineteenth century. The escape probability method effectively breaks down our transfer problem into separate emission ($\mu'_\nu B_\nu(T)$) and absorption ($P_0^{(\nu)}$) factors, averaged separately over the entire body. This factorization is very convenient mathematically for a number of problems. By definition, the optically thin limit corresponds to $P_0^{(\nu)} \cong 1$. Hence, if we pass to this limit for the total emissivity above, we can perform the frequency integration:

$$\varepsilon\,(\text{optically thin}) = \frac{4V}{S}\,\bar{\mu}_P \tag{2.7.34}$$

where $\bar{\mu}_P$ is again the Planck mean absorption coefficient. This result shows that the geometric numerical factor that relates the Planck mean, in the thin limit, to the emissivity of a general geometric sample shape is the mean chord length. For a slab, this mean chord is $2L$, where L is the thickness of the slab. This value, of course, makes eqn. (2.7.34) agree with eqn. (2.4.16). Similarly, the mean chord length for a sphere, $(4/3)\,a$, produces agreement with eqn. (2.7.31). For an infinite cylinder, a geometry which we have not heretofore considered, $4V/S = 2a$, where a is the radius of the cylinder. Because of the simplicity of the formula for the mean chord length, values for it can be obtained immediately for a large variety of regular shapes from the surface and volume formulas in most mathematical handbooks.

Let us now return to the problem of intermediate optical depths and the calculation of the average escape probabilities $P_0^{(\nu)}$. Case, de Hoffman and Placzek (1953) discuss the calculation of these quantities by the method of chord distributions due to Dirac (see also Weinberg and Wigner, 1958). Escape probabilities for a variety of shapes including the hemisphere, oblate spheroid, infinite cylinder, and oblate hemispheroid have been calculated by this method. Briefly, it proceeds as follows. Let $\hat{\boldsymbol{\Omega}}$ be a direction from a surface element $d\boldsymbol{S}$ of a convex body for which the chord length is L' and $d\Omega$ is a solid angle element about $\hat{\boldsymbol{\Omega}}$ such that all chords in this element of solid angle have length between L' and $L' + dL'$. Then the number of chords in this solid angle element having lengths in this range is $\hat{\boldsymbol{\Omega}} \cdot d\boldsymbol{S}\, d\Omega$. By summing over all directions $\hat{\boldsymbol{\Omega}}$ and all elements of surface $d\boldsymbol{S}$ having chords of length L' one obtains the total number of chords of length L'. We reduce this number to a probability $\varphi(L')\, dL'$ that a chord has length between L' and $L + dL'$ by dividing the sum for length L' by the sum over all chords. Hence, we obtain for this probability

$$\varphi(L')\, dL' = \frac{\int_{S(L')} \int_{\hat{\Omega}(L')} \hat{\boldsymbol{\Omega}} \cdot d\boldsymbol{S}\, d\Omega}{\int \int \hat{\boldsymbol{\Omega}} \cdot d\boldsymbol{S}\, d\Omega}.$$

The denominator can be integrated immediately:

$$\int \int \hat{\boldsymbol{\Omega}} \cdot d\boldsymbol{S}\, d\Omega = 2\pi S \int \cos\theta\, d\,[\cos\theta] = \pi S$$

where S is the total surface area and θ is the angle between $d\boldsymbol{S}$ and $\hat{\boldsymbol{\Omega}}$. By definition, the average chord length is given by

$$L\,(\text{average}) = \int L'\, \varphi(L')\, dL' = \frac{\int L' \hat{\boldsymbol{\Omega}} \cdot d\boldsymbol{S}\, d\Omega}{\pi S}.$$

But $\int L'\widehat{\boldsymbol{\Omega}} \cdot d\boldsymbol{S} = V$, the volume of the body; therefore, the above formula yields the result L (average) $= 4V/S$, as we stated earlier. Case, de Hoffman and Placzek also show that the average escape probability is related to this chord distribution function by

$$P_0^{(\nu)} = \left(\frac{S}{4V}\right) \mu_\nu'^{-1} \int (1 - e^{-L'\mu_\nu'})\, \varphi(L')\, dL'$$

where we have substituted $\mu_\nu'^{-1}$ for the (neutron) mean free path used by these authors. They discuss chord distributions for the various shapes mentioned above. For example, for a slab of thickness L, they show that

$$\varphi(L') = \frac{2L^2}{(L')^3}$$

and for a sphere of radius a,

$$\varphi(L') = \frac{L'}{2a^2}.$$

For the infinite cylinder and other geometries, the result is considerably more complicated. The appropriate chord distributions, when inserted into the above formula for the average escape probability yield the following results for a plane slab, a sphere, and an infinite cylinder.

(plane slab of thickness L, $\tau \equiv \mu_\nu' L$) $P_0^{(\nu)} = \tau^{-1}\,[\tfrac{1}{2} - E_3(\tau)]$

(sphere of radius a, $\tau \equiv \mu_\nu' a$) $P_0^{(\nu)} = \dfrac{3}{8\tau^3}\,[2\tau^2 - 1 + (1 + 2\tau)\, e^{-2\tau}]$

(infinitive cylinder of radius a, $\tau \equiv \mu_\nu' a$) (2.7.35)

$$P_0^{(\nu)} = \frac{\tau}{3}\{2\,[\tau K_1(\tau)\, I_1(\tau) + K_0(\tau)\, I_0(\tau) - 1] + \tau^{-1} K_1(\tau)\, I_1(\tau) - K_0(\tau)\, I_1(\tau) + K_1(\tau)\, I_0(\tau)\}.$$

In the first of these expressions, E_3 is the third exponential integral of eqn. (2.4.9), and in the last, K_n and I_n are modified Bessel functions (Abramowitz and Stegun, 1964). Case, de Hoffman and Placzek also give expansions of these formulas for small and large τ. Since such expansions are readily available for the exponential integral, we cite their results only for the sphere and cylinder formulas above.

$$\text{Sphere,}\quad \tau \ll 1 \quad P_0^{(\nu)} \cong 1 - \tfrac{3}{4}\tau,$$

$$\tau \gg 1 \quad \cong \frac{3}{4\tau}\left[1 - \frac{1}{2\tau^2}\right].$$

$$\text{Cylinder,}\quad \tau \ll 1 \quad P_0^{(\nu)} \cong 1 - \frac{4}{3}\tau + \frac{\tau^2}{2}\ln\frac{2}{\tau} + \frac{1}{2}\tau^2\left(\frac{5}{4} - \gamma\right)$$

$$\tau \gg 1 \quad \cong \frac{1}{2\tau} - \frac{3}{32\tau^3}$$

where γ = Euler's constant [eqn. (2.4.10)], which has the value 0.577216...

The Equivalent Width

Spectrum lines may constitute a large fraction of the detail which must be accounted for in a radiation transport or absorption coefficient calculation. Since they often exhibit common features and regularities which can be used to simplify their treatment, it is worthwhile to consider briefly the transport of radiation within the profile of an individual spectrum line. Furthermore, since lines, as opposed to more continuous spectral features (such as ionization and dissociation), occur in virtually infinite number and with a wide variety of shapes and features, it is preferable to group such lines whenever possible into classes which can be treated as units. For this reason, we will also touch upon the subject of band models. By this is meant a relatively simple analytic or stochastic representation of a large number of actual lines. Goody (1964, chap. 4) gives an extensive discussion of band models.

Before considering band models *per se*, we give a brief exposition of absorption or emission for a single line designated by the index α. The absorption coefficient for a line be written as

$$\mu'_{\nu\alpha} = \varrho S_\alpha b_\alpha(\nu - \nu_0) \tag{2.7.36}$$

where S_α is the "integrated line strength", $b_\alpha(\nu - \nu_0)$ is a shape factor as introduced in connection with eqn. (2.2.4a) and in eqn. (2.6.20a), and ϱ is the mass density. A similar quantity to S has been given in eqn. (2.6.22); different definitions of the integrated line strength often contain different density factors. The line shape factor is to be normalized to unity, as in eqn. (2.6.20b), over the profile of a line, so that the frequency integral of the absorption coefficient is equal to ϱS_α. We can apply the definition of the absorption function $A(\mu_\nu s)$ or the column emissivity $\varepsilon_{\nu c}$ to a single line using eqns. (2.7.9a, b). These definitions for the line α, become

$$\varepsilon_{\nu c\alpha}(s) = 1 - \mathrm{Tr}_\alpha(\mu'_\nu s), \tag{2.7.37}$$

$$A_\alpha(\mu_\nu s) \doteq 1 - \mathrm{Tr}_\alpha(\mu'_\nu s) = \varepsilon_{\nu c\alpha}(s). \tag{2.7.38}$$

The approximate equality symbol $\doteq$ has been used for the absorption function to remind us that this quantity represents an observable net attenuation only approximately, as was discussed in connection with eqns. (2.7.9). The approximation is, of course, very good at low temperatures where the re-emission into the incident beam is negligible.

Another useful definition is that of *average absorption* $\bar{A}$, of a single line, which, for the line α that is a member of a group of lines, is given by

$$\bar{A}_\alpha = \frac{1}{d}\int_{-\infty}^{+\infty} \{1 - \exp(-\mu_{\nu\alpha} s)\}\, d\nu, \tag{2.7.39}$$

where d is the average spacing between lines. This is related to the so-called *equivalent width* of a line α, $W_\alpha(s)$, by

$$W_\alpha(s) \doteq \bar{A}_\alpha d \tag{2.7.40}$$

where the basic definition of $W_\alpha(s)$ is given by

$$W_\alpha(s) \equiv \int_{-\infty}^{+\infty} [1 - \exp(-\mu'_{\nu\alpha} s)]\, d\nu \doteq \int_{-\infty}^{+\infty} A_\alpha(\mu_\nu s)\, d\nu. \tag{2.7.41}$$

By use of eqn. (2.7.37), the equivalent width is seen to satisfy the exact relation

$$W_\alpha(s) = \int_{-\infty}^{+\infty} \varepsilon_{\nu c\alpha}(s)\, d\nu. \tag{2.7.42}$$

The origin of the name "equivalent width" lies in the fact that $W_\alpha(s)$ is the integrated absorption of the line, or equivalently, the width of a totally absorbing line ($\mu_{\nu\alpha} = \infty$) having the same integrated absorption function, viz. the same value of $\int A_\alpha (\mu_\nu s)\, d\nu$ as the given line. The relation between $W(s)$ and s is known, for the historical reasons in astrophysics, as the "curve of growth". The equations above demonstrate that $W(s)$ is equal to the frequency integral of the column emissivity, or to one minus the transmission function. We note, however, that this is not the total column emissivity as we defined this latter quantity in eqn. (2.7.10), since the Planck function appears as a weighting factor in the definition of total emissivity.

Experimental observations of the intensity of a line profile as a function of absorbing thickness are often reduced to curves of growth for comparison with theory. The theoretical prediction for such a curve for a homogeneous atmosphere (constant pressure and temperature) and the Lorentz shape of eqn. (2.6.20a),

$$b(\nu - \nu_0) = \frac{1}{\pi} \frac{w}{(\nu - \nu_0)^2 + w^2}, \qquad (2.7.43)$$

where w is the Lorentz half-width, can be obtained by inserting eqn. (2.7.43) into the definition of $W_\alpha(s)$ as given in eqn. (2.7.41) and performing the integration. The result one obtains (due originally to Ladenburg and Reiche, 1913) is

$$W(x) = 2\pi w x\, e^{-x} \{I_0(x) + I_1(x)\} \qquad (2.7.44)$$

where $x \equiv \varrho S_\alpha s / 2\pi w$, and I_0 and I_1 are Bessel functions, for imaginary argument, of the zeroth and first order, respectively, as appeared in eqn. (2.7.35). Results are available in the literature for Doppler and Voigt line shapes. These have been reviewed by Goody, and some convenient graphs have been presented by Yamada (1968). Calculations for non-isothermal paths for Lorentz lines have been carried out by Simmons (1967) and by Cogley (1970) and for Doppler lines by Yamada (1967). Kyle (1968) has given results for Voigt line profiles in an isothermal atmosphere with an exponentially decreasing pressure, and Jansson and Korb (1968) give tables of the equivalent width for Voigt profiles in a homogeneous atmosphere as a function of optical depth and the ratio of Lorentz to Doppler half-widths [parameter y of eqn. (3.7.6)]. An analytic solution has been available for some time for the equivalent width of a Lorentz line in a hydrostatic, isothermal atmosphere under certain restrictions (Goody, 1964, p. 233). This has been generalized recently by Yamamoto and Aida (1970) who also applied their results approximately to the non-isothermal case.

Band Models

The above considerations of a single line are applicable to a band of isolated (non-overlapping) lines of equal strength either by summation over the band or by reduction of the band to a single average line. For bands of isolated lines of unequal strengths, a generalization can readily be made to N lines if a distribution function can be specified for the N line strengths. The average equivalent width of the lines is

$$\overline{W}(s) = \frac{1}{N} \sum_{\alpha=1}^{N} W_{\alpha}(s). \tag{2.7.45}$$

In order to compute $\overline{W}$, some assumption must be made regarding the distribution of line strengths. Two such distributions sometimes employed (Goody, 1964) are the exponential

$$P(S) = \frac{1}{S} e^{-S/\overline{S}} \tag{2.7.46}$$

and the inverse first-power

$$P(S) = kS^{-1} \tag{2.7.47}$$

distributions. $P(S)\,dS$ is the probability for a line in a band to lie in the strength interval dS, and k is the normalization constant.

For proper normalization the second of these distributions must be cut off at some upper limit S' on S, and, depending on its use, a lower cutoff may be needed as well. These cutoffs determine the normalization constant k. These two distributions lead, for the Lorentz line shape, to

$$\overline{W}(\text{exp}) = \frac{2\pi wx}{(2x+1)^{1/2}} \tag{2.7.48}$$

and

$$\overline{W}\,(\text{inverse}) = 2\pi w\,[e^{-x}I_0\,(x) + 2x\,e^{-x}\,\{I_0(x) + I_1(x)\} - 1] \quad (2.7.49)$$

respectively. The definition of x as given following eqn. (2.7.44) must be modified in each case by replacing S by the average line strength $\bar{S}$ for the exponential distribution, and by the maximum line strength S' for the inverse first-power distribution.

To illustrate how the effect of line overlap may be accounted for, we will describe briefly the Elsasser "regular model" and Goody's "general random model".

The use of a band model was first suggested by Elsasser (1938). His model consisted of an infinite number of lines of equal strength with equal spacing between lines, which may be allowed to overlap. The absorption coefficient for Lorentz lines may then be written as [cf. eqns. (2.6.20a) and (2.7.43)]

$$\mu'_\nu = \sum_{j=-\infty}^{j=+\infty} \left(\frac{\varrho S}{\pi}\right) \frac{w}{(\nu - j\delta)^2 + w^2} \quad (2.7.50)$$

where δ is the frequency spacing between lines and the average absorption can then be shown to be

$$\bar{A} = 1 - E\,(y, x) \quad (2.7.51)$$

where $y = w/\delta$ and

$$E\,(y, x) = \int_{-1/2}^{+1/2} \exp\left(-2\pi xy\,\frac{\sinh 2\pi y}{\cosh 2\pi y - \cos 2\pi t}\right) dt\,. \quad (2.7.52)$$

The function $E\,(y, x)$ cannot be written in closed form and must be numerically integrated, treated in limiting cases, or by means of a series expansion (Seitz and Lundholm, 1964).

The random model was first considered by Mayer (1947) and Goody (1952). In its most general form, it refers to N overlapping lines of unequal intensity randomly spaced in an interval $\Delta\nu = Nd$. In view of the excellent discussion by Goody (1964)† we will not go into the details

† See also Bond, Watson and Welch (1965), sect. 11.3.

here, but merely point out that the result, in this general case, for the average transmission function is

$$\bar{T} = \exp\left[-\overline{W}/d\right] \tag{2.7.53}$$

where d in this case is the *mean* spacing of the lines and $\overline{W}$ is the average equivalent width for an array of independent, or non-overlapping lines. Hence, $\overline{W}$ may be evaluated for a given line shape by such line-strength distributions as given in eqns. (2.7.46) and (2.7.47).

The Elsasser band model has been extended to the Doppler line shape by Kyle (1967) and by Golden (1967; 1968), and to the Voigt profile by Golden (1969). Malkmus (1967) has demonstrated the plausibility of a strength-distribution model which is a superposition of the models of eqns. (2.7.46) and (2.7.47); this is further discussed by Rodgers (1968).

2.8. The Diffusivity Factor

In the foregoing discussion, we have seen in a number of instances (stratified atmospheres) how the transmission function $e^{-\mu' x/\cos\theta}$ leads to exponential integrals when integrated over angle to obtain a flux. In large-scale detailed calculations which might involve thousands of spectrum lines, extensive frequency integrations, and spatial (length) integrations, the angular integral needed to obtain a flux from a transmission function is often considered a nuisance. The appearance of the exponential integrals was considered a handicap even before large-scale problems were attacked. This, of course, is because it is only the increase in computational power that has come about with the development of computers in the past two decades that made the large-scale problems tractable. Even with computers one may wish to avoid an integration, such as the above-mentioned angular integration, if this is possible, in order to include more spectrum lines or other detail, with a given amount of computing power. Because of a reluctance to evaluate the exponential integral [e.g. eqns. (2.4.8) and (2.4.9)] that arises in radiation flux calculations, it has been customary for many years, particularly in problems of atmospheric radiative transfer, to approximate $E_3(\tau)$ by a simple exponential

$$E_3(\tau) \cong \tfrac{1}{2}e^{-r\tau}, \tag{2.8.1}$$

where the factor r is usually called the diffusivity factor. The first suggestion of the use of this approximation appears to have been made by Roberts (1930). Many comments and a number of exploratory calculations regarding this form of approximation have since been made, but most of these have been of an empirical nature specific to particular problems. Among the more noteworthy discussions are those given by Elsasser (1942), Kondrat'yev (1965, chaps. 3 and 5), Kaplan (1952), by Hitschfeld and Houghton (1961), Plass (1952), by Yamamoto (1951) and by Rodgers and Walshaw (1966). The value selected for the factor r has generally been between 1.5 and 1.7 with 1.66 appearing as the most favored value.

A complete theory for the determination of this factor has been given by Armstrong (1968b) and we follow that discussion closely. Figure 2.13 illustrates qualitatively the difference between the two functions of eqn. (2.8.1). The crossover point where the two are equal moves in or out (towards small or large τ) as r increases or decreases, respectively. Since most physically observable quantities correspond to integrals over τ, one might surmise from this figure that it would be simpler and more reasonable to require the integrated difference between E_3 and $\frac{1}{2}\exp(-r\tau)$ to vanish than to attempt to match the two functions in some point-wise ashion. For example, the simple mathematical requirement that

$$\int_0^\infty [E_3(\tau) - \tfrac{1}{2}\exp(-r\tau)]\, d\tau = 0 \tag{2.8.2}$$

can be easily shown to hold if $r = \frac{3}{2}$ as was first pointed out by Roberts (1930). For this value of r the positive and negative areas shown in Fig. 2.13 exactly cancel. However, eqn. (2.8.2) does not have any physical content, so it is of no quantitative value in a practical application.

A quantitative physical basis for determining such an exact r by cancellation was suggested implicitly by Elsasser (1942). He noted that in the regime of "square-root" absorption† the flux transmission function could be obtained from the intensity transmission function, *after performing the frequency integration*, by simply multiplying the normal optical depth

† For large optical depth, the equivalent width $W(x)$ as given by eqn. (2.7.44) behaves as $x^{1/2}$. This can be seen most clearly from the asymptotic expansion for W given later by eqn. (2.8.23a), and is discussed by Goody (1964, p. 127).

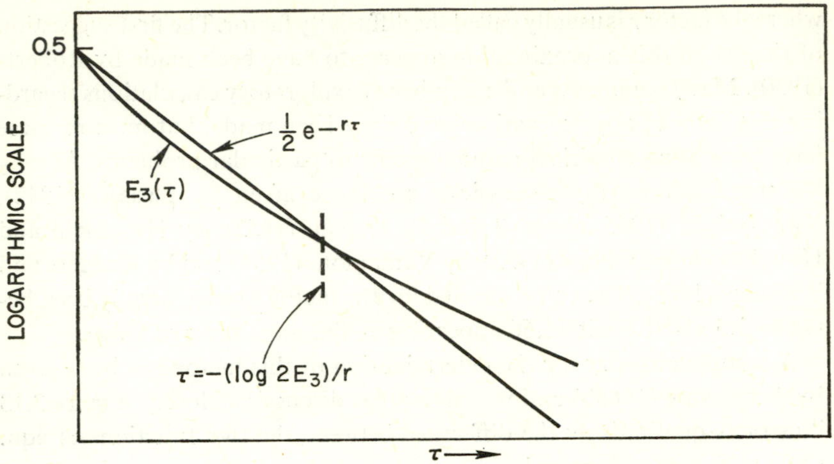

FIG. 2.13. A qualitative comparison of $\frac{1}{2}e^{-r\tau}$ and $E_3(\tau)$ as functions of τ.

by $\frac{16}{9}$. In order to show this we rewrite the intensity transmission function of eqn. (2.7.8) as

$$T_I(\tau') = e^{-\tau'} \tag{2.8.3}$$

where τ' is any optical depth. The flux and intensity transmission functions are then related through

$$T_f(\tau) = 2\int_1^\infty T_I(\tau t)\, t^{-3}\, dt \tag{2.8.4}$$

[compare, for example, eqns. (2.7.4) and (2.7.5)]. Since the optical depth in general depends strongly upon the frequency we make use of an average transmission function as in eqn. (2.7.12b), except that we employ a more general definition,

$$T(u) \equiv (\Delta\nu)^{-1}\int_{\Delta\nu} T(\tau_\nu)\, d\nu, \tag{2.8.5}$$

where $\Delta\nu$ is some spectral region of interest, $u = \int \varrho\, dz$ is the mass of absorbing gas per cm^2, ϱ is the density, z is the height variable of a stratified atmosphere, and T is a flux or intensity transmission function. This frequency average provides the means of effecting the cancellation mentioned above.

In the regime of "square-root" absorption by Lorentz lines in a homogeneous atmosphere, the average intensity transmission function can be written (Elsasser, 1942)

$$\bar{T}_I\,(u \sec\theta) = 1 - (4w\,Su \sec\theta)^{1/2}\,(\delta)^{-1} \tag{2.8.6}$$

where θ is the zenith angle, or the angle with respect to the z-axis. By use of this formula the angular integration can be carried out (the frequency integration has, of course, already been performed) quite readily. In eqn. (2.8.6), S is the integrated line strength, as defined in eqn. (2.7.36). The Lorentz half-width is w as defined earlier and δ is the line spacing. By the use of eqn. (2.8.5), eqn. (2.8.4) can be written in terms of the average transmission:

$$\bar{T}_f(u) = 2\int_1^{\infty} \bar{T}_I\,(ut)\,t^{-3}\,dt. \tag{2.8.7}$$

Inserting eqn. (2.8.6) into eqn. (2.8.7) with $t = \sec\theta$ we obtain

$$\bar{T}_f(u) = 1 - \tfrac{4}{3}\,(4w\,Su)^{1/2}\,(\delta)^{-1} = \bar{T}_1\,(\tfrac{16}{9}u). \tag{2.8.8}$$

Thus, once the frequency integration is performed, the diffusivity factor, $r = \frac{16}{9}$ yields an exact result in place of the angular integration in this limiting case of large optical depth.

This suggested to Yamamoto (1951) and to Plass (1952) that the requirement

$$\bar{T}_I\,(ru) = \bar{T}_f(u) \tag{2.8.9}$$

should be used to define r in general, both for non-homogeneous atmospheres and arbitrary optical depth. It provides, in principle, an exact result for the angular integration by means of a diffusivity factor. However, only a few specific values of r were provided, so that a general application of the method was not exploited until the work of Armstrong (1968b). As we will show, this definition has relatively simple analytic solutions for r in the cases of some band models for the Lorentz line shape.

The existence of these analytic solutions adds considerable insight into the nature and use of the diffusivity factor method. Armstrong's (1968b) results which include for the first time the effects of superimposing lines of different strength, indicate that the spread of values of r encountered in practical problems is not as broad as was concluded by Plass (1952) on the basis of considerations involving the Elsasser band alone.

In the cases for which we obtain simple results, this possibility arises from inverting the order of integration over the line strength and the frequency so that the frequency integration [implied in eqn. (2.8.9)] does not have to be carried out first as it does, of course, in the case of single lines.

From the definitions of the appropriate average transmission functions, eqn. (2.8.9) can be rewritten as

$$\int_{\Delta\tau(\Delta\nu)} [e^{-r\tau_\nu} - 2E_3(\tau_\nu)] \left(\frac{d\nu}{d\tau_\nu}\right) d\tau_\nu = 0. \qquad (2.8.10)$$

For a given frequency range $\Delta\nu$, this equation determines a value of r which produces a cancellation of the integrated difference between $\exp(-r\tau)$ and $2E_3(\tau)$. This value of r can be used in the intensity transmission function averaged over $\Delta\nu$ to convert it, according to eqn. (2.8.9), to the average flux transmission function. In this form, the relation of this definition to that of Roberts [eqn. (2.8.2)] can be easily seen, since they are the same except for the "weighting function" $d\nu/d\tau_\nu$, and the limits of integration. It is just these factors that provide the physical content missing from the qualitative definition, eqn. (2.8.2).

Application to Individual Lines and to Non-overlapping Rfgularly Spaced Lines

If we assume a regular band of non-overlapping lines of identical shape but of differing strengths, the total frequency interval which the band spans can be broken up into equal subintervals $\Delta\nu$ each of which contains just one line, and the contribution of each line can be considered independently of the others. In view of the assumptions of regular spacing and of no overlap, the individual frequency subintervals can all be taken equal. After dividing up the band in this fashion, since the lines all have the same shape, the individual contributions can be regrouped as a function of their strengths and then integrated over a strength distribution function $P(S)$ and over a single line profile, instead of performing the original frequency integration. This procedure and the kinds of distribution functions encountered in practice have been discussed by Goody (1964). Equation (2.8.10) then becomes

$$\int_{\Delta\nu} \int_{S=0}^{\infty} \{E_3 [\tau_\nu(S)] - \tfrac{1}{2} \exp [-r\tau_\nu (S)]\} P(S)\, dS\, d\nu = 0. \qquad (2.8.11)$$

As stated above, the frequency integral in eqn. (2.8.11) is to be taken across a single line profile. This result is a general equation for determining r for a band of non-overlapping, regularly spaced lines. Such bands are arithmetically simpler than the individual lines; this simplicity is also represented by the fact that some of the regular band results can be integrated analytically on the angle θ. Thus, in those cases where a regular band offers a realistic representation of a molecular spectrum [small optical depth and rapid fall-off of intensity in the line wing, as, for example, Doppler lines in the upper atmosphere (Rogers and Walshaw, 1966)] there may be no need of a diffusivity factor method.

The inverse first-power distribution, eqn. (2.7.47), has been applied to the water vapor spectrum which is often the most important absorber in the terrestrial troposphere (Godson, 1955). By means of eqns. (2.1.7) and (2.7.36), we can express S as a function of τ by means of the relation

$$\tau_\nu = Sub(\nu). \qquad (2.8.12)$$

This enables us to write eqn. (2.8.11) as

$$\int_{\Delta\nu} \int_{\tau_\nu=\tau_\nu(S_1)}^{\tau_\nu(S_2)} [E_3(\tau_\nu) - \tfrac{1}{2} \exp (-r\tau_\nu)] \frac{d\tau_\nu}{\tau_\nu}\, d\nu = 0. \qquad (2.8.13)$$

We can now consider τ_ν and ν as independent variables, and because of the strict proportionalities between $P(S)$, S, and τ in this model, the major physical variables of the problem, viz. amount of absorber and frequency, appear only in the limits of the inner integral and not in the integrand itself. This enables us to obtain an estimate of r immediately, without performing the frequency integral, as follows. If we restrict ourselves to that portion of the line profile for which $b(\nu)$ is neither very large nor very small (on the basis of the physical assumption, say, that regions of small $b(\nu)$ will lie under other lines and regions of large $b(\nu)$ will be black), the assumption can be made that a small value of S, say S_1, corresponds to small τ_1 and that a large value of S, say S_2, corresponds to a large τ_2. On this basis, the limits appearing in eqn. (2.8.13) can be replaced by zero and infinity, since the integrand vanishes at these limits, irrespective of the value of r. (The error in making this assumption is even further reduced by the fact that the integrand of eqn. (2.8.13) has opposite signs for small and for large τ.) In order to determine an approximate value for r it is therefore sufficient to set

$$\int_0^\infty [E_3(\tau) - \tfrac{1}{2}\exp(-r\tau)] \frac{d\tau}{\tau} = 0 \tag{2.8.14}$$

without regard to the frequency integration.† The value of this integral can be obtained analytically (Armstrong, 1968b) and turns out to be $\frac{1}{2}\log r - \frac{1}{4}$. Hence eqn. (2.8.14) yields

$$r = \sqrt{e} = 1.649 \tag{2.8.15}$$

which agrees nicely with the value of r most favored in practice. This result, as derived above, is independent of the amount of water vapor present which is consistent with the empirically determined uniqueness of r. It is, of course, a "lowest-order" result, since, in principle, corrections should be obtained from a more accurate consideration of the integration limits as dependent upon the problem variables. These, however, should be small in view of the behavior of the integrand at the limits. The major deficiency lies in the crudity of the model. It does, however, provide an indication of how a parameter-independent value of r can be obtained in principle.

The next most important contributor to radiative cooling, after water vapor, is CO_2. We assume here that its spectrum can be described by the exponential distribution of line strength (Goody, 1964) as given by eqn. (2.7.46). If we substitute τ for S by means of eqn. (2.8.12), specialize to

† Another way to obtain this same result—eqn. (2.8.14)—would be to assume a rectangular line profile and carry out the resulting trivial frequency integration.

the Lorentz shape of eqn. (2.7.43) and define $\bar{q} \equiv \bar{S}u/\pi w$ with $\Delta\nu = \delta$ as the line spacing, eqn. (2.8.11) yields

$$\int_{x=0}^{\delta/2\omega} \int_{\tau=0}^{\infty} [E_3(\tau) - \tfrac{1}{2} \exp(r\tau)]\, e^{-\tau(1+x^2)/\bar{q}} (1 + x^2)\, d\tau\, dx = 0. \quad (2.8.16)$$

The integration variable x is $(\nu - \nu_0)/w$. For our present purposes it is not a bad approximation to replace $\delta/2w$ by ∞ in the upper limit of the x-integration of eqn. (2.8.16). At sea level pressure the actual value of $\delta/2w$ for the 15 micron CO_2 spectrum (Walshaw and Rodgers, 1963) is ~ 6 and, of course, it increases with altitude. We adopt this approximation here as part of the model with the understanding that we will include the effect of line overlap (which is inextricably connected with any line shape cutoff) later by use of the Goody random model or the Elsasser model.

With this replacement, eqn. (2.8.16) can be integrated on x immediately, leading to the result (after an integration by parts)

$$\int_0^{\infty} \left[E_2(\tau) - \frac{r}{2} \exp(-r\tau) \right] e^{-\tau/\bar{q}} \tau^{-1/2}\, d\tau = 0. \quad (2.8.17)$$

By use of the integral definition $E_2 = \int_0^1 e^{-\tau/t} dt$, the integrations over τ and t can be performed in that order. The final result is

$$\frac{r}{2\,(r\bar{q} + 1)^{1/2}} = (1 + \bar{q})^{1/2} - \bar{q} \log \bar{q}^{-1/2} + (1 + \bar{q}^{-1})^{1/2}. \quad (2.8.18)$$

Figure 2.14 shows a graph of r as a function of $\bar{q}$ determined by this equation. Also shown in the figure is the value of r as determined for a single Lorentz line. This case can be recovered from eqn. (2.8.11) by setting $P(S) = \delta(S)$ where δ is the Dirac delta function, or directly from eqn. (2.8.10) by using the relation

$$\frac{d\nu}{d\tau} = \pm \frac{w q^{1/2}}{2\tau^{3/2} (1 - \tau/q)^{1/2}} \quad (2.8.19)$$

which holds in this case. We have set $q \equiv Su/\pi w$ and note that the plus sign in eqn. (2.8.19) pertains to one side of the line center, and the minus sign to the other. If we make the customary approximation of extending

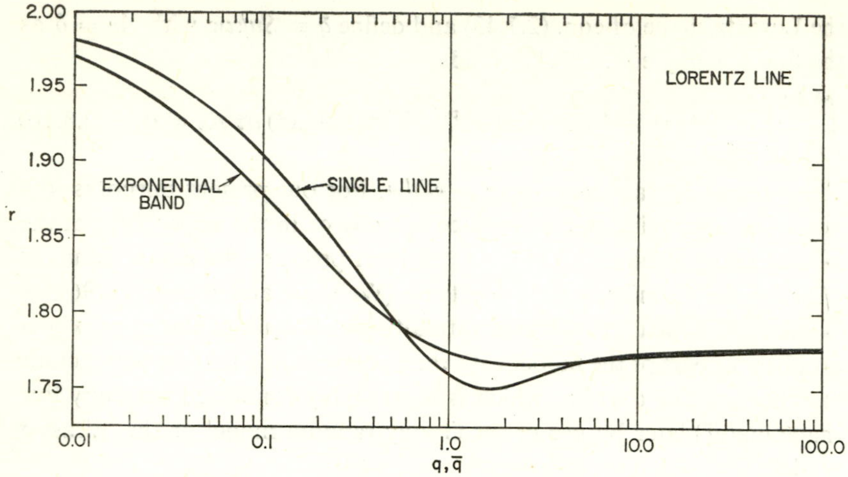

FIG. 2.14. The diffusivity factor r for a single Lorentz line and for an exponential distribution of non-overlapping Lorentz lines. The abscissa is the optical depth scale parameter $q = Su/\pi w$ for the single line and $\bar{q} = \bar{S}u/\pi w$ for the exponential band.

the frequency integration on each side of the line center to infinity, the result for a single Lorentz line is

$$\int_0^q [E_3(\tau) - \tfrac{1}{2}\exp(-r\tau)] \frac{d\tau}{\tau^{3/2}(1 - \tau/q)^{1/2}} = 0. \qquad (2.8.20)$$

This integral is not well suited to numerical evaluation as it stands on account of the singularities. However, the limiting values of r determined as $q \to 0$ and $q \to \infty$ can be readily evaluated from it and are 2 and $\frac{16}{9}$, respectively. These limits have been adequately discussed in the literature (see, for example, Goody, 1964† and Plass, 1952) so our interest here is

† It should be noted that in Goody's discussion of the $r = 2$ limit, an error has occurred in his use of Gaussian integration following eqn. (6.38) on p. 245. Gaussian integration properly yields $E_3(0) = 0.5$ and $dE_3(0)/dx = -1$ *in all orders*, as must be true since the integrands of $dE_3(0)/dx$ and $E_3(0)$ are zeroth and first-order polynomials, respectively. The different values of $E_3(0)$ stated as corresponding to different orders of Gaussian integration by Goody result from using half the Gaussian points for the standard interval $(-1, +1)$ to integrate the interval $(0, 1)$ rather than using the customary linear transformation between these intervals (Abramowitz and Stegun, 1964).

to verify that the formulas we derive correctly reproduce the known limits.

The exponential band of Lorentz lines has the same limits as can be readily verified from eqn. (2.8.18). In order to compute the diffusivity factor for the isolated Lorentz line case, we can convert the equation to a more familiar and non-singular form. Unity can be added and subtracted to the integrand in eqn. (2.8.10) to yield

$$\int_{\Delta\nu} (e^{-r\tau_\nu} - 1)\, d\nu = \int_{\Delta\nu} [2E_3(\tau_\nu) - 1]\, d\nu \tag{2.8.21}$$

and this expression can be evaluated for a Lorentz line shape in a homogeneous atmosphere. For such a case, by use of eqns. (2.8.12) and (2.8.19), eqn. (2.8.21) becomes, after some manipulation (Armstrong, 1968b),

$$W(rq/2) = \int_0^1 W(q/2y^{1/2})\, dy \tag{2.8.22}$$

where W denotes the Ladenburg and Reiche function of eqn. (2.7.44).

If the functional relationship $r = r(q)$ which this equation determines is slowly varying over some useful range of q, then it may be sensible to approximate it by $r \cong$ constant in a problem involving this range of q. One is then, in effect, making a point-wise fit to $r(q)$. In this type of equation for r, the mean value theorem (Franklin, 1940) can be used to obtain an interpretation for r. This theorem states that for an integral of the form

$$\int_b^a f(x)\, dx$$

there exists a value of x, $x = \xi$, say, such that

$$(b - a) f(\xi) = \int_a^b f(x)\, dx.$$

Thus, the equation in W, eqn. (2.8.22) states that $r = 1/\sqrt{\xi}$ where ξ is the mean value† of y in the integral $\int_0^1 W(q/2y^{1/2})\, dy$, and lies between 0 and 1. Yamamoto (1951) first suggested this interpretation. The diffusivity factor method is, therefore, tantamount to expressing the angular integral

† This is to say, ξ is a mean value of $\cos^2\theta$ where θ is the zenith angle.

in terms of the integrand evaluated at a mean value of the angular variable. It is only an approximation if an approximate mean value is used. By use of the asymptotic expansion (Kondrat'yev, 1965)

$$(2\pi w)^{-1}\,W(u) \cong \sqrt{\left(\frac{2u}{\pi}\right)}\left\{1 - \frac{1}{8u} - \frac{3}{128u} + \cdots\right\} \qquad (2.8.23a)$$

for the Ladenburg and Reiche function,† an asymptotic solution to eqn. (2.8.22) can be obtained. This is

$$r \cong \frac{16}{9} - \frac{1}{30q} - \frac{729}{22{,}400q^2} \qquad (2.8.23b)$$

to second order in q.

Armstrong (1968b) performed a numerical integration of eqn. (2.8.22) to determine $r(q)$ and the result is graphed in Fig. 2.14 as the "single Lorentz line" case. The asymptotic formula given above fits quite nicely onto the numerical results as q begins to get large compared to one. A detailed numerical comparison of the asymptotic and exact results is given by Armstrong (1968b).

We note the remarkably flat behavior of the "exponential band" curve of Fig. 2.14. The exponential distribution of line strengths smooths the already fairly flat behavior of the single line so that there is less than 1 per cent variation in r from $q = 0.8$ to $q = \infty$. The single Lorentz line curve varies approximately 2 per cent over this interval. This indicates that the diffusivity factor method would be quite good for the CO_2 15-micron band with $r \cong \frac{16}{9}$ as long as $q \sim 1$ or greater, if the regular exponential band were an adequate model. As we will see later, the values of r begin to drop below $\frac{16}{9}$ and to spread as we include line overlap.

For Doppler-broadened lines, we can set [see eqns. (3.7.2) and (3.7.6)]

$$b(\nu) = \exp\left[-(\nu - \nu_0)^2/\alpha_D^2\right]/\alpha_D\sqrt{\pi} \qquad (2.8.24)$$

where

$$\alpha_D\sqrt{(\ln 2)} = \nu_0\left(\frac{2kT\ln 2}{Mc^2}\right)^{1/2}$$

† Note that the asymptotic formula given by Goody (1964) is in error.

is the customary Doppler half-width (k is Boltzmann's constant, T the absolute temperature, M the mass of the radiating atom and c is the velocity of light). For this line shape, eqn. (2.8.11) can be transformed to (Armstrong, 1968b)

$$\int_0^{\bar{q}_D} \left\{ \frac{r}{2(1 + rv)} + v \log(1 + v^{-1}) - 1 \right\} [\log(\bar{q}_D/v)]^{-1/2} \, dv = 0 \quad (2.8.25)$$

where $\bar{q}_D \equiv u\bar{S}/\sqrt{\pi}\alpha_D$, and the integration variable v has arisen by setting

$$\bar{q}_D \exp[-(\nu - \nu_0)^2/\alpha_D^2] \equiv v.$$

Values of r obtained by numerically integrating this equation (Armstrong, 1968b) are plotted in Fig. 2.15. This figure also includes the result for a

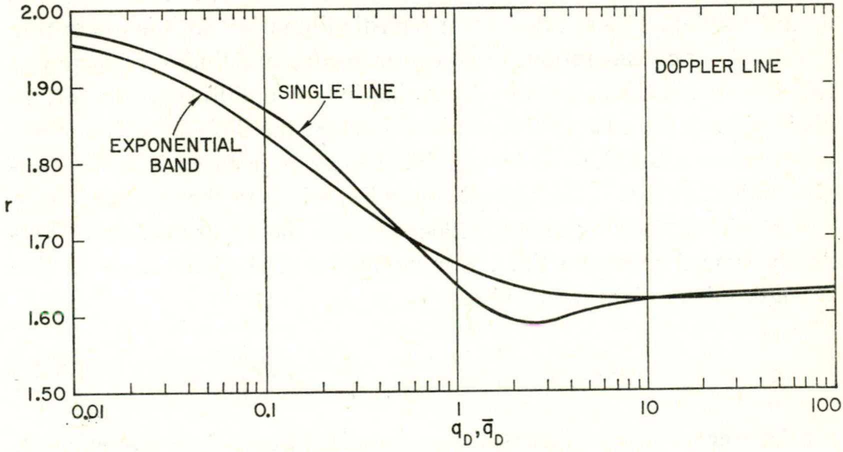

FIG. 2.15. The diffusivity factor r for a single Doppler line and for an exponential distribution of such lines. The abscissa is the optical depth scale parameter $q_D = Su/\sqrt{\pi}\alpha_D$ for the single line and $\bar{q}_D = \bar{S}u/\sqrt{\pi}\alpha_D$ for the band.

single Doppler line. The appropriate equation for this latter case can be obtained from eqns. (2.8.11) and (2.8.12) with $P(S) = \delta(S)$, and is

$$\int_0^1 \left[E_2(q_D t) - \frac{r}{2} \exp(-r q_D t) \right] [\log(1/t)]^{1/2} \, dt = 0 \quad (2.8.26)$$

where $q_D \equiv Su/\sqrt{\pi}\alpha_D$. The usual limit $r = 2.0$ is readily obtainable as $q \to 0$ for both the single line and the exponential distribution. As $q_D \to \infty$, eqn. (2.8.26) will be satisfied independently of the value of q_D if

$$\int_0^\infty [E_3(\tau) - \tfrac{1}{2}\exp(-r\tau)]\frac{d\tau}{\tau} = 0$$

a result which we have met before in eqn. (2.8.14). Thus, the limit in the case of a single Doppler line is $r = \sqrt{e} = 1.649$. The exponential band also has this limit, but it is somewhat harder to show (Armstrong, 1968b).

Application to Band Models of Overlapping Lines

Our previous considerations have been limited to spectrum lines that do not overlap. This is often not a realistic limitation so it is interesting to extend our considerations to allow for overlap and find out what effect this feature produces on the determination of a diffusivity factor. In distinction to the case of the previous models, the diffusivity factor will now become a function of two variables instead of just one. The new variable which appears is the ratio R of the line width to the line spacing.

A model devised by Goody (1964) permits the superposition of randomly spaced spectrum lines. For an exponential distribution in line strength this random model yields

$$\bar{T}_I = \exp\left\{-d^{-1}\int_{-\infty}^{+\infty}\frac{\bar{\tau}_\nu}{1+\bar{\tau}_\nu}\,d\nu\right\} \tag{2.8.27}$$

for the average intensity transmission function, where $\bar{\tau}_\nu = Sub(\nu)/\cos\theta$. The integral in eqn. (2.8.27) specialized to the Lorentz line shape in a homogeneous atmosphere becomes (Goody, 1964)

$$\frac{1}{d}\int\frac{\bar{\tau}_\nu}{1+\bar{\tau}_\nu}\,d\nu = \frac{\bar{S}u/(d\cos\theta)}{(1+\bar{S}u/w\pi\cos\theta)^{1/2}} \equiv \frac{\bar{q}R}{\cos\theta\,(1+\bar{q}/\cos\theta)^{1/2}}. \tag{2.8.28}$$

We have set

$$R \equiv \pi w/d \tag{2.8.29}$$

where d is the average line spacing and $\bar{q}$ has been defined earlier. Use of these formulas in the defining eqn. (2.8.9) yields

$$\exp\left\{\frac{-r\bar{q}R}{(1+r\bar{q})^{1/2}}\right\} = \int_0^1 \exp\left\{\frac{-\bar{q}R}{(y+\bar{q}\sqrt{y})^{1/2}}\right\} dy \qquad (2.8.30)$$

for the equation that determines the diffusivity factor r for this model. For small R:

$$R \ll \bar{q}^{-1/2},$$

the exponential in eqn. (2.8.30) can be expanded leading to

$$\frac{r}{(1+r\bar{q})^{1/2}} \cong \int_0^1 \frac{dy}{(y+\sqrt{y\bar{q}})^{1/2}}.$$

It is not difficult to show that this limit is identical to eqn. (2.8.18) for the non-overlapping case. Hence, in the limit $R = 0$ the random model yields back the limits on r of 2 and $\frac{16}{9}$ for small and large $\bar{q}$, respectively, obtained in the non-overlapping case. One can also show that as $\bar{q} \to 0$, $r \to 2$ irrespective of the value of R. One the other hand, for finite R and large $\bar{q}$, a different limit is obtained. In this case for $\bar{q} \gg 1$, the right-hand side of eqn. (2.8.30) becomes

$$\int_0^1 \exp\left\{-\frac{\bar{q}R}{(y+\bar{q}\sqrt{y})^{1/2}}\right\} dy \cong 4E_5\,[\sqrt{(\bar{q}R^2)}]$$

where E_5 has been defined in eqn. (2.4.9). This approximation yields, when inserted into eqn. (2.8.30),

$$r \cong \frac{1}{\bar{q}R^2}\{-\log 4E_5\,[\sqrt{(\bar{q}R^2)}]\}^2, \quad \bar{q} \to \infty \qquad (2.8.31)$$

which shows that $r \to 1$ as $\bar{q} \to \infty$ for finite R.

Figure 2.16 presents a graph of r as a function of $\bar{q}$ as determined by numerical integration of eqn. (2.8.30) for various values of R; and Fig. 2.17 shows r as a function of R for various values of $\bar{q}$. Since r becomes a function $\bar{q}R^2$ only as $\bar{q} \to \infty$, a graph of eqn. (2.8.31) against this variable is also presented in Fig. 2.18. Results for the inverse first-power distribution in the random model have been given by Armstrong (1968b).

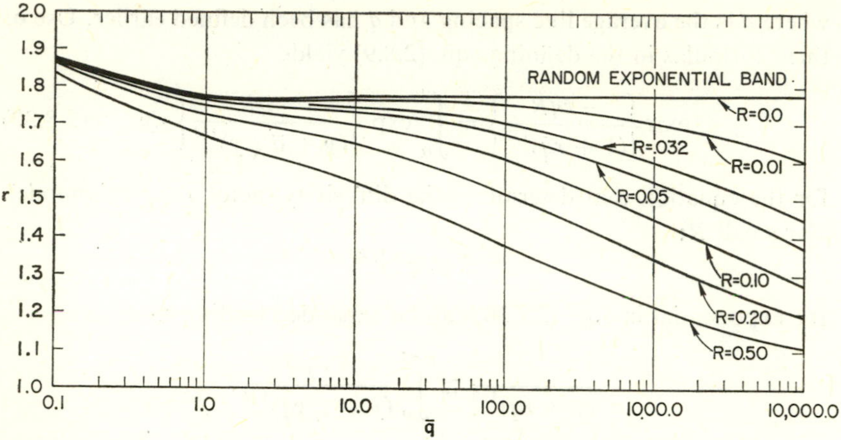

FIG. 2.16. The diffusivity factor r for the random exponential band of Lorentz lines for various values of the width-to-average-spacing ratio R. The abscissa is the optical depth scale factor $\bar{q} = \bar{S}u/\pi w$.

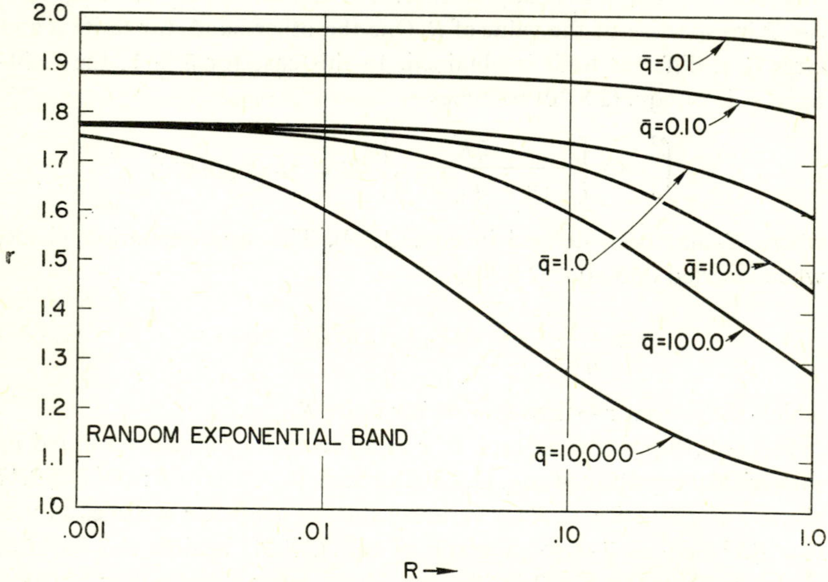

FIG. 2.17. A plot of the diffusivity factor r for the random exponential band as a function of the width-to-spacing ratio R for constant values of $\bar{q}$.

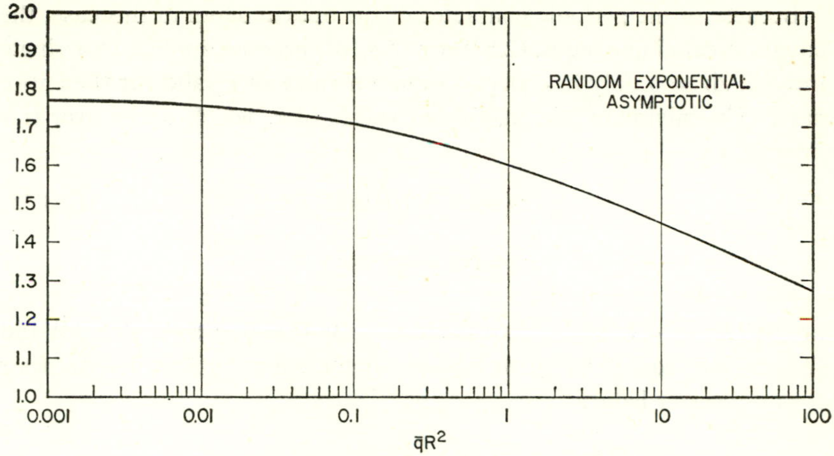

FIG. 2.18. The asymptotic diffusivity factor r for the random exponential band as a function of the single variable $\overline{q}R^2 = \bar{S}uR^2/\pi w$.

In the case of an Elsasser band of Lorentz lines, the optical depth can be written (Goody, 1964)

$$\tau_x = \frac{qR \sinh 2R}{\cosh 2R - \cos 2Rx} \tag{2.8.32}$$

by use of the definitions

$$q \equiv Su/\pi w,$$
$$R \equiv \pi w/\delta, \tag{2.8.33}$$
$$x \equiv \nu/w.$$

The characteristic line spacing, or period, in the Elsasser model has been denoted by δ, and the other quantities, including the half-width w, are as previously defined. For eqn. (2.8.32) the relation

$$\frac{d\tau_x}{dx} = \pm 2\tau_x (2qR\tau_x \coth 2R - \tau_x^2 - q^2R^2)^{1/2}/q \tag{2.8.34}$$

can be readily obtained.

Because of the periodic character of the optical depth in this case, the equation defining r [eqns. (2.8.9) or (2.8.10)] need be applied to a single interval $\Delta\nu = \delta$ only in order to obtain a value of r valid for the entire band. The minimum and maximum values of τ_x which occur over this interval are

$$\tau\,(\min) = qR \tanh R, \tag{2.8.35}$$

$$\tau\,(\max) = qR \coth R$$

at $\cos 2Rx = -1$, and $+1$, respectively. By inserting eqns. (2.8.34) and (2.8.35) into eqn. (2.8.10), we obtain for the equation that determines r for an Elsasser band:

$$\int_{qR\tanh R}^{qR\coth R} \frac{[E_3(\tau) - \frac{1}{2}\exp(-r\tau)]\,d\tau}{[2qR\tau\coth 2R - \tau^2 - q^2R^2]^{1/2}\tau} = 0. \tag{2.8.36}$$

Let us consider first two limiting cases. The first of these is the case when both q and R are small. In this event, we can set

$$qR \tanh R \ll 1 \tag{2.8.37}$$

and

$$\coth R \approx \frac{1}{R}$$

so that eqn. (2.8.36) becomes, approximately,

$$\int_0^q \frac{[E_3(\tau) - \frac{1}{2}\exp(-r\tau)]\,d\tau}{\tau^{3/2}(1 - \tau/q)^{1/2}} = 0. \tag{2.8.38}$$

This is the same as eqn. (2.8.20) for a single Lorentz line as one would expect for $R \to 0$. The second case we consider here is that of the limit of small R and large q. It is convenient in this case to set

$$s \equiv qR \tanh R \tag{2.8.39}$$

and transform eqn. (2.8.36) into

$$\int_1^{\coth^2 R} \frac{[E_3(sx) - \frac{1}{2}\exp(-rsx)]\,dx}{x(x-1)^{1/2}(\coth^2 R - x)^{1/2}} = 0. \tag{2.8.40}$$

So far we have made no approximations so that eqn. (2.8.40) is still exact. Now, for

$$R \ll 1$$

and

$$q \cong \frac{s}{R^2} \gg 1$$

the numerator in eqn. (2.8.40) vanishes strongly at the upper limit. We can, therefore, neglect x compared to $\coth^2 R$, since there is no significant contribution to the integral when x is near $\coth^2 R$. On this basis, eqn. (2.8.40) becomes, approximately,

$$\int_1^{\infty} \frac{E_3(sx) - \frac{1}{2}\exp(-rsx)}{x(x-1)^{1/2}}\,dx = 0. \tag{2.8.41}$$

This is the case of the Elsasser band with lines that are black in the centers (Plass, 1952) and it can be integrated exactly (Elsasser, 1942; Plass, 1952; Yamamoto, 1951) with the result that r is determined by

$$\operatorname{erf}[\sqrt{(rs)}] = \frac{2}{3}\left(\frac{s}{\pi}\right)^{1/2}(1-2s)\,e^{-s} + \frac{4}{3}s^2 + \left(1 - \frac{4}{3}s^2\right)\operatorname{erf}(\sqrt{s}) \tag{2.8.42}$$

where

$$\operatorname{erf}(x) = \frac{2}{\sqrt{\pi}}\int_0^x e^{-t^2}\,dt \tag{2.8.43}$$

is the customary error function. Approximations to eqn. (2.8.42) for small and large s have been given by Plass (1952).

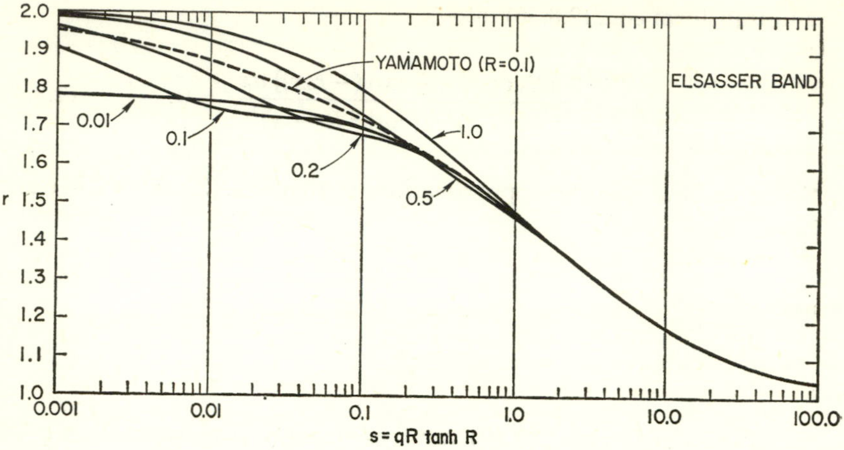

FIG. 2.19. The diffusivity factor r as a function of $s = qR \tanh R$ for various values of the width-to-spacing ratio R, for the Elsasser band of Lorentz lines. The dotted line is Yamamoto's result for $R = 0{\cdot}1$.

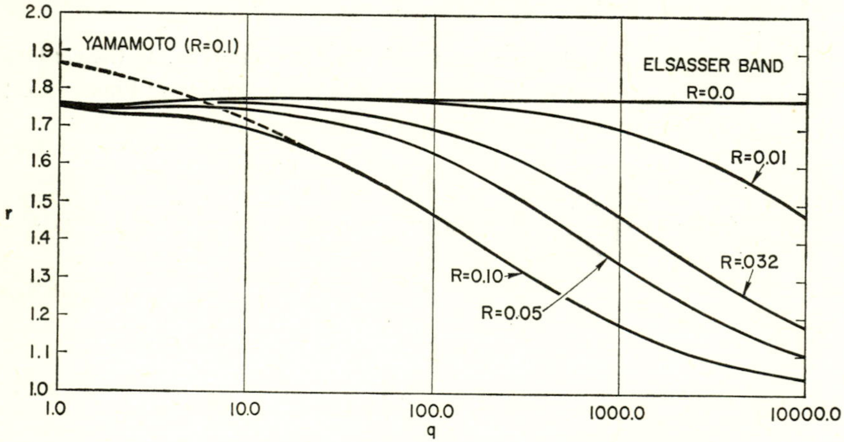

FIG. 2.20. The Elsasser band diffusivity factor r as a function of the optical depth scale parameter $q = Su/\pi w$ for various values of the width-to-spacing ratio R. The dotted line is Yamamoto's result for $R = 0{\cdot}1$.

Equation (2.8.40) can be integrated numerically as it stands as the singularities are not very strong; however, this is not an efficient procedure. A form better suited for numerical computation has been suggested by Armstrong (1968b) and the necessary calculations carried out. Values of r as determined numerically are presented in Fig. 2.19 as a function of s for width-to-spacing ratios $R = 0.01, 0.1, 0.2, 0.5$ and 1.0. The asymptotic formula, eqn. (2.8.42), yields a result which is essentially coincident with the $R = 0.01$ curve shown in Fig. 2.19 over the region given (the two deviate at still smaller s than shown on the figure). Therefore, we do not present it explicitly. The curves of r as a function of the variable s as given in Fig. 2.19 (following Yamamoto, 1951) have a somewhat complicated behavior. From small s values each curve comes in above the asymptotic limit (as given by eqn. (2.8.42)) dips below it and then towards still larger s rises up and meets the asymptotic curve. This results in an overlapping which makes it hard to discriminate between the curves for moderate s when graphed in this fashion. A plot of r against q rather than s can be employed as in Fig. 2.20 to disentangle the curves; they do not overlap when presented in this manner. We have also plotted Yamamoto's result, which was computed for $R = 0.1$, on these figures, and as can be noted his results are at variance with ours towards small q (or s) due to some computational errors that occurred during his calculation.

In summary, the diffusivity-factor definition of Yamamoto (1951) and Plass (1952), which is tantamount to an exact angular integration for the radiative flux in a plane-parallel atmosphere, has been applied in the above discussion to some of the line shapes and band models of interest in atmospheric radiative transfer. Although the results tend to corroborate the value $r \sim 1.65$ as perhaps the best single value for the Earth's atmosphere, it is now possible to select a more accurate value specific to the problem, line shape, and band model under consideration. Only a small amount of additional work is required to use these more accurate values in a practical calculation. In fact, approximate values of r can be determined from the graphs presented herein and by Armstrong (1968b) which will be more accurate than the use of the usual single value of $r \sim 1.65$. As suggested by Yamamoto, the diffusivity factor can be regarded as a bona fide physical parameter—an inverse mean value of the zenith angle—determined by the distribution of line strength, the line shape, the distribution of the absorber, and the thickness of the atmospheric layer. In the case of an inverse-first-power-of-the-line-strength distribution, as has been employed for water vapor, a single approximate value $r = \sqrt{e} = 1.649$ is derived which is independent of the optical parameters. Further discussion, including results for other band models, have been given by Armstrong (1968b, 1969b).

Chapter 3

Applied Radiation Theory for Atoms

IN CHAPTER 2 we reviewed the role of absorption coefficients and related experimentally measurable quantities in the transport of radiation through gases. The absorption coefficients and related quantities were introduced as atomic constants and no discussion was made of the theories which determine them. In this chapter we therefore review the classical and quantum models of radiation theory which can be used to calculate the absorption coefficients and other transition parameters of atoms and molecules.

3.1. Classical Lorentz Radiation Theory

The Lorentz theory, based on the classical model of interaction between electromagnetic radiation and matter, is well known. Excellent discussions have been given by Heitler (1954), by Stone (1963), and Kramers (1958) has provided a rigorous analysis of exceptional thoroughness. This theory involves the examination of the behavior of an ensemble of damped linear, simple harmonic electron oscillators (as models of atoms) driven by the electric vector of the electromagnetic wave. The theory has been spectacularly successful in discussing dispersion, and many other phenomena of physical optics. In spite of its limitations, and the apparently arbitrary assumptions which have to be applied to overcome them, much physical insight can be obtained by study of the Lorentz model. Its features are still useful in the correlation and interpretation of experimental measurements, and in providing perspective and intuition for the quantum theory of radiation. For these reasons, we include a brief summary of these classical results. The reader is referred to the references above for a more thorough account.

In the discussion of anomalous dispersion, the equation of motion of a typical damped, driven, oscillating electron is

$$m\ddot{x} + b\dot{x} + kx = eE_0 \exp(i\omega t) \tag{3.1.1}$$

where

$$\gamma = b/m \tag{3.1.2}$$

is called the damping constant. The characteristic angular frequency of the electron is

$$\omega_0 = (k/m)^{1/2} \tag{3.1.3a}$$

$$\omega_0 = 2\pi\nu_0 \tag{3.1.3b}$$

and the angular frequency of the driving E-vector is $\omega = 2\pi\nu$. Straightforward (and well-known) analysis of this model leads to the following expression for the total cross section σ and for the energy that is extracted from the incident field E_0 per unit time in the steady state:

$$\sigma = \frac{8\pi}{3} r_0^2 \frac{\nu^4}{(\nu_0^2 - \nu^2)^2 + \nu^2 (\gamma/2\pi)^2} \tag{3.1.4}$$

where

$$r_0 = e^2/mc^2 \tag{3.1.5}$$

is called the "classical electron radius". That is to say, if ε_ω is the energy per second extracted from a beam of intensity $I_0 = cE^2/8\pi$ (energy per cm^2 per sec), then the cross section σ is an area such that when multiplied into I_0 yields ε_ω:

$$\varepsilon_\omega = \sigma I_0. \tag{3.1.6}$$

In the classical model this energy ε_ω is re-radiated or scattered out of the beam direction, and the damping force arises from the radiation reaction (Heitler, 1954). The theory of the radiation reaction yields the result

$$\gamma = \frac{8\pi^2}{3} \frac{\nu_0^2}{c} r_0 = \frac{0.233}{\lambda_0^2} \sec^{-1} \tag{3.1.7}$$

where λ_0 is the wavelength in cm corresponding to the frequency ν_0. The classical scattering cross section of eqn. (3.1.4) contains three significant limiting cases:

1. for $\nu_0 \ll \nu$ and $\gamma \ll \nu$, we obtain the Thomson scattering cross section

$$\sigma = \frac{8\pi}{3} r_0^2, \tag{3.1.8}$$

2. when $\nu_0 \gg \nu$, we have the Rayleigh scattering cross section,

$$\sigma = \frac{8\pi}{3} r_0^2 \frac{\nu^4}{\nu_0^4}, \tag{3.1.9}$$

3. with $\nu \sim \nu_0$, the Lorentz cross section results,

$$\sigma = \frac{2\pi}{3} r_0^2 \frac{\nu^2}{(\nu_0 - \nu)^2 + (\gamma/4\pi)^2}. \tag{3.1.10}$$

When $\nu \sim \nu_0$ in case (3) we have the phenomenon of coherent scattering at resonance or resonance fluorescence, which from a quantum viewpoint can also be considered as a *single* process equivalent to absorption and re-emission of a photon. If the driving field in eqn. (3.1.1) is turned off, the transient solution that governs the decay of the oscillator motion is given by (Heitler, 1954):

$$x = x_0 \, e^{-\gamma t/2} \, e^{-i\omega_0 t} \tag{3.1.11}$$

(for $\gamma \ll \omega_0$) and the radiation emitted by the oscillator during its decay will have an amplitude given by the same formula:

$$E = E_0 \, e^{-\gamma t/2} \, e^{-i\omega_0 t}. \tag{3.1.12}$$

From this expression we see that the damping constant γ is the reciprocal of the time τ for the power ($\sim |E|^2$) to decay to $1/e$ of its initial value. From eqn. (3.1.7), we note that

$$\tau = \gamma^{-1} = 4.30\lambda_0^2 \quad \text{sec}$$

with the wavelength λ_0, again, in cm.

If the field amplitude of eqn. (3.1.12) is expanded in a Fourier integral to find its frequency distribution and the result squared to obtain an intensity frequency distribution, the result is

$$I(\nu) \sim |E(\nu)|^2 = \frac{I_0 \gamma}{4\pi^2} \frac{1}{(\nu - \nu_0)^2 + (\gamma/4\pi)^2}. \tag{3.1.13}$$

Thus, we again find the Lorentz shape. In probability theory this ubiquitous function is called the "Cauchy distribution". The same shape is obtained for an oscillator radiating continuously at the frequency ν_0 except for randomly occurring collisions that terminate the oscillations. The finite lifetime of the oscillations produce an analogous spread in frequency to that given above. In this case the "damping coefficient" corresponding to γ becomes $1/(\pi\tau_0)$ where τ_0 is the average collision time. This average time between collisions is given by l/v where l is the collision mean free path and v is the velocity of the radiating "oscillator" (White, 1934). τ_0 can also be expressed as $n\sigma v$ where v again is the velocity, n is the number density of atoms ("radiating oscillators") and σ is the collision cross section.

The line shape given by eqn. (3.1.13), called the "natural shape", is, as we have seen, the classical shape for the radiation intensity *emitted* by an atom. Now in view of Kirchoff's law which states that (Milne, 1930)

$$\frac{j_\nu}{\varkappa_\nu} = I_\nu \tag{3.1.14}$$

we see that, whenever I_ν is constant over the width of a line, the shape of an absorption line will also be given by eqn. (3.1.13). This constancy of I_ν will be satisfied under most conditions of interest to us. (As we will see later, it is also required in quantum theory for a constant transition probability per unit time to exist.) The natural line shape therefore implies the existence of a "natural absorption coefficient" proportional to the same frequency function as given in eqn. (3.1.13). The constant of proportionality can be determined by calculating the absorption by the classical oscillator of eqn. (3.1.1) (Heitler, 1954), and the result for the classical "natural absorption coefficient" is

$$\mu_\nu = \frac{N_v e^2}{mc} \frac{(\gamma/4\pi)}{(\nu - \nu_0)^2 + (\gamma/4\pi)^2} \tag{3.1.15}$$

in terms of the number of absorbing atoms per unit volume N_v.

This expression yields the absorption coefficient in cm^{-1} as defined in eqn. (2.1.1). The other absorption coefficients as introduced in eqns. (2.1.6a) and (2.1.6b), are again related to μ_ν by factors of the particle number density or mass density as given in Chapter 2.

The above equations illustrate the Lorentz radiation-damping natural line profile and apart from a redefinition of γ in the quantum formulation retain this same form of a driven tuned, damped oscillator in the quantum theory (of independent isolated radiating atoms). A typical Lorentz line is illustrated in Fig. 3.1, where the full width $2w$ at "half-power point" is seen to be the distance between the two frequencies (symmetric

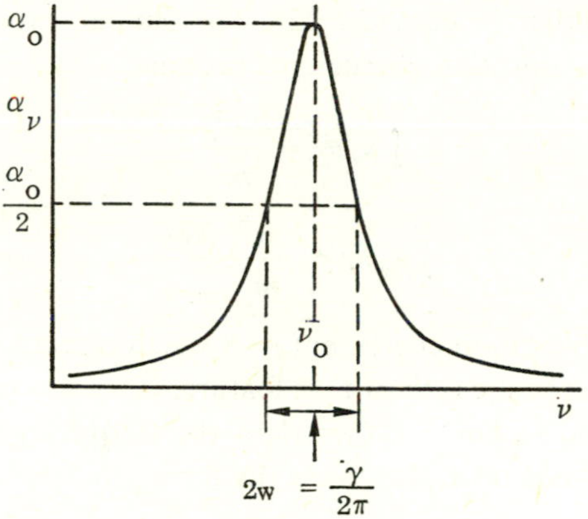

FIG. 3.1. Lorentz profile.

about ν_0) at which the two terms in the denominator of eqn. (3.1.15) are equal. Thus, we have

$$2w = \frac{\gamma}{2\pi}. \tag{3.1.16}$$

Integration by elementary methods of eqn. (3.1.15) with resepect to ν over a spectral line leads to the important result (see Aller, 1963)

$$\int \mu_\nu \, d\nu = \frac{\pi e^2 N_v}{mc}. \tag{3.1.17a}$$

In order to compare such an equation with experiment the model has to be refined somewhat. Atoms exhibit more than one spectrum line and it is thus assumed that the electron is so bound that it can oscillate at one of a number of characteristic frequencies ν_0. The "fraction" of the electron associated with any one characteristic frequency is designated f and is called the classical oscillator strength at that frequency. This requires that the right-hand side of eqn. (3.1.17a) be multiplied by f, and thus we find

$$\int \mu_\nu \, d\nu = \frac{\pi e^2 N_v}{mc} f \tag{3.1.17b}$$

as anticipated earlier in eqn. (2.6.22). The similar expressions for the other integrated absorption coefficients become

$$\int \alpha_\nu \, d\nu = \frac{\pi e^2 f}{mc} \tag{3.1.17c}$$

$$\int \varkappa_\nu \, d\nu = \frac{\pi e^2}{mc} \frac{N_v f}{\varrho} \tag{3.1.17d}$$

The integrated absorption coefficient of a spectral feature is thus a measure of the oscillator strength of that feature. If we use the wave number $\bar{\nu} = \lambda^{-1}$ measured in cm^{-1} rather than the frequency in sec^{-1} as the spectral variable, eqn. (3.1.17c) takes the form

$$\int \alpha_{\bar{\nu}} \, d\bar{\nu} = \frac{\pi e^2}{mc^2} f \tag{3.1.17e}$$

and a similar change occurs in the denominator of eqn. (3.1.17b and d). A basic explanation of this quantity f was, of course, not provided by classical physics. It must be found from quantum mechanics as we will demonstrate shortly. The above integrated absorption coefficients can be related directly to the equivalent width W of a spectrum line in the optically thin case. Recall that the equivalent width was defined and discussed in Section 2.7, and its historical interpretation was cited there as the width of a line of infinite absorption possessing the same integrated absorption function as the given line. If we expand the Beer's law expression

$$I_\nu = I_{0\nu} e^{-\mu_\nu s} \tag{3.1.18}$$

[cf. eqn. (2.1.2a) in the case of small optical depth] and keep only the first term, we find

$$I_{0\nu} - I_\nu \cong I_{0\nu} \mu_\nu s. \tag{3.1.19}$$

Then, since W is defined as the integral of the absorption function $(I_{0\nu} - I_\nu)/I_{0\nu}$ [see eqns. (2.7.9b) and (2.7.41)] which in the present case is just $\mu_\nu s$, we have:

$$WI_{0\bar{\nu}} = \int (I_{0\bar{\nu}} - I_\nu)\, d\nu = sI_{0\bar{\nu}} \int \mu_\nu d\nu. \tag{3.1.20a}$$

In this definition $I_{0\bar{\nu}}$ is written for the maximum spectral intensity across the line (viz. the incident intensity, which is assumed to be slowly varying compared to the intensity variation across the line).

In terms of the optical depth τ_ν or the absorption coefficient per atom α_ν, this definition becomes

$$W = \int \tau_\nu\, d\nu = N_v s \int \alpha_\nu d\nu = \frac{\pi e^2}{mc} N_v f s \tag{3.1.20b}$$

where the last relation follows from the integrated expression, eqn. (3.1.17c). Thus for optically thin lines, W can be used to measure N_v, f or s. The application of eqn. (3.1.20b) to the optically thick case involves a discussion of the curve of growth (see Aller, 1963, and Section 2.7).

A few properties of classical oscillator strengths are noted briefly.

(a) The "sum rule"

$$\sum f = 1 \tag{3.1.21}$$

is obeyed.

(b) For continuous spectra eqn. (3.1.17c) is written

$$\int \alpha_\nu\, d\nu = \frac{\pi e^2}{mc} \int \frac{df}{d\nu}\, d\nu \tag{3.1.22}$$

where df is the element of oscillator strength associated with the frequency increment $d\nu$.

(c) In the classical formulation, the oscillator strength of a line is a good parameter with which to specify the integrated absorption coefficient of the line. This feature, plus the fact that the sum rule, eqn. (3.1.21), has a rigorous basis in the quantum theory, have led to the persistence of its use in quantum theory in spite of its *ad hoc* origin.

Relations Between the Various Transition Parameters

We have now introduced altogether quite a variety of atomic parameters such that it is worthwhile to explicitly note their interconnecting relations. We recall the definition of the Einstein A_{nm} and B_{nm} coefficients of Section 2.2. In a transition of spontaneous emission from U(pper) to L(ower) states, the number of radiative transitions per second is $N_v(U)\,A_{UL}$, where $N_v(U)$ is the number density of excited atoms in the state U. Thus the energy $E_S^{(UL)}$ radiated per second from unit volume into all 4π of solid angle is given by this number of transitions per second times the energy $h\nu_{UL}$ emitted per transition:

$$E_S^{(UL)} = N_v(U)\,A_{UL} h\nu_{UL}. \tag{3.1.23}$$

The energy $h\nu_{UL}$ is, of course, equal to the difference in energy between the two states as in eqn. (2.2.3).

The same result can be obtained by integration of eqn. (2.2.4a) over solid angle $\widehat{\Omega}$ and over the frequency range about the line profile.

Similarly, in the L–U absorption transition stimulated by a beam of radiation with specific intensity I_ν, and hence, energy density per unit solid angle I_ν/c, the number of absorption transitions per second is†

$$N_v(L)\,B_{LU}\,\frac{I_\nu}{c}\,d\Omega \tag{3.1.24}$$

per unit volume.

The coefficient B_{LU} is called the Einstein B-coefficient for absorption or for stimulated (induced) emission. Thus, the energy per unit area per unit time absorbed out of the beam I_ν within a path length dx due to this transition is

$$dE_A^{(UL)} = dx\,N_v(L)\int B_{LU} I_\nu\,\frac{h\nu_{LU}}{c}\,d\Omega. \tag{3.1.25}$$

However, from the definition of the absorption coefficient α_ν, we know that this absorbed energy should be, as stated following eqn. (3.1.23):

$$dE_A^{(UL)} = N_v(L)\,dx\int_{\hat{\Omega}}\int_{\nu} \alpha_\nu I_\nu\,d\nu\,d\Omega \tag{3.1.26}$$

where the frequency integration is carried out over the width of the line U–L. By comparing these two formulas for the case of a line over whose

† The B-coefficient is sometimes defined relative to the intensity alone, rather than relative to the energy density. In this event, no factor c appears in this expression.

width I_ν does not vary appreciably, we find the relationship between α_ν, B_{LU}, and the line oscillator strength:

$$\int \alpha_\nu \, d\nu = B_{LU} \frac{h\nu_{LU}}{c} = \frac{\pi e^2}{mc} f_{LU}. \tag{3.1.27}$$

The last equality follows from eqn. (3.1.17c).

We can therefore write

$$f_{LU} = \frac{m}{\pi e^2} h\nu_{LU} B_{LU}. \tag{3.1.28}$$

When U is g_U-fold degenerate and L is g_L-fold degenerate, the B-coefficients of the component transitions may be summed to give an overall B-coefficient which will apply when the levels all lie at the same energy (Davidson, 1962):

$$\sum_{U_i L_i} B_{U_i L_i}(\nu) = g_U B_{UL} = g_L B_{LU}. \tag{3.1.29}$$

Similarly, by a detailed balancing argument the following relation can be shown to exist between the A- and B-coefficients in the general case of degeneracy (Davidson, 1962):

$$g_U A_{UL} = g_L B_{LU} \frac{8\pi h\nu^3}{c^3} = g_U B_{UL} \frac{8\pi h\nu^3}{c^3}. \tag{3.1.30}$$

The intensities are, however, still controlled by eqns. (3.1.23) and (3.1.26).

The energy absorbed per unit time, E_{cl}, by a (single) classical electron oscillator [corresponding to eqn. (3.1.25)] is given by

$$E_{cl} = \frac{\pi e^2}{mc} \int_{\hat{\Omega}} I_\nu \, d\Omega \tag{3.1.31}$$

[cf. Heitler, 1954, p. 38, eqn. (19)] in terms of the specific intensity I_ν in which the oscillator is immersed. The quantum-mechanical result (which we will derive in the next section)† is

$$E_{\text{Q.M.}} = \frac{4\pi^2 e^2}{3\hbar c} \nu_{\alpha\beta} |\langle\alpha| \boldsymbol{r} |\beta\rangle|^2 \int_{\hat{\Omega}} I_\nu \, d\Omega \tag{3.1.32}$$

† See eqn. (3.3.47). By multiplying $w_{\beta\alpha}$ of this equation by $\hbar\omega$, the result used here can be obtained after integration over the line profile. $I_0(\omega)$ has been re-expressed to indicate the solid-angle integration.

in terms of the matrix element $\langle\alpha|\, \boldsymbol{r}\, |\beta\rangle$ between the quantum states α and β and the frequency $\nu_{\alpha\beta} = (E_\beta - E_\alpha)/h$ associated with the transition between the two states whose energies are E_β and E_α. The two results can be connected by the definition of the *quantum-mechanical f*-number:

$$f_{\alpha\beta} \equiv \frac{4\pi m}{3\hbar} \nu_{\alpha\beta} \, |\langle\alpha|\, \boldsymbol{r}\, |\beta\rangle|^2 \tag{3.1.33}$$

[see eqn. (3.3.58), following]. With this definition, eqn. (3.1.32) yields

$$E_{\text{Q.M.}} = \frac{\pi e^2}{mc} f_{\alpha\beta} \int_{\hat{\Omega}} I_\nu \, d\Omega \tag{3.1.34}$$

in analogy to eqn. (3.1.31). If we compare these two expressions for the absorbed energy, we obtain

$$E_{\text{Q.M.}} = f_{\alpha\beta} E_{cl}. \tag{3.1.35}$$

Thus, the oscillator strength, which in classical terms is associated with the fraction of the electron responsible for the radiative transition, may also be thought of as the fraction or factor that converts the value of the classical energy absorbed or radiated per unit time to the correct quantal value.

A classical derivation of photoelectric absorption has been given by Thompson and an improved semiclassical derivation has been given by Kramers. These are described by Compton (1926) in chapters 6 and 12, respectively.

3.2. Formal Quantum Formulation of Radiation Theory

We will not attempt to expound or review the basic principles of the quantum theory of radiation. Rather, in the spirit of the applied investigator, we shall derive the working formulas needed for applications of the theory from its basic formal results. We will also attempt, where it appears useful, to correlate the results of several more or less standard texts or references (Bates and Damgaard, 1949; Bethe and Salpeter, 1957; Dirac, 1958; Griem, 1964; Heitler, 1954; Mayer, 1947; Schiff, 1968; Slater, 1960) where their authors employ different conventions, and to include some of the details of the reduction of the general formulas that are normally glossed over or left to the reader in these texts. Blatt and Weisskopf (1952, chap. XII) also give an excellent treatment of radiative transitions.

The basic formula that we need, called "Golden Rule no. 2" by Fermi (1950), is

$$w_{ij} = \frac{2\pi}{\hbar} |\langle j| H' |i\rangle|^2 \frac{dn}{dE}. \tag{3.2.1}$$

This formula, from first-order perturbation theory, gives the transition probability per unit time for a transition from state i to state j induced by a perturbation H' with matrix element $\langle j| H' |i\rangle$. The analogous formula for second-order perturbation theory. called "Golden Rule no. 1" by Fermi, is

$$w_{ij} = \frac{2\pi}{\hbar} \left| \sum_k \frac{\langle j| H' |k\rangle \langle k| H' |i\rangle}{E_j - E_k} \right|^2 \frac{dn}{dE}. \tag{3.2.2}$$

but we will not make use of it in this volume.

The indices i and j are symbolic for all the quantum numbers needed to define the system (atom + radiation) before and after the transition, respectively. The factor dn/dE is the number density of final states in the cases where there is a continuous range of final states (e.g. spontaneous emission, photoelectric absorption), or the density of initial states in the cases where there is a single final state but a continuous range of initial states (e.g. line absorption). For the existence of a constant transition probability per unit time, a continuous range of states must exist in either the initial or final state. For a discussion of this point see Heitler (1954) or Schiff (1968, p. 401). These formulas are derived by Schiff (1955; 1968) and are derived and justified in detail by Heitler (1954). Although they can be derived heuristically for radiative transitions within the framework of ordinary quantum mechanics and semiclassical radiation theory, they cannot be rigorously justified except by appeal to quantum field theory or to the rules of Quantum Electrodynamics (see Feynman, 1962, p. 4). As is shown in the derivations cited, these transition-probability formulas contain implicitly the principle of conservation of energy and the uncertainty principle. Thus, all of our radiation formulas are dependent on the requirement that the change in energy of the atomic or molecular system is equal to the energy of the photon that is absorbed or emitted. The conservation of energy (and of momentum) can be displayed explicitly in the transition probability formulas to a conceptual advantage, as we will indicate later in Section 3.6. Since the original formulation of time-dependent perturbation theory and the derivation of eqn. (3.2.1) by Dirac, there has been considerable clarification and generalization of the ideas involved, but no basic change in the principles. This clarification and increased sophistication have come from scattering theory (e.g. Goldberger and Watson, 1964). In Section 3.6 we will make a limited use of this more general formalism when we take up free–free transitions, where it is of great advantage. An excellent review of the application of modern scattering theory to atomic radiation problems has been given by Shore (1967).

Although in principle one is interested in all processes which result in the absorption or emission of a photon from or into an incident beam, in practice one usually confines oneself to the processes which dominate the absorption and emission in the temperature-density regime of interest. In this spirit, we will limit our attention to discrete atomic transitions, photoelectric and free–free transitions. We will not concern ourselves with scattering, radiative corrections or relativistic effects.

These three processes which we address, viz. discrete atomic transitions, photoelectric, and free–free transitions, termed "simple absorption" by Mayer (1947), are often the most important in heated gases and we will emphasize them accordingly. We will concern ourselves with the reduction of eqn. (3.2.1) to the specific transitions listed above that we wish to treat. This can be done without reference to the type of atomic system because the nature of the atomic system is concealed in the wave functions used to compute the matrix elements $\langle j|\, H'\, |i\rangle$. After deriving the formulas specific to the various transitions, the user can then proceed to specify the atomic system. In applying eqns. (3.2.1) and (3.2.2) for the transition probability to different situations, one usually must deal with *degenerate* states (i.e. states with the same energy, but differing in some other quantum number or numbers). In view of the principle of equal *a priori* probabilities in phase space one should, in the absence of additional information, average over degenerate initial states and sum over degenerate final states (Tolman, 1938, sects. 98 and 99). This degeneracy must arise from different spatial orientations or from degeneracy on quantum numbers other than the energy, as the sum over the energy states lying in a small region ΔE allowed by the uncertainty principle (and therefore effectively degenerate) has already been carried out in the derivation of eqn. (3.2.1). It is this sum that leads to the appearance of the "number density of states" dn/dE. With this prescription, the transition probability per unit time becomes

$$w_{ij} = \frac{2\pi}{\hbar} \frac{1}{g_i} \sum_{\alpha(i)} \sum_{\beta(j)} |\langle\beta| H' |\alpha\rangle|^2 \frac{dn}{dE} \tag{3.2.3}$$

for first-order processes, where the index $\alpha(i)$ labels the degenerate family of g_i states (i) and the index $\beta(j)$ labels the family of g_j states (j). That is to say, i and j uniquely specify the energy of two degenerate states and α and β run over the remaining quantum numbers. For simplicity in the derivations which we now undertake, we will compute transition probabilities, cross sections, etc., for a pair of distinct, non-degenerate substates α and β of a *single* atomic electron, and after doing this return to the general situation.

As implied above, we will be concerned in the following sections of this chapter with the reduction of the basic equation (3.2.1) into the product of an angular factor and a radial factor for our processes of interest. For the radial factors we will use the notation

$$\mathscr{R}_{\beta\alpha} \equiv \mathscr{R}(\beta, \alpha) \equiv \int_0^\infty R_\beta^* R_\alpha r^3 \, dr \tag{3.2.4}$$

where the R's under the integral are bound or free state radial wave functions, $r^2\,dr$ is the radial part of the volume element, and the additional factor r arises from the electric field perturbation in the dipole-length approximation for radiative transitions. We will usually express our results in terms of this approximation. Since we do not concern ourselves herein with the actual calculation of wave functions, the dipole velocity and acceleration forms of the matrix elements will be only briefly alluded to in Section 3.6, eqn. (3.6.19). For the subject of the calculation of the radial function, the reader is referred to the literature, e.g. Bates and Damgaard (1949), Hartree (1957), Stewart and Rotenberg (1965) and Kelly (1964a,

1964b). Since the radial wave functions usually depend on the orbital quantum number 1 and not on the magnetic quantum number, we will also use the notation

$$\mathscr{R}(l_i, l_j) \equiv \int_0^\infty R_{l_i}^* R_{l_j} r^3 \, dr \tag{3.2.5}$$

for our radial factor. In the study of the angular factors, which we take up in Section 3.5, it is convenient to define the radial dipole integral σ_{ij}^2:

$$\sigma_{ij}^2 \equiv (4l_>^2 - 1)^{-1} \left[\int_0^\infty R_{l_i}^* R_{l_j} r^3 \, dr\right]^2 = (4l_>^2 - 1)^{-1} \mathscr{R}^2(l_i, l_j). \tag{3.2.6}$$

The subscript $>$ signifies the greater of the two angular momenta involved in the transition. This particular normalization of the radial factor is convenient for the definition of the angular factor.

It is presumed that the form of the radial wave function, viz. whether it describes a bound or a free state, will be clear from the context and does not need to be explicitly incorporated into the notation.

3.3. Reduction of the General Quantum Formulas for Bound–Bound Processes

For radiative processes, the perturbation term that enters the Hamiltonian for the interaction between the radiation field and an atomic electron is

$$H' = -\frac{e}{mc} \boldsymbol{p} \cdot \boldsymbol{A} \tag{3.3.1}$$

where $\boldsymbol{p}$ is the momentum operator of the atomic electron and $\boldsymbol{A}$ is the electromagnetic vector potential. This arises, of course, from expanding the term

$$\frac{1}{2m}\left(\boldsymbol{p} - \frac{e}{c}\boldsymbol{A}\right)^2$$

in the Hamiltonian; see Schiff (1968). It also can be interpreted as the interaction energy

$$-\frac{1}{c}\boldsymbol{j} \cdot \boldsymbol{A}$$

between the atomic electron current $\boldsymbol{j} = e\boldsymbol{v}$ and the vector potential $\boldsymbol{A}$. (See also Power, 1964, p. 101.) The normalization of $\boldsymbol{A}$ and the density of states of the radiation field dn_R/dE are related so that a consistent choice of the two must be made. For emission of a photon by an atomic system which makes a transition between two bound states, the necessary continuous range of states is provided by the radiation field which is produced in the final state. Similarly for absorption accompanied by a transition between two bound states, the continuous range of states is provided by the presence of the radiation field in the initial state. A consistent choice of dn_R/dE and the matrix element of the interaction term given by eqn. (3.3.1) is given by Heitler (1954) as

$$\frac{dn_R}{dE} = \frac{\omega^2\, d\Omega}{(2\pi c)^3\, \hbar}, \tag{3.3.2}$$

$$\begin{aligned} \langle\beta|\, H'\, |\alpha\rangle &= \langle\alpha|\, H'^*\, |\beta\rangle \\ &= -\frac{e}{m}\left(\frac{2\pi\hbar}{\omega}\right)^{1/2} (n_\omega + 1)^{1/2} \int \Psi_\beta^*\, \boldsymbol{p}\cdot\hat{\varepsilon}\, e^{-i\boldsymbol{k}_\gamma\cdot\boldsymbol{r}}\, \Psi_\alpha\, dV. \end{aligned} \tag{3.3.3}$$

In these equations dn_R is the number of photon states per unit volume of a given polarization in the energy range E to $E + dE$ and in the solid angle element $d\Omega$ about the propagation vector $\boldsymbol{k}_\gamma$. The polarization is specified by the unit vector $\hat{\varepsilon}$ and there are two independent directions of polarization. The matrix element given in eqn. (3.3.3) is appropriate for the state α to contain n_ω photons per radiation oscillator (cell in the phase space of the radiation field) and the state β to contain $n_\omega + 1$ photons per oscillator. Hence, we can use α and β as initial and final states, respectively, for emission, or β and α as initial and final states, respectively, for absorption. In this latter instance, however, it will be necessary to replace n_ω by $n_\omega - 1$ and $n_\omega + 1$ by n_ω. We also note that $|\boldsymbol{k}_\gamma| = \omega/c$ and that $\omega = \omega_{\alpha\beta} = (E_\beta - E_\alpha)/\hbar$, where E_α and E_β are the energies of the two atomic states involved. We will not always display this fact explicitly because of the notational complications.

The matrix element of eqn. (3.3.3) depends on a particular radiation oscillator within the infinitesimal range $dn_R = \omega^2\, d\Omega\, dE/(2\pi c)^3\hbar$ only through the "occupation number" of the cell, n_ω. Since, as stated previously, the square of the matrix element appearing in eqn. (3.2.1) or eqn. (3.2.3) has been summed over this infinitesimal volume of phase space, when we make use of eqn. (3.3.2) as the density of final states and eqn. (3.3.3) as the matrix element in eqn. (3.2.1), n_ω should be replaced by $\bar{n}_\omega$, the *average number* of photons per radiation oscillator in the range dn_R.

Bound-state Emission

We now insert eqns. (3.3.2) and (3.3.3) into eqn. (3.2.1) ($\alpha \equiv i$, $\beta \equiv j$) to obtain

$$dw_{\alpha\beta} = \frac{e^2\omega_{\alpha\beta}}{2\pi\hbar m^2 c^3} |\langle\beta|\, \boldsymbol{p}\cdot\hat{\boldsymbol{\varepsilon}}\, e^{-i\boldsymbol{k}_\gamma\cdot\boldsymbol{r}}\, |\alpha\rangle|^2 (\bar{n}_\omega + 1)\, d\Omega \tag{3.3.4}$$

for the transition probability per unit time for *emission* into the solid angle element $d\Omega$. The matrix element of $\boldsymbol{p}\cdot\hat{\boldsymbol{\varepsilon}}\, e^{-i\boldsymbol{k}_\gamma\cdot\boldsymbol{r}}$ indicated by the symbols α, β within the brackets is defined by the integral appearing in eqn. (3.3.3).

We now need to relate the occupation number n_ω of the radiation oscillators to the intensity of a beam of photons. Since the number of radiation oscillators (cells in phase space) per unit volume in an energy range dE is given by eqn. (3.3.2), we can from this equation compute the intensity of radiation corresponding to the presence of some average number $\bar{n}_\omega$, of photons per radiation oscillator. The physical properties of photons within the infinitesimal range

$$dn_R = \frac{\omega^2\, d\Omega\, dE}{(2\pi c)^3\, \hbar}$$

of oscillator states are the same; hence, a measurement of the intensity would not discriminate among them and would therefore constitute a measure of the *average* occupation of these dn_R states. Since

$$\frac{1}{\hbar\omega c} I_\omega(\widehat{\boldsymbol{\Omega}})\, d\Omega\, d\omega$$

is the number of photons per cm^3 in the (angular) frequency range $d\omega$ and in the element of solid angle $d\Omega$, it must be equal to $\bar{n}_\omega$ times the number of radiation oscillators per cm^3 in the same differential ranges $d\Omega$ and $d\omega$. Hence, we have

$$\frac{1}{\hbar\omega c} I_\omega(\hat{\Omega})\, d\omega\, d\Omega = \bar{n}_\omega \left\{\frac{\omega^2\, d\Omega}{(2\pi c)^3 \hbar}\right\} \hbar\, d\omega \qquad (3.3.5)$$

which can be solved for $I_\omega(\hat{\Omega})$ in terms of $\bar{n}_\omega$:

$$I_\omega(\hat{\Omega}) = \bar{n}_\omega \left\{\frac{\hbar\omega^3}{(2\pi)^3 c^2}\right\} \qquad (3.3.6)$$

or vice versa. The intensity function $I_\omega(\hat{\Omega})$ has been defined previously in Section 2.1, with respect to frequency $\nu = \omega/2\pi$.

We will usually make use of the dipole-length approximation wherein the exponential in the matrix element is taken to be unity and the electron momentum operator $\boldsymbol{p}$ is replaced by the coordinate operator $\boldsymbol{r}$ through use of the relation (Schiff, 1968, p. 404)

$$\langle\alpha|\, \boldsymbol{p}\, |\beta\rangle = im\omega_{\alpha\beta} \langle\alpha|\, \boldsymbol{r}\, |\beta\rangle \qquad (3.3.7)$$

(where $i = \sqrt{-1}$).

Making use of eqn. (3.3.7), we obtain from eqn. (3.3.4)

$$\frac{dw_{\alpha\beta}}{d\Omega} = \frac{e^2\omega_{\alpha\beta}^3}{2\pi\hbar c^3} |\hat{\varepsilon} \cdot \langle\beta|\, \boldsymbol{r}\, |\alpha\rangle|^2 (\bar{n}_\omega + 1) \qquad (3.3.8)$$

which by use of eqn. (3.3.6) becomes:

$$\frac{dw_{\alpha\beta}}{d\Omega} = \frac{e^2\omega_{\alpha\beta}^3}{2\pi\hbar c^3} |\hat{\varepsilon} \cdot \langle\beta|\, \boldsymbol{r}\, |\alpha\rangle|^2 \left(1 + \frac{I_\omega(\hat{\Omega})\, 8\pi^3 c^2}{\hbar\omega^3}\right) \qquad (3.3.9)$$

The first term in this expression is the *spontaneous* transition prob ability and the second is the *induced* transition probability. The tota transition probability (with emission into any solid angle element $d\Omega$) is of course, obtained by integrating eqn. (3.3.9) over all angles. However, except for isotropic radiation, this is not interesting as an atomic parameter, as it will depend on the intensity I_ω. The total *spontaneous* transition probability is an atomic parameter in any event, and can be obtained from eqn. (3.3.9) by integrating the first term over all angles. This parameter is also known as the Einstein A-coefficient $A_{\alpha\beta}$ (see also Sections 2.2 and 3.1):

$$w_{\alpha\beta}\,(\text{spont}) \equiv A_{\alpha\beta} = \int \frac{dw\,(\text{spont})}{d\Omega}\, d\Omega = \frac{e^2\omega_{\alpha\beta}^3}{2\pi\hbar c^3} \sum_l \int |\hat{\varepsilon}_l \cdot \boldsymbol{r}_{\beta\alpha}|^2\, d\Omega. \tag{3.3.10}$$

The indicated summation (l) over the two independent polarization directions must be carried out before following the usual procedure of taking an average over the random orientations of the dipole moment of the atom, since the contribution of a given polarization depends on the dipole-moment orientation which is specified by the vector

$$\boldsymbol{r}_{\beta\alpha} = \langle\beta|\, x\, |\alpha\rangle\, \hat{\imath} + \langle\beta|\, y\, |\alpha\rangle\, \hat{\jmath} + \langle\beta|\, z\, |\alpha\rangle\, \hat{\boldsymbol{k}}.$$

The summation can be carried out with reference to Fig. 3.2 by taking $\boldsymbol{r}_{\beta\alpha}$ as the direction of the z-axis. The angle between this axis and the propagation vector $\boldsymbol{k}_\gamma$ is denoted as θ. We further take $\hat{\varepsilon}_1$ and $\hat{\varepsilon}_2$ as two mutually perpendicular but otherwise arbitrary unit polarization vectors whose orientation is specified relative to a fixed set of axes in the plane perpendicular to $\boldsymbol{k}_\gamma$. This fixed set of axes is further chosen (Fig. 3.2) so

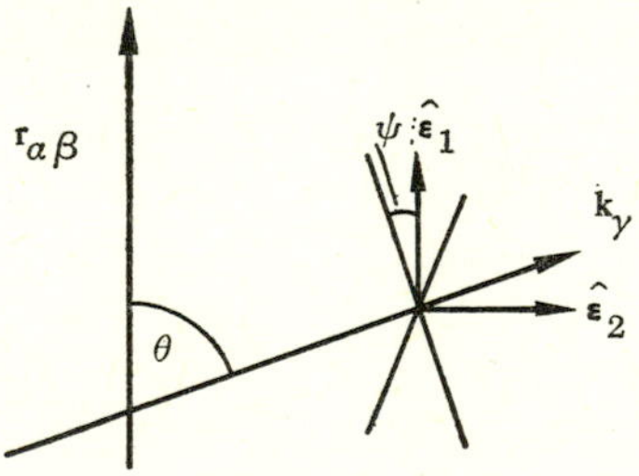

FIG. 3.2. Directional relationships between the dipole matrix element vector $r_{\alpha\beta}$, the propagation vector $\boldsymbol{k}$, and the unit polarization vectors $\hat{\varepsilon}_1$ and $\hat{\varepsilon}_2$.

that one of them lies in the plane of $\boldsymbol{r}_{\alpha\beta}$ and $\boldsymbol{k}_\gamma$ and the other then is perpendicular to this latter plane. With these conventions, we obtain, since $d\Omega$ is an element of solid angle about the direction $\boldsymbol{k}_\gamma$ and since

$$\int_0^\pi \sin^2\theta\, d\Omega = 2\pi \int_0^\pi \sin^3\theta\, d\theta = \frac{8\pi}{3},$$

the result (valid for a fixed direction $\boldsymbol{r}_{\beta\alpha}$):

$$A_{\alpha\beta} = \frac{e^2\omega_{\alpha\beta}^3}{2\pi\hbar c^3}\int |\boldsymbol{r}_{\beta\alpha}|^2 \sin^2\theta\,[\cos^2\psi + \sin^2\psi]\,d\Omega$$

$$= \frac{4}{3}\frac{e^2\omega_{\alpha\beta}^3}{\hbar c^3}|\boldsymbol{r}_{\beta\alpha}|^2 \quad \text{(a)}, \qquad (3.3.11)$$

or

$$= \frac{64\pi^4 e^2}{3h\lambda_{\alpha\beta}^3}|\boldsymbol{r}_{\beta\alpha}|^2 \quad \text{(b)}.$$

Note that we might as well have taken a single unit polarization vector lying in the plane of $\boldsymbol{r}_{\beta\alpha}$ and $\boldsymbol{k}_\gamma$, since the components perpendicular to this plane do not contribute. It is not necessary now to average over orientations of the dipole moment, as no directional dependence on $\boldsymbol{r}_{\beta\alpha}$ remains, due to the integration over solid angle. Equation (3.3.11a) is in the form given by Bethe and Salpeter (1957, eqn. 59.11) and eqn. (3.3.11b) is in the form given by Bates and Damgaard (1949), with the added provision that $|\boldsymbol{r}_{\beta\alpha}|^2$ be summed and averaged over final and initial states.

Because of the isotropic nature† of the emission given by eqn. (3.3.11), the emission per unit solid angle is just $A_{\alpha\beta}/4\pi$. We now compute the transition probability for induced emission from the second term of eqn. (3.3.9). The result is

$$\frac{dw_{\alpha\beta}}{d\Omega}(\text{ind}) = \frac{e^2\omega_{\alpha\beta}^3}{2\pi\hbar c^3} \times \frac{8\pi^3c^2}{\hbar\omega^3}\,|\hat{\boldsymbol{\varepsilon}}_1\cdot\boldsymbol{r}_{\beta\alpha}|^2 I_\omega^{(1)}(\widehat{\boldsymbol{\Omega}}) + |\hat{\boldsymbol{\varepsilon}}_2\cdot\boldsymbol{r}_{\beta\alpha}|^2 I_\omega^{(2)}(\widehat{\boldsymbol{\Omega}}) \qquad (3.3.12)$$

where the superscripts on the two $I_\omega(\widehat{\boldsymbol{\Omega}})$ factors indicate the polarization. We now assume that the radiation is unpolarized so that $I_\omega^{(1)}(\widehat{\boldsymbol{\Omega}}) = I_\omega^{(2)}(\widehat{\boldsymbol{\Omega}}) = \frac{1}{2}I_\omega(\widehat{\boldsymbol{\Omega}})$ and obtain

$$\frac{dw_{\alpha\beta}}{d\Omega}(\text{ind}) = \frac{2\pi^2e^2}{\hbar^2c}|\boldsymbol{r}_{\beta\alpha}|^2 I_\omega(\widehat{\boldsymbol{\Omega}})\sin^2\theta. \qquad (3.3.13)$$

As stated above, θ is the angle between the propagation vector $\boldsymbol{k}_\gamma$ and the atom dipole moment vector $\boldsymbol{r}_{\beta\alpha}$. Thus, we can simplify eqn. (3.3.13)

† On account of our dipole approximation, $|r_{\beta\alpha}|$ does not depend on the propagation vector k_γ.

by now averaging over random orientations of the dipole moment. To do this we multiply eqn. (3.3.13) by $d\Omega'/4\pi$ where $d\Omega'$ is an element of solid angle about $\boldsymbol{r}_{\beta\alpha}$. Since $(1/4\pi)\int \sin^2\theta\, d\Omega' = \frac{2}{3}$, we obtain

$$\frac{dw_{\alpha\beta}}{d\Omega}\,(\text{ind}) = \frac{4\pi^2 e^2}{3\hbar^2 c}\,|\boldsymbol{r}_{\beta\alpha}|^2\, I_\omega(\widehat{\boldsymbol{\Omega}}). \tag{3.3.14}$$

The total induced or stimulated emission can now be calculated from this result by integration over all angles of emission:

$$w_{\alpha\beta}\,(\text{ind}) = \frac{4\pi^2 e^2}{3\hbar^2 c}\,|\boldsymbol{r}_{\beta\alpha}|^2 \int I_\omega(\widehat{\boldsymbol{\Omega}})\, d\boldsymbol{\Omega}. \tag{3.3.15}$$

The induced or stimulated transition probability given by eqn. (3.3.15) is related to the Einstein B-coefficient. The B-coefficient is defined here so that [cf. eqn. (3.1.24)]

$$\frac{dw_{\alpha\beta}}{d\Omega}\,(\text{ind}) = B_{\alpha\beta}\left(\frac{1}{c}\, I_\nu(\widehat{\boldsymbol{\Omega}})\right) \tag{3.3.16}$$

which is the radiation density (ϱ_ν) rather than specific intensity (I_ν) definition since $\varrho_\nu = I_\nu/c$. [Cf. eqns. (2.2.4b) and (2.2.7a, b, c).] Comparing this with eqn. (3.3.14) and noting that

$$I_\nu(\widehat{\boldsymbol{\Omega}}) = I_\omega(\widehat{\boldsymbol{\Omega}})\,\frac{d\omega}{d\nu} = 2\pi\, I_\omega(\widehat{\boldsymbol{\Omega}})$$

we obtain

$$B_{\alpha\beta} = \frac{2\pi e^2}{3\hbar^2}\,|\boldsymbol{r}_{\beta\alpha}|^2 = \frac{8\pi^3 e^2}{3h^2}\,|\boldsymbol{r}_{\beta\alpha}|^2. \tag{3.3.17}$$

The last form agrees with the result given by Slater (1960, p. 142), whereas eqn. (3.3.15) is the form given by Schiff (1968, p. 405). Griem (1964, p. 28) employs a quantity $\mathscr{B}_{\alpha\beta}$ defined such that the total transition probability $w_{\alpha\beta}$ is expressed in terms of the average intensity $\bar{I}_\omega = (1/4\pi)\int I_\omega(\widehat{\boldsymbol{\Omega}})\, d\boldsymbol{\Omega}$:

$$w_{\alpha\beta} = A_{\alpha\beta} + B_{\alpha\beta}\,\frac{2\pi}{c}\int I_\omega\, d\Omega = A_{\alpha\beta} + \frac{8\pi^2 B_{\alpha\beta}}{c}\,\bar{I}_\omega = A_{\alpha\beta} + \mathscr{B}_{\alpha\beta}\bar{I}_\omega.$$

Thus, his formula for $A_{\alpha\beta}$ is the same as ours (recalling that for the time being we are assuming our states to be non-degenerate, so that the statistical weights are all unity) but his formula for $\mathscr{B}_{\alpha\beta}$ differs from ours by the factor $8\pi^2/c$. Griem also uses mks units; hence, to convert to cgs units the factor $4\pi\varepsilon_0$ should be removed from his formulas [specifically his eqns. (2.48) and (2.54)].

It is perhaps appropriate to observe at this point that the ratio $A_{\alpha\beta}/B_{\alpha\beta}$ as given by eqns. (3.3.11) and (3.3.17) has the value

$$\frac{A_{\alpha\beta}}{B_{\alpha\beta}} = \frac{8\pi h\nu_{\alpha\beta}^3}{c^3} \tag{3.3.18}$$

which, as stated prior to eqn. (3.1.30) of Section 3.1, is required by general considerations of detailed balance (see, for example, Slater, 1960, p. 21). Although the more or less heuristic method of "averaging over atom orientations" (performing an average over the angle between $\boldsymbol{r}_{\alpha\beta}$ and $\boldsymbol{k}_\gamma$) is the simplest method of eliminating the direct dependence of the transition probability on the coordinate system chosen, it is not necessarily the most perspicuous, and is, in fact, somewhat redundant. In the absence of an external field, different orientations of the atom correspond to different superpositions over degenerate sublevels of the magnetic quantum numbers which are to be averaged and summed over according to the fundamental prescription. The final result must still be summed and averaged over the magnetic quantum numbers of the final and initial states, although this procedure is simpler once the average over atom orientations is carried out. This problem can be avoided by carrying out the average and sum over quantum substates *ab initio*, and then demonstrating that this is equivalent to the usual procedure of first "averaging over atom orientations" followed by a simpler average and sum over substates. We therefore proceed to demonstrate the equivalence of these two procedures in a one-electron approximation as follows.

We can write $\boldsymbol{r} \cdot \hat{\boldsymbol{\varepsilon}} = rP_1(\cos\theta)$, where $\hat{\boldsymbol{\varepsilon}}$ is a unit vector and where the Legendre polynomial P_1 can be expressed in terms of the normalized spherical harmonics $Y_{lm}(\theta, \varphi)$ and the orbital angular momentum l by the relation

$$P_l(\cos\theta) = \left(\frac{4\pi}{2l+1}\right)^{1/2} Y_{l0}(\theta, \varphi).$$

By use of the spherical harmonics addition theorem (Rose, 1957, eqn. 4.27),

$$Y_{l0}(\theta, \varphi) = \sum_{m'} Y^*_{lm'}(\Theta, \Phi)\, Y_{lm'}(\theta_1, \varphi_1) \left(\frac{4\pi}{2l+1}\right)^{1/2}, \qquad (3.3.19)$$

we can relate $\boldsymbol{r} \cdot \hat{\varepsilon}$ to the angles $\Theta, \Phi, \theta_1, \varphi_1$ which specify the orientation of $\boldsymbol{r}$ and $\hat{\varepsilon}$, respectively, with respect to some fixed axis. We can then write for the relevant matrix element

$$\int \psi^*_\beta \boldsymbol{r} \cdot \hat{\varepsilon}\, \psi_\alpha\, dV = \left(\frac{4\pi}{3}\right) \int R^*_\beta r R_\alpha Y^*_\beta(\Theta, \Phi) \times \sum_{m'} Y^*_{lm'}(\Theta, \Phi)\, Y_{lm'}(\theta_1, \varphi_1)\, Y_\alpha(\Theta, \Phi)\, dV.$$

Now the integral over the product of three spherical harmonics has the value (Rose, 1957, pp. 61–62)

$$\int d\Omega\, Y^*_{l_3 m_3} Y_{l_2 m_2} Y_{l_1 m_1} = \left[\frac{(2l_1+1)(2l_2+1)}{4\pi(2l_3+1)}\right]^{1/2} C(l_1 l_2 l_3; m_1 m_2 m_3)\, C(l_1 l_2 l_3; 000) \qquad (3.3.20)$$

where the $C(l_1 l_2 l_3; m_1 m_2 m_3)$ are Clebsch–Gordan coefficients† (we follow Rose's notation). Using this result and the fact that $m_3 = m_1 + m_2$, we obtain for the matrix element above:

$$\int \psi^*_\beta \boldsymbol{r} \cdot \hat{\varepsilon} \psi_\alpha\, dV \equiv \boldsymbol{r}_{\beta\alpha} \cdot \hat{\varepsilon} = \left(\frac{4\pi}{3}\right)^{1/2} \int R^*_\beta r\, R_\alpha r^2\, dr \left[\frac{(2l_\alpha+1)}{(2l_\beta+1)}\right]^{1/2} \times C(l_\alpha 1 l_\beta, m_\alpha, m_\beta = m_\alpha, m_\beta)\, C(1 l_\alpha l_\beta, 000)\, Y_{1 m_\beta - m_\alpha}(\theta_1, \varphi_1)$$

$$= \left(\frac{4\pi}{3}\right)^{1/2} \mathcal{R}_{\alpha\beta} \left(\frac{2l_\alpha+1}{2l_\beta+1}\right)^{1/2} C(l_\alpha 1 l_\beta, m_\alpha, m_\beta - m_\alpha, m_\beta) \times C(l_\alpha 1 l_\beta, 000)\, Y_{1 m_\beta - m_\alpha}(\theta_1, \varphi_1) \qquad (3.3.21)$$

where we have defined the radial factor of the matrix element as in eqn. (3.2.4), $\int R^*_\beta R_\alpha r^3\, dr \equiv \mathcal{R}_{\beta\alpha}$. We now wish to sum this expression over final

† For the definition of these quantities, see Rose (1957) and Section 3.4 which follows.

states (β) and average it over initial states (α). First we carry out the sum over final states by summing over m_β and noting that $m_\beta - m_\alpha$ can take on only the values 0, ± 1. Denoting the matrix element $\int \psi_\beta^* \boldsymbol{r} \cdot \hat{\varepsilon} \psi_\alpha \, dV$ by $M_{\beta\alpha}$, we can write

$$\sum_{m_\beta} |M_{\beta\alpha}|^2 = \left(\frac{4\pi}{3}\right) \mathscr{R}_{\beta\alpha}^2 \left(\frac{2l_\alpha + 1}{2l_\beta + 1}\right) C^2(l_\alpha 1 l_\beta; 000) [C^2(l_\alpha 1 l_\beta; m_\alpha, 0, m_\alpha) Y_{1,0}^2 + C^2(l_\alpha 1 l_\beta; m_\alpha, 1, m_\alpha + 1) Y_{1,1}^2 + C^2(l_\alpha 1 l_\beta; m_\alpha, -1, m_\alpha - 1) Y_{1,-1}^2].$$

In order to perform the average over m_α, we make use of eqn. (3.16c) of Rose (1957):

$$C^2(j_1 j_2 j_3; m_1 m_2 m_3) = \left(\frac{2j_3 + 1}{2j_2 + 1}\right) C^2(j_1 j_2 j_3; m_1, -m_3, -m_2). \tag{3.3.22}$$

By interchanging the indices according to this relation, the sum and average needed becomes

$$\frac{1}{2l_\alpha + 1} \sum_{m_\alpha} \sum_{m_\beta} |M_{\beta\alpha}|^2 = \left(\frac{4\pi}{3}\right) \mathscr{R}_{\beta\alpha}^2 \left(\frac{1}{2l_\beta + 1}\right) C^2(l_\alpha 1 l_\beta; 000) \times \sum_{m_\alpha} \left[\frac{2l_\beta + 1}{3} C^2(l_\alpha l_\beta 1; m_\alpha, -m_\alpha, 0) Y_{1,0}^2 + \frac{2l_\beta + 1}{3} \times C^2(l_\alpha l_\beta 1; m_\alpha, -1 - m_\alpha, -1) Y_{11}^2 + \frac{2l_\beta + 1}{3} \times C^2(l_\alpha l_\beta 1; m_\alpha, 1 - m_\alpha, 1) Y_{1,-1}^2\right]. \tag{3.3.23}$$

Each of the sums over the squares of the C-coefficients can now be carried out by means of the orthonormality relations satisfied by these coefficients (Rose, 1957, eqn. 3.7),

$$\sum_{m_1} C(j_1 j_2 j; m_1, m - m_1, m)\, C(j_1 j_2 j'; m_1, m - m_1, m) = \delta_{jj'} \tag{3.3.24}$$

with the result that each sum is unity. Equation (3.3.23) thus becomes

$$\frac{1}{2l_\alpha + 1} \sum_{m_\alpha} \sum_{m_\beta} |M_{\beta\alpha}|^2 = \frac{4\pi}{9} \mathscr{R}^2_{\alpha\beta} C^2 (l_\alpha 1 l_\beta ; 000) \left[\sum_m Y^2_{1m} (\theta_1, \varphi_1)\right].$$

By use of the formula (Bethe and Salpeter, 1957, eqn. A.42)

$$\sum_m |Y_{lm} (\theta, \varphi)|^2 = \frac{2l + 1}{4\pi}, \tag{3.3.25}$$

we can finally write this result for *one polarization vector* $\hat{\varepsilon}_1$ as

$$\frac{1}{2l_\alpha + 1} \sum_{m_\alpha} \sum_{m_\beta} |M_{\beta\alpha}|^2 = \frac{1}{3} \mathscr{R}^2_{\beta\alpha} C^2 (l_\alpha 1 l_\beta ; 000). \tag{3.3.26}$$

Since this result no longer depends on the polarization vector we need only multiply by 2 to account for the two independent polarization vectors. For spontaneous emission, we found previously

$$w_{\alpha\beta} \text{(spont)} = \frac{e^2 \omega^3}{2\pi \hbar c^3} \int \{|\boldsymbol{r}_{\beta\alpha} \cdot \hat{\varepsilon}_1|^2 + |\boldsymbol{r}_{\beta\alpha} \cdot \hat{\varepsilon}_2|^2\} \, d\Omega .$$

If we make use of the sum and average we have just found, we can evaluate this as

$$\begin{aligned} w \text{(spont)} &= \frac{1}{2l_\alpha + 1} \sum_{m_\alpha} \sum_{m_\beta} w_{\alpha\beta} \text{(spont)} = \frac{e^2 \omega^3}{3\pi \hbar c^3} \mathscr{R}^2 (l_\alpha l_\beta) C^2 (l_\alpha 1 l_\beta ; 000) \int d\Omega \\ &= \frac{4 e^2 \omega^3}{3 \hbar c^3} \mathscr{R}^2 (l_\alpha l_\beta) C^2 (l_\alpha 1 l_\beta ; 000) \end{aligned} \tag{3.3.27}$$

with an analogous result for induced emission. We have changed notation on $\mathscr{R}$ from $\mathscr{R}_{\beta\alpha}$ to $\mathscr{R} (l_\alpha l_\beta)$ since this quantity does not depend on m_α or m_β. From the properties of the Clebsch–Gordan coefficients we know that $C (l_\alpha 1 l_\beta ; 000)$ vanishes unless $l_\beta = l_\alpha \pm 1$. The values of the particular Clebsch–Gordan coefficients $C (l, 1, l \pm 1 ; 000)$ may be found in a number of places. Condon and Shortley (1935, table 2^3), for example, give

$$
\begin{aligned}
C^2(l, 1, l+1; 000) &= \frac{(l+1)}{2l+1}, \\
C^2(l, 1, l-1; 000) &= \frac{l}{2l+1}.
\end{aligned}
\qquad (3.3.28)
$$

These are the coefficients (for the single, non-equivalent electron case) called C_{l+1}, and C_{l-1} by Bates (1946, see table I for s^2S and p^2P electron), and called $C(l \to l+1)$ and $C(l \to l-1)$ by Burgess and Seaton (1960). With these values of the C-coefficients we now obtain for the spontaneous-emission transition probability

$$
\begin{aligned}
w_{l,l+1}(\text{spont}) &= \frac{4e^2\omega^3}{3\hbar c^3} \frac{l+1}{2l+1} \mathscr{R}^2(l, l+1), \\
w_{l,l-1}(\text{spont}) &= \frac{4e^2\omega^3}{3\hbar c^3} \frac{l}{2l+1} \mathscr{R}^2(l, l-1).
\end{aligned}
\qquad (3.3.29)
$$

In order to demonstrate that this result is indeed the same as the result obtained by averaging over "atom orientations" and then averaging and summing over initial and final states, we must show that

$$
\frac{1}{2l_\alpha + 1} \sum_{m_\alpha} \sum_{m_\beta} |\boldsymbol{r}_{\beta\alpha}|^2 = C^2(l_\alpha 1 l_\beta; 000)\, \mathscr{R}^2(l_\alpha l_\beta) \qquad (3.3.30)
$$

since if this is true, eqn. (3.3.27) and eqn. (3.3.11) will agree. To show this, first we note that (cf., for example, Rose, 1957, eq. III.22)

$$
\begin{aligned}
x &= \left(\frac{4\pi}{6}\right)^{1/2} r\,[Y_{1,-1} - Y_{1,1}], \\
y &= i\left(\frac{4\pi}{6}\right)^{1/2} r\,[Y_{1,-1} + Y_{1,1}], \\
z &= \left(\frac{4\pi}{3}\right)^{1/2} r Y_{1,0}.
\end{aligned}
\qquad (3.3.31)
$$

With these relations and our previous definition of $\mathscr{R}_{\beta\alpha} \equiv \int \mathscr{R}_\beta^* \mathscr{R}_\alpha r^3 dr$ we can write the matrix elements of x, y and z as

$$\langle\beta|\, x\, |\alpha\rangle \equiv \int \psi_\beta^* x \psi_\alpha \, dV$$

$$= \left(\frac{2\pi}{3}\right)^{1/2} \mathscr{R}_{\beta\alpha} \int Y_\beta^* \left(Y_{1,-1} - Y_{1,1}\right) Y_\alpha \, d\Omega,$$

$$\langle\beta|\, y\, |\alpha\rangle = i\left(\frac{2\pi}{3}\right)^{1/2} \mathscr{R}_{\beta\alpha} \int Y_\beta^* \left(Y_{1,1} + Y_{1,-1}\right) Y_\alpha \, d\Omega,$$

$$\langle\beta|\, z\, |\alpha\rangle = \left(\frac{4\pi}{3}\right)^{1/2} \mathscr{R}_{\beta\alpha} \int Y_\beta^* Y_{10} Y_\alpha \, d\Omega. \tag{3.3.32}$$

We will abbreviate these three matrix elements by $x_{\beta\alpha}$, $y_{\beta\alpha}$ and $z_{\beta\alpha}$, respectively.

By eqn. (3.3.20), the angular integration over the three spherical harmonics can be performed, yielding the result

$$x_{\beta\alpha} = \left(\frac{2\pi}{3}\right)^{1/2} \mathscr{R}_{\beta\alpha} \left[\frac{3\,(2l_\alpha+1)}{4\pi\,(2l_\beta+1)}\right]^{1/2} C\,(l_\alpha 1 l_\beta;\, 000)$$
$$\times\, [C\,(l_\alpha 1 l_\beta;\, m_\alpha, -1, m_\beta) - C\,(l_\alpha 1 l_\beta;\, m_\alpha 1 m_\beta)],$$

$$y_{\beta\alpha} = \left(\frac{2\pi}{3}\right)^{1/2} \mathscr{R}_{\beta\alpha} \left[\frac{3\,(2l_\alpha+1)}{4\pi\,(2l_\beta+1)}\right]^{1/2} i\, C\,(l_\alpha 1 l_\beta;\, 000)$$
$$\times\, [C\,(l_\alpha 1 l_\beta;\, m_\alpha, -1, m_\beta) + C\,(l_\alpha 1 l_\beta;\, m_\alpha 1 m_\beta)], \tag{3.3.33}$$

$$z_{\beta\alpha} = \left(\frac{4\pi}{3}\right)^{1/2} \mathscr{R}_{\beta\alpha} \left[\frac{3\,(2l_\alpha+1)}{4\pi\,(2l_\beta+1)}\right]^{1/2} C\,(l_\alpha 1 l_\beta;\, 000)$$
$$\times\, C\,(l_\alpha 1 l_\beta;\, m_\alpha 0 m_\beta).$$

If now we square and add these components, the cross terms in $|x_{\beta\alpha}|^2$ and $|y_{\beta\alpha}|^2$ cancel yielding for the sum of the squares of the matrix elements the value,

$$|\boldsymbol{r}_{\beta\alpha}|^2 = |x_{\beta\alpha}|^2 + |y_{\beta\alpha}|^2 + |z_{\beta\alpha}|^2$$

$$= \mathscr{R}^2_{\beta\alpha} \left(\frac{2l_\alpha+1}{2l_\beta+1}\right) C^2\,(l_\alpha 1 l_\beta;\,000)\,[C^2\,(l_\alpha 1 l_\beta;\,m_\alpha,\,-1,\,m_\beta)$$

$$+\; C^2\,(l_\alpha 1 l_\beta;\,m_\alpha 1 m_\beta) + C^2\,(l_\alpha 1 l_\beta;\,m_\alpha 0 m_\beta)].$$

But we have that (Rose, 1957, eq. 3.16b)

$$C(j_1 j_2 j_3;\,m_1 m_2 m_3) = (-1)^{j_1+j_2-j_3}\,C(j_2 j_1 j_3;\,m_2 m_1 m_3) \qquad (3.3.34)$$

which allows us to permute the quantum numbers in the above expression to obtain

$$|\boldsymbol{r}_{\beta\alpha}|^2 = \left(\frac{2l_\alpha+1}{2l_\beta+1}\right) \mathscr{R}^2_{\beta\alpha}\,C^2\,(l_\alpha 1 l_\beta;\,000)\,[C^2\,(1 l_\alpha l_\beta;\,-1,\,m_\alpha,\,m_\beta)$$

$$+\; C^2\,(1 l_\alpha l_\beta;\,1 m_\alpha m_\beta) + C^2\,(1 l_\alpha l_\beta;\,0 m_\alpha m_\beta)]. \qquad (3.3.35)$$

The term in brackets is now the orthonormal sum on the first m-index and hence equals unity by eqn. (3.3.24). Thus, since the radial factor again does not depend on the m's we finally arrive at

$$\frac{1}{2l_\alpha+1}\sum_{m_\alpha}\sum_{m_\beta}|\boldsymbol{r}_{\beta\alpha}|^2 = \mathscr{R}^2\,(l_\alpha l_\beta)\,C^2\,(l_\alpha 1 l_\beta;\,000)$$

which is eqn. (3.3.30) as was to be demonstrated. In what follows, we will continue the tradition of averaging over "atom orientations" because of its simplicity, with the understanding, however, that it can be replaced by the foregoing more detailed considerations if necessary.

Bound-state Absorption

To obtain the transition probability per unit time for absorption we want the initial state β to contain $\bar{n}_\omega$ photons per radiation oscillator [to correspond to the specific intensity $I_\omega(\widehat{\boldsymbol{\Omega}})$ through the relation, eqn. (3.3.6)], and the final state α to contain $\bar{n}_\omega - 1$ photons per oscillator. In addition, we need to specify that the density of states given by eqn. (3.3.2) now pertains to the *initial* state. The matrix element given by eqn. (3.3.3) becomes

$$\langle\alpha|\, H'\,|\beta\rangle = -\frac{e}{m}\sqrt{\left(\frac{2\pi\hbar\bar{n}_\omega}{\omega}\right)}\,\langle\alpha|\,\boldsymbol{p}\cdot\hat{\boldsymbol{\varepsilon}}\,e^{+ik_\gamma\cdot \boldsymbol{r}}|\beta\rangle. \tag{3.3.36}$$

With dn_R/dE as given by eqn. (3.3.2), the transition probability per unit time as given by eqn. (3.2.1) for absorption of radiation from an element of solid angle $d\Omega$ becomes, using eqn. (3.3.7) and the dipole approximation,

$$dw_{\beta\alpha} = \frac{e^2\omega_{\alpha\beta}^3}{2\pi c^3\hbar}\,|\hat{\boldsymbol{\varepsilon}}\cdot\langle\alpha|\,\boldsymbol{r}\,|\beta\rangle|^2\, n_\omega^{(\varepsilon)}\, d\Omega,$$

for a given polarization $\hat{\boldsymbol{\varepsilon}}$. By use of eqn. (3.3.6) this can also be expressed as

$$dw_{\beta\alpha} = \frac{4\pi^2 e^2}{\hbar^2 c}\,|\hat{\boldsymbol{\varepsilon}}\cdot\langle\alpha|\,\boldsymbol{r}|\beta\rangle^2\, I_\omega^{(\varepsilon)}\,(\widehat{\boldsymbol{\Omega}})\, d\Omega. \tag{3.3.37}$$

If now we average this over polarizations, we obtain in the same manner as before

$$dw_{\beta\alpha} = \frac{2\pi^2 e^2}{\hbar^2 c}\,|\boldsymbol{r}_{\alpha\beta}|^2\, I_\omega(\widehat{\boldsymbol{\Omega}})\sin^2\theta\, d\Omega \tag{3.3.38}$$

where $\boldsymbol{r}_{\alpha\beta}$ was defined following eqn. (3.3.10). Again we average over atom dipole-moment orientations

$$\left(\int\frac{d\Omega'}{4\pi}\right)$$

and this produces a factor $\frac{2}{3}$. The net result is

$$dw_{\beta\alpha} = \frac{4\pi^2 e^2}{3\hbar^2 c}\,|\boldsymbol{r}_{\alpha\beta}|^2\, I_\omega(\widehat{\boldsymbol{\Omega}})\, d\Omega, \tag{3.3.39}$$

or, the total transition probability per unit time for absorption becomes

$$w_{\beta\alpha} = \frac{4\pi^2 e^2}{3\hbar^2 c}\,|\boldsymbol{r}_{\alpha\beta}|^2 \int I_\omega(\widehat{\boldsymbol{\Omega}})\, d\Omega. \tag{3.3.40}$$

These results [eqns. (3.3.39) and (3.3.40)] are seen to be identical with the differential and total transition probabilities for induced or stimulated emission†, eqns. (3.3.14) and (3.3.15), respectively, since $\boldsymbol{r}_{\alpha\beta}$ is simply the complex conjugate of $\boldsymbol{r}_{\beta\alpha}$. These results can be correlated with those of Schiff (1968, eqn. 44.23) by noting that the intensity $I(\omega)$ as defined by Schiff is the total intensity integrated over all angles; viz. our $\int I_\omega(\widehat{\Omega})\, d\Omega$.

As in the case of induced emission the total transition probability per unit time given by eqn. (3.3.40) is not solely an atomic parameter except in the case of isotropic radiation. Since we do not, in general, wish to be restricted to isotropic radiation it is convenient for us to define a purely atomic parameter which can be called an absorption coefficient per atom which is a cross section. In order to define such a parameter, we first must take into account the fact that a line is not infinitesimally sharp. In any actual physical system this is true, as the absorption will cover a finite (even if small) range of frequencies (Heitler, 1954, pp. 181–6; Mayer, 1947, p. 8) about the resonance frequency $\omega_{\alpha\beta} = (E_\beta - E_\alpha)/\hbar$. We can define the transition probability per unit frequency interval w_ω by the relation

$$w_\omega = \lim_{\Delta\omega \to 0} \frac{\Delta w}{\Delta\omega}$$

where Δw is that contribution to the total transition probability which occurs in the angular frequency range $\Delta\omega$. Then we must have

$$w = \int w_\omega \, d\omega \tag{3.3.41}$$

where the integration is taken over the entire profile of the line in order that the total transition probability be equal to the sum of its parts. In view of eqn. (3.3.41) it is convenient to account for the line shape‡ phenomenologically by assigning a line-shape factor $b(\omega)$ to the line as we did

† The absorption and emission transition probabilities are equal for the exact matrix element, as well as in the dipole approximation as shown here; see Schiff (1968).

‡ For a very entertaining quantum mechanical derivation of a line shape similar to that of eqn. (3.1.4), the reader is referred to Franken (1969).

in Chapter 2 [e.g. eqn. (2.2.4a)], such that

$$w_\omega = wb(\omega). \tag{3.3.42}$$

Equation (3.3.42) will be valid if we normalize $b(\omega)$ according to

$$\int_0^\infty b(\omega)\, d\omega = 1. \tag{3.3.43}$$

Thus, w_ω as given by eqn. (3.3.42) is the number of transitions per unit time per unit frequency interval. For the idealized infinitely sharp transition, we have

$$b(\omega) = \delta(\omega - \omega_{\alpha\beta}), \tag{3.3.44}$$

whereas, for real transitions, $b(\omega)$ will be a finite (but usually still sharp) function.

We now recall that $I_\omega(\hat{\boldsymbol{\Omega}})\, d\Omega$ refers to the incident beam intensity in the direction $\hat{\boldsymbol{\Omega}}$. A cross section for absorption can be defined by reducing this angularly distributed beam intensity to a plane-parallel incident beam through

$$d\Omega\, I_\omega(\hat{\boldsymbol{\Omega}}) = I_0(\omega)\, \delta(\hat{\boldsymbol{\Omega}})\, d\Omega \tag{3.3.45}$$

where $I_0(\omega)$ (energy per cm^2/sec per unit frequency interval) is the parallel spectral incident beam intensity. Since $\int \delta(\hat{\boldsymbol{\Omega}})\, d\Omega = 1$, we can substitute eqn. (3.3.45) into eqn. (3.3.40) together with eqns. (3.3.41) and (3.3.42). The result after carrying out the integration over $\hat{\boldsymbol{\Omega}}$ is

$$w_{\beta\alpha} = \frac{4\pi^2 e^2}{3\hbar^2 c} |\boldsymbol{r}_{\alpha\beta}|^2 \int_\omega I_0(\omega)\, b(\omega)\, d\omega \tag{3.3.46}$$

for the total transition probability per unit time, and by dropping the ω-integration, we obtain

$$w_{\beta\alpha}(\omega) = \frac{4\pi^2 e^2}{3\hbar^2 c} |\boldsymbol{r}_{\alpha\beta}|^2 I_0(\omega)\, b(\omega) \tag{3.3.47}$$

for the transition probability per unit time *per unit frequency interval.*

A cross section can now be defined by

$$\sigma_{\beta\alpha}(\omega) = \frac{w_{\beta\alpha}(\omega)\, \hbar\omega}{I_0(\omega)} \tag{3.3.48}$$

which, according to eqn. (3.3.47), will have the value

$$\sigma_{\beta\alpha}(\omega) = \frac{4\pi^2 e^2 \omega}{3\hbar c} |\boldsymbol{r}_{\alpha\beta}|^2 \, b(\omega). \tag{3.3.49}$$

The frequency integral of this quantity

$$\int \sigma_{\beta\alpha}(\omega)\, d\omega = \frac{4\pi^2 e^2 \omega}{3\hbar c} |\boldsymbol{r}_{\alpha\beta}|^2 \tag{3.3.50}$$

is just the ratio $S/I_0(\omega)$ as given by Heitler (1954, p. 180, eqn. 19) where S is the energy absorbed per unit time as he defines it. Since, as noted after eqn. (3.3.40), the transition probabilities per unit time for absorption and induced emission are identical, it can be seen that the cross sections for these two processes are also equal. This requires that the shape of an absorption line be the same as an emission line† which must be true from general equilibrium considerations (Heitler, 1954, p. 186).

Just as in the discussion by Bethe and Salpeter (1957, p. 296) of photoelectric absorption, the cross section we have defined here has the physical interpretation that

$$\int \sigma(\omega) \frac{I_0(\omega)}{\hbar\omega}\, d\omega \equiv w$$

is the probability per second that an atom upon which the flux $I_0(\omega)$ is incident will be excited. Also, the inverse mean free path or absorption coefficient $N_v \sigma(\omega)$ is the probability per centimeter that a photon of angular frequency ω will be absorbed in a medium consisting of N_v absorbing atoms per cm^3. Thus, we can interpret the product $\sigma N_v I_0(\omega)\, dx$ as:

$$\begin{aligned}(\sigma N_v\, dx)\, I_0(\omega) &= (\text{probability of photon absorption in } dx) \\ &\quad \times \frac{\text{Number of photons} \times \hbar\omega}{\text{cm}^2\text{-sec-frequency interval}} \\ &= \text{energy absorbed in } dx \text{ per cm}^2 \text{ per sec} \\ &\quad\ \text{per unit frequency interval} \\ &= dI_0(\omega).\end{aligned} \tag{3.3.51}$$

† When emitted or absorbed at the atom—not after transfer through a medium.

Alternatively, we can interpret this product as:

$$\frac{\sigma(\omega)\, I_0(\omega)}{\hbar\omega}\,(N_v\,dx)\,\hbar\omega = \frac{\text{Probability of excitation per atom}}{\text{sec} \times \text{frequency interval}}$$

$$\times\ (\text{number of atoms/cm}^2) \times \text{energy of excitation}$$

$$= \text{energy absorbed in } dx \text{ per cm}^2 \text{ per sec}$$

$$\text{per unit frequency interval}$$

$$= dI_0\,(\omega). \qquad (3.3.52)$$

As far as the transfer equation (Section 2.1) is concerned, it is not really necessary to define a cross section. This was done in Section 2.1 for convenience of interpretation, but all that is needed is the energy that is absorbed (and/or emitted) from the beam. The absorption can be obtained directly from the transition probability (per unit frequency interval) since

$$dI_\omega(\widehat{\boldsymbol{\Omega}}) = -N_v\,\hbar\omega\,dw_{\alpha\beta}(\omega)\,ds \qquad (3.3.53)$$

for the beam intensity decrement $dI_\omega\,(\widehat{\boldsymbol{\Omega}})$ in a distance ds. The phenomenological definition of the absorption coefficient is (Section 2.1)

$$\mu(\omega) = -\frac{dI_\omega(\widehat{\boldsymbol{\Omega}})}{ds}\,\frac{1}{I_\omega(\widehat{\boldsymbol{\Omega}})}. \qquad (3.3.54)$$

Inserting eqn. (3.3.39) into eqn. (3.3.53) with the definitions eqn. (3.3.41) and eqn. (3.3.42) now leads by the use of eqn. (3.3.54) to†

$$\mu(\omega) = N_v \cdot \frac{4\pi^2 e^2 \omega}{3\hbar c}\,|\boldsymbol{r}_{\alpha\beta}|^2\, b(\omega). \qquad (3.3.55)$$

Thus, the quantity

$$\frac{4\pi^2 e^2 \omega}{3\hbar c}\,|\boldsymbol{r}_{\alpha\beta}|^2\, b(\omega)$$

can also be referred to as the absorption coefficient per atom, $\mu(\omega)/N_v$.

† This result implicitly assumes that the absorption is due entirely to the single transition $\alpha \rightarrow \beta$. This has been done for simplicity, to avoid cumbersome summation signs.

Dirac (1958, p. 245) gives

$$\frac{8\pi^3 \nu e^2}{hc} |\langle\alpha| \boldsymbol{r} \cdot \hat{\boldsymbol{\varepsilon}} |\beta\rangle|^2 \tag{3.3.56}$$

as an absorption coefficient. Referring to p. 245 of this book, however, we see that this is the "probability per unit time of an absorption taking place with an incident beam of one particle per unit area per unit time per unit frequency range". Hence, it is obtainable from our eqn. (3.3.37). First we set $I_\omega(\widehat{\boldsymbol{\Omega}}) = \hbar\omega N_0^{(\varepsilon)}(\omega)\,\delta(\widehat{\boldsymbol{\Omega}})$ where $N_0^{(\varepsilon)}(\omega)\,d\omega$ is the number of photons of a given polarization $\hat{\boldsymbol{\varepsilon}}$ in the angular frequency interval $d\omega$ crossing unit area per second, and then integrate over $\widehat{\boldsymbol{\Omega}}$. We obtain

$$w_{\beta\alpha}^{(\varepsilon)}(\sec^{-1}) = \frac{4\pi^2 e^2 \omega}{\hbar c} |\langle\alpha| \hat{\boldsymbol{\varepsilon}} \cdot \boldsymbol{r} |\beta\rangle|^2 N_0^{(\varepsilon)}(\omega). \tag{3.3.57}$$

Now we convert to frequency ν rather than angular frequency ω by $N_0(\omega)\,d\omega = n_0(\nu)\,d\nu$, or $N_0(\omega) = n_0(\nu)/2\pi$. Dirac's absorption coefficient, eqn. (3.3.56), is then equal to $w_{\beta\alpha}^{(\varepsilon)}$ for $n_0(\nu) = 1$.

Another quantity commonly defined is the oscillator strength or f-number (Section 3.1; Bethe and Salpeter, 1957, p. 256; Condon and Shortley, 1935, p. 108) for a transition from a (single) state β to a (single) state α:

$$f_{\alpha\beta} = \frac{2m}{3\hbar} \omega_{\beta\alpha} |\boldsymbol{r}_{\alpha\beta}|^2. \tag{3.3.58}$$

In terms of this quantity our cross section $\sigma_{\beta\alpha}(\omega)$ of eqn. (3.3.49) becomes

$$\sigma_{\beta\alpha}(\omega) = \frac{2\pi^2 e^2}{mc} f_{\alpha\beta} b(\omega), \tag{3.3.59}$$

or, in terms of $b(\omega)\,d\omega = b(2\pi\nu) \times 2\pi d\nu = b(\nu)\,d\nu$ and $\sigma(\omega)\,d\omega = \sigma(\nu)\,d\nu$, this can be written as

$$\sigma(\nu) = \frac{\pi e^2}{mc} f b(\nu) \tag{3.3.60}$$

where we have dropped the subscripts for simplicity. The f-number definition is made, of course, because of the classical correspondences noted in

Section 3.1. That is to say, if we multiply the above definition by the number density N_v and integrate over frequency ν, we obtain

$$\int \mu_\nu \, d\nu = N_v \int \sigma(\nu) \, d\nu = \frac{\pi e^2 N_v f}{mc}. \tag{3.3.61}$$

Thus, we recover the classical expression, eqn. (3.1.17b) of Section 3.1, and complete the quantal derivation of that important formula, first introduced by eqn. (2.6.22).

3.4. Reduction of the General Quantum Formulas for Bound–Free Processes

There are relatively few pedagogical discussions in the literature of the details of the calculation of photoelectric cross sections. Most discussions give the final formulas without the details of the reduction from the formal theory, or else give a very specific calculation, such as the transition from the *K*-shell, which avoids the general angular momentum reduction. The most comprehensive modern theoretical discussion is the advanced treatment of Bethe and Salpeter (1957). Marr (1967) gives an excellent modern review of both theory and experiment, but does not treat the mathematical details of the theory. The present section is intended to supplement these discussions from the pedagogical standpoint by illustrating some of the manipulations useful in the application of the theory to a variety of (non-relativistic) processes and systems. The details of the photoelectric process provide excellent exercises in applied quantum mechanics of considerable heuristic value to the student. The level of difficulty we address is very similar in spirit to that of the book by Bethe and Jackiw (1968). We also derive some of the final formulas often quoted in the literature without derivation and this should be of value both to the student and to the applied investigator. We make a one-electron, central-field approximation throughout the section, wherein we assume it is valid to factor a single electron wave function out of the total wave function, and we neglect any explicit consideration of exchange. Table 3.1 gives a bibliography of selected references on the calculation of non-relativistic photoelectric cross sections. The early references are included for their very considerable historical interest. For large-scale practical calculations of absorption coefficients for heated gases, the paper of Burgess and Seaton is of particular importance. Their results were first applied to opacity calculations by Peach (1962), and later by Armstrong *et al.* (1967). This work of Burgess and Seaton was further extended and corrected in later papers by Peach (1965; 1967b) and applied by her to a large variety of atomic systems (Peach, 1967c; 1970). Travis and Matsushima (1968) have made similar calculations for stellar atmosphere studies.

In the case of photoelectric absorption and emission we must deal with a free electron in the presence of a residual ion. Hence, we must specify the type of wave function (and its normalization) that is to be used to describe the free electron. The simplest description, if applicable, is to use a plane-wave momentum eigenfunction normalized to the volume L^3. This is the method illustrated by Schiff (1968) and by Heitler (1954), and is an appropriate approximation for high-energy absorption, and is mathematically and conceptually convenient as well. For the limitations of this approximation and improvements thereto, see, for example, Bethe and Salpeter (1957, section 70), Mott and Massey (1965, chap. XXI) and Marr (1967, chap. 3).

TABLE 3.1 Selected references on photoelectric absorption cross sections

Kramers (1923)	Bethe and Salpeter (1957)
Milne (1924)	Seaton (1958)
Wentzel (1926)	Armstrong (1959)
Oppenheimer (1928, 1929)	Burgess and Seaton (1960)
Gaunt (1930)	Dalgarno and Parkinson (1960)
Stobbe (1930)	Karzas and Latter (1961)
Maue (1932)	Cooper and Martin (1962)
Menzel and Pekeris (1935)	Cooper (1962)
Massey and Smith (1936)	Ditchburn and Opik (1962)
Hall (1936)	Altick and Glassgold (1964)
Sommerfeld (1939)	Dalgarno, Henry, and Stewart (1964)
Bates, Buckingham, Massey and Unwin (1939)	Johnston (1964)
Bates and Massey (1943)	Burgess (1964)
Bates (1946)	Peach (1962)
Bates and Seaton (1949)	Armstrong *et al.* (1967)
Seaton (1951)	Marr (1967)
Weissler (1956)	Peach (1967b)

However, at low energies the effect of the potential in the final state is important, the dipole approximation is usually valid, and one is usually interested in a final state that is an angular momentum eigenfunction rather than a linear momentum eigenfunction. Since the questions involved in passing from a free-state eigenfunction of one type to one another type as well as the free-state normalization methods are not trivial, we shall attempt to give a fairly complete discussion of these aspects in order to facilitate the use and understanding of the fundamental formulas.

Differential Cross Section for Linear Momentum Final Eigenstate

Although in the calculation of the photoelectric transition probability our basic equation is still eqn. (3.2.1), the calculation is somewhat more subtle on account of the free electron which appears in the final state. In order to obtain a transition probability per unit time in the case of bound–bound absorption, it was necessary to sum over the continuous range of initial (photon) states. Hence in that case, the density of states dn/dE was taken to be the density of initial photon states per unit volume, and was expressed in terms of the incident intensity through eqn. (3.3.6). For photoelectric transitions, there is a continuous range of states in the final state of the system provided by the free electron, so that the derivation employed in the case of bound–bound transitions is no longer appropriate. The final state sum employed in arriving at eqn. (3.2.1) should now be over a differential element of the free electron phase space, and dn/dE taken as the density of final free electron *momentum eigenstates* in this phase space. If $\boldsymbol{p}$ is the electron momentum, and we denote the element of solid angle about its direction as $d\Omega_e$, this density becomes:

$$\frac{dn_e}{dE} = \frac{d^3\boldsymbol{p}\, L^3}{(2\pi\hbar)^3\, dE} = \frac{mp\, d\Omega_e\, L^3}{(2\pi\hbar)^3} \tag{3.4.1}$$

since $dE = (p/m)\, dp$. The normalization volume for the free electron has been denoted by L^3, and the result is for electrons of a *given* spin, not both spins. We will discuss the sums and averages over degenerate states later. The matrix element is obtained from eqn. (3.3.3) just as for bound–bound absorption except that n_ω is not now *required* to be the average value $\bar{n}_\omega$ (we are not summing over radiation oscillators) but we could take it to be $\bar{n}_\omega$ if convenient, since the rest of the matrix element and dn_e/dE is independent of the radiation oscillator (within the infinitesimal element dn_R).

As in the case of eqn. (3.3.36), for an absorption transition we want the occupation numbers to go from n_ω to $n_\omega - 1$ so that our matrix element for photoelectric absorption becomes

$$\langle\alpha|\, H'\,|\beta\rangle = \frac{1}{V^{1/2}}\frac{e}{m}\sqrt{\left(\frac{2\pi\hbar n_\omega}{\omega}\right)}\,\langle\alpha|\,\boldsymbol{p}\cdot\hat{\varepsilon}\,e^{+ik_\gamma\cdot r}|\beta\rangle \tag{3.4.2}$$

and we must bear in mind that we have assumed, in specifying dn_e/dE by eqn. (3.4.1), that the free electron is normalized within a volume L^3. We will return to this point later. We have also explicitly inserted the photon normalization volume V. This was set equal to unity previously since dn_R/dE was expressed as the number density per unit volume, and the two factors if included explicitly would cancel against each other (see Heitler, 1954, p. 57).

Since dn_R/dE is not used as before, it seems advisable to display it explicitly now. Heitler (1954) evaluates a transition probability and cross section in the following manner. He assumes an initial system of just the atom plus one photon.† Then n_ω in eqn. (3.4.2) is unity for some one radiation oscillator (in the initial state) and vanishes for all others. In the final state all $n_\omega = 0$. Thus, with no photons in the final state, the density of final states per unit energy interval (of the free electron) is just that given by eqn. (3.4.1). Inserting eqns. (3.4.1) and (3.4.2) with $n_\omega = 1$ into eqn. (3.2.1), we obtain (initial state β, final state α):

$$dw_{\beta\alpha} = \frac{1}{2\pi}\frac{e^2}{V\omega_{\beta\alpha}}\,|\langle\alpha|\,\boldsymbol{p}\cdot\hat{\varepsilon}\,e^{+ik_\gamma\cdot r}|\beta\rangle|^2\,\frac{v\,d\Omega_e L^3}{\hbar^3} \tag{3.4.3}$$

for the transition probability per unit time for the absorption of a photon of angular frequency ω with emission of an electron of velocity v into the element of solid angle $d\Omega_e$. If again, we make use of the dipole approxima-

† That this procedure is not possible in the case of discrete absorption follows from the requirement of a continuous range of initial photon states in order for a constant transition probability per unit time to exist for discrete absorption.

tion and eqn. (3.3.7), this result can be written as

$$dw_{\beta\alpha} = \frac{m^2 e^2 \omega_{\beta\alpha}}{2\pi\hbar^3 V} |\langle\alpha| \boldsymbol{r} \cdot \hat{\varepsilon}| |\beta\rangle|^2 v \, d\Omega_e L^3. \tag{3.4.4}$$

We can obtain a cross section from this result by noting that the intensity corresponding to one photon in the volume V is (*not* per unit frequency interval):

$$I_0 = \frac{\hbar\omega c}{V}. \tag{3.4.5}$$

With this value of the intensity, the formula

$$d\sigma_{\beta\alpha} = dw_{\beta\alpha}\hbar\omega/I_0 \tag{3.4.6}$$

for the cross section (cf. eqn. (3.3.48)) becomes

$$d\sigma_{\beta\alpha} = \frac{dw_{\beta\alpha}V}{c} \tag{3.4.7}$$

so that, from eqn. (3.4.4)

$$d\sigma_{\beta\alpha} = \frac{m^2 e^2 \omega_{\beta\alpha}}{2\pi\hbar^3 c} |\langle\alpha| \boldsymbol{r} \cdot \hat{\varepsilon} |\beta\rangle|^2 vL^3 \, d\Omega_e. \tag{3.4.8}$$

For unpolarized light, with reference to Fig. 3.2, we can now average over the two independent polarization directions:

$$\begin{aligned} d\sigma_{\beta\alpha} &= \frac{m^2 e^2 \omega_{\beta\alpha}}{2\pi\hbar^3 c} vL^3 \, d\Omega_e \left(\frac{1}{2}\right) (|\langle\alpha| \boldsymbol{r} \cdot \hat{\boldsymbol{\varepsilon}}_1| \beta\rangle|^2 + |\langle\alpha| \boldsymbol{r} \cdot \hat{\boldsymbol{\varepsilon}}_2 |\beta\rangle|^2) \\ &= \frac{m^2 e^2 \omega_{\beta\alpha}}{4\pi\hbar^3 c} vL^3 \, d\Omega_e \{|\boldsymbol{r}_{\alpha\beta}|^2 \sin^2\theta\} \end{aligned} \tag{3.4.9}$$

where θ, as noted in Fig. 3.2, is the angle between the photon propagation direction $\boldsymbol{k}_\gamma$ and the vector $\boldsymbol{r}_{\alpha\beta}$, which was defined following eqn. (3.3.10).

Before proceeding to obtain the total cross section, it is interesting to rederive the above result without making the assumption $n_\omega = 1$.

If we wish to use the matrix element eqn. (3.4.2) as it stands with $n_\omega \neq 1$, then there will be $n_\omega - 1$ photons present in the final state. For each free electron state there will now be (cf. eqn. (3.3.2))

$$dn_R = V\omega^2 \, d\Omega_\gamma \, d\omega/(2\pi c)^3$$

photon states (containing photons with the same physical properties). For clarity, we have appended a subscript γ on the solid angle element about the photon (radiation) direction. We must now average the square of the matrix element eqn. (3.4.2) over the initial photon states which requires setting $n_\omega = \bar{n}_\omega$, and using the factor $dn_R \times dn_e/dE$ for the final state density. The transition probability now becomes

$$\delta w_{\beta\alpha} = \frac{e^2 \bar{n}_\omega}{2\pi V \omega_{\beta\alpha}} |\langle\alpha| \boldsymbol{p} \cdot \hat{\boldsymbol{\varepsilon}} \, e^{+i\boldsymbol{k}_\gamma \cdot \boldsymbol{r}} |\beta\rangle|^2 \, \frac{v \, d\Omega_e L^3}{\hbar^3} \times \frac{V\omega^2 \, d\Omega_\gamma \, d\omega}{(2\pi c)^3}. \quad (3.4.10)$$

But, by eqn. (3.3.5),

$$\frac{\bar{n}_\omega \omega^2 \, d\Omega_R \, d\omega}{(2\pi c)^3} = \frac{I_\omega(\widehat{\boldsymbol{\Omega}}) \, d\omega \, d\Omega_\gamma}{\hbar \omega c}$$

whereby we obtain from eqn. (3.4.10)

$$\delta w_{\beta\alpha} = \frac{e^2}{2\pi \omega_{\beta\alpha}} |\langle\alpha| \boldsymbol{p} \cdot \hat{\boldsymbol{\varepsilon}} \, e^{+i\boldsymbol{k}_\gamma \cdot \boldsymbol{r}} |\beta\rangle|^2 \, \frac{v \, d\Omega_e L^3}{\hbar^3} \times \frac{I_\omega(\widehat{\boldsymbol{\Omega}}) \, d\omega \, d\Omega_\gamma}{\hbar \omega c}.$$

From this result, which can be written in the dipole approximation as

$$\delta w_{\beta\alpha} = \left\{ \frac{e^2 m^2 \omega_{\beta\alpha}}{2\pi \hbar^3} |\langle\alpha| \boldsymbol{r} \cdot \hat{\boldsymbol{\varepsilon}} |\beta\rangle|^2 \, v L^3 \, d\Omega_e \right\} \frac{I_\omega(\widehat{\boldsymbol{\Omega}}) \, d\omega \, d\Omega_\gamma}{\hbar \omega c} \quad (3.4.11)$$

it is easy to see greater the flexibility of the initial state description in the case of photoelectric absorption as opposed to the case of discrete absorption. For a plane-parallel, monochromatic beam

$$I_\omega(\widehat{\boldsymbol{\Omega}}) = I_0 \, \delta(\widehat{\boldsymbol{\Omega}}) \, \delta(\omega - \omega_{\beta\alpha}).$$

Equation (3.4.10) reduces to

$$dw_{\beta\alpha} = \iint dw_{\beta\alpha} \, d\omega \, d\Omega = \frac{e^2 m^2 \omega_{\beta\alpha}}{2\pi \hbar^3} |\langle\alpha| \boldsymbol{r} \cdot \hat{\boldsymbol{\varepsilon}} |\beta\rangle|^2 \, v L^3 \, d\Omega_e \, \frac{I_0}{\hbar \omega_{\beta\alpha} c}. \quad (3.4.12)$$

A differential cross section is usually defined on the basis of these definitions and is [cf. eqn. (3.4.6)]:

$$\frac{d\sigma_{\beta\alpha}}{d\Omega_e} = \frac{dw/d\Omega_e \times \hbar\omega_{\beta\alpha}}{I_0} \tag{3.4.13}$$

This formula yields, using eqn. (3.4.12),

$$\frac{d\sigma_{\beta\alpha}}{d\Omega_e} = \frac{e^2 m^2 \omega_{\beta\alpha}}{2\pi\hbar^3 c} |\langle\alpha| \boldsymbol{r} \cdot \hat{\boldsymbol{\varepsilon}} |\beta\rangle|^2 \, vL^3.$$

This is, of course, identical to eqn. (3.4.8). However, it can also be defined directly from eqn. (3.4.11) by means of

$$\frac{d\sigma}{d\Omega_e} = \frac{\delta w \times \hbar\omega}{I_\omega(\widehat{\Omega}) \, d\omega \, d\Omega_\gamma \, d\Omega_e}. \tag{3.4.14}$$

This leads to the same result, of course, but is explicitly independent of the nature of the incident beam. We can also derive the photoionization cross section semiclassically†—that is, without the use of the field-theoretic matrix element eqn. (3.3.3). We follow the treatment of Schiff (1968). In this case, we have the perturbation as given by eqn. (3.3.1), but the radiation field is not quantized. Thus, the final state density does not include photon states and is just as given by eqn. (3.4.1). The transition probability is still

$$w_{\beta\alpha} = \frac{2\pi}{\hbar} |\langle\alpha| H' |\beta\rangle|^2 \frac{dn}{dE}$$

since this formula comes from ordinary quantum mechanics if the radiation field is not quantized. Thus for a transition with emission of an electron into $d\Omega_e$ we obtain

$$dw_{\beta\alpha} = \frac{2\pi}{\hbar} \frac{e^2}{m^2 c^2} |\langle\alpha| \boldsymbol{p} \cdot \boldsymbol{A} |\beta\rangle|^2 \frac{mpL^3 \, d\Omega_e}{(2\pi\hbar)^3}. \tag{3.4.15}$$

† This can also be done, of course, for bound–bound transitions (Schiff, 1968). We include both methods for heuristic purposes.

Now classically we can take†

$$A = \hat{\varepsilon} A_0 \, e^{+i(k_\gamma \cdot r - \omega t)} \tag{3.4.16}$$

for the vector potential. The Poynting vector $c/4\pi\,(\boldsymbol{E} \times \boldsymbol{H})$ averaged over a period $2\pi/\omega$ of the oscillation has the magnitude $\omega^2/2\pi c\,|A_0|^2$. This is the intensity I_0 of the classical wave given by eqn. (3.4.16). (I_0 is not defined per unit frequency interval.)

The cross section is again given by

$$d\sigma = \frac{dw\,\hbar\omega}{I_0}.$$

With the above expression for the intensity and eqn. (3.4.16), we obtain from eqn. (3.4.15) by use of eqn. (3.3.7),

$$d\sigma_{\beta\alpha} = \frac{1}{2\pi}\frac{e^2 m^2 \omega}{\hbar^3 c}\,|\langle\alpha|\,\boldsymbol{r}\cdot\hat{\varepsilon}\,|\beta\rangle|^2\, vL^3\, d\Omega_e$$

in agreement with our previous result, eqn. (3.4.8).

Unfortunately this simple form of the differential cross section [or the more explicit form averaged over polarizations, eqn. (3.4.9)] is not a particularly useful formula, except at high energies where a plane-wave final state is a good approximation. Except in this latter approximation the angle θ which appears in eqn. (3.4.9) is not the angle of physical interest. In radiation absorption studies of the type we are concerned with here, this is not a disadvantage, as one is not usually concerned with the differential cross section *per se*‡, but rather with the total cross section obtainable from it by a suitable integration over angles.

Recombination and Photoelectric Total Cross Sections by Integration of Differential Cross Sections

It is evident from the form of eqn. (3.4.9) that it would be conceptually simpler to integrate this equation over photon propagation directions

† The conditions under which this is valid and the matrix element eqn. (3.3.1) is the only required perturbing term are discussed by Schiff (1968) in sect. 44.

‡ An extensive discussion of the differential cross section and the angular distribution of the photoejected electrons is given by Bethe and Salpeter (1957); see especially Sects. 69, 70, and 72.

(for a fixed $|\boldsymbol{r}_{\alpha\beta}|$) than over photoelectron directions. We would thereby avoid having to determine directly the dependence of $|\boldsymbol{r}_{\alpha\beta}|$ on the direction of the photoelectron in the final state. We can perform the integration of eqn. (3.4.9) over photon propagation directions in lieu of photoelectron directions by relating the basic photoelectric cross section of eqn. (3.4.8) to the cross section for the inverse process of radiative recombination. In this latter process, a free electron is incident, in the initial state, on an atom or ion. It is captured with the emission of a photon in the transition to the final state, which therefore contains a photon and an atom or ion of charge one unit less than that of the initial bound system.

If we equate the squared matrix elements for the photoelectric process and for the radiative recombination process by means of eqn. (3.2.1) we obtain, in an obvious notation,

$$\frac{dw_{\beta\alpha}\,(\text{P.E.})}{(mpL^3 d\Omega_e)/(2\pi\hbar)^3} = \frac{dw_{\alpha\beta}(R)}{(\omega_{\beta\alpha}^2 d\Omega_\gamma V)/(2\pi c)^3\,\hbar} \tag{3.4.17}$$

where we have written dw in each case instead of w as we wish to consider transitions into an angular element of phase space. The denominators in this expression are the final state densities for photons (right-hand side) and for electrons (left-hand side), discussed previously [cf. eqns. (3.3.2) and (3.4.1)]. For the photoelectric process, eqn. (3.4.7) relates the photoelectric transition probability to the differential cross section. For recombination, we have from the definition of cross section

$$d\sigma = \frac{dw}{|\boldsymbol{S}|} \tag{3.4.18}$$

where $\boldsymbol{S}$ (particles per cm^2 per sec) is the electron flux. In the photoelectric case we have previously specified that the free electron is normalized to a volume L^3; since $\boldsymbol{S}$ is given by

$$\boldsymbol{S} = \frac{\hbar}{2im}(\psi^*\,\nabla\psi - \psi\,\nabla\psi^*), \tag{3.4.19}$$

($i = \sqrt{-1}$) we can use a plane wave $\psi(\boldsymbol{r}) = e^{i\boldsymbol{k}_e\cdot\boldsymbol{r}}L^{-3/2}$ so normalized to keep track explicitly of the proper factors†. Equation (3.4.19) yields, for

† k_e is the electron propagation vector $\boldsymbol{p}/\hbar$.

the free electron plane wave ($\boldsymbol{v}$ is the electron velocity),

$$\boldsymbol{S} = \frac{\boldsymbol{v}}{L^3}, \tag{3.4.20}$$

so that we can write eqn. (3.4.18) as

$$d\sigma_R = \frac{L^3\, dw_R}{v}. \tag{3.4.21}$$

where the R subscript here signifies recombination. From this result and from eqns. (3.4.7) and (3.4.17) we obtain the relation

$$\frac{d\sigma_{\beta\alpha}\,(\text{P.E.})}{d\Omega_e} = \frac{m^2v^2c^2}{\hbar^2\omega_{\alpha\beta}^2}\,\frac{d\sigma_{\alpha\beta}\,(R)}{d\Omega_\gamma}. \tag{3.4.22}$$

We now refer both differential cross sections to the angle between $\boldsymbol{p}$ and $\boldsymbol{k}_\gamma$, since both cross sections can be defined physically only in terms of this angle. Equation (3.4.22) can be multiplied by an element of solid angle $d\Omega$ taken about $\boldsymbol{p}$ and referred to an axis along $\boldsymbol{k}_\gamma$ on the left-hand side of eqn. (3.4.22), and taken about $\boldsymbol{k}_\gamma$ and referred to an axis along $\boldsymbol{p}$ on the right-hand side of eqn. (3.4.22). Both sides can then be integrated over the full solid angle to yield

$$\sigma_{\beta\alpha}\,(\text{P.E.}) = \frac{m^2v^2c^2}{\hbar^2\omega_{\alpha\beta}^2}\,\sigma_{\alpha\beta}(R). \tag{3.4.23}$$

Since the cross sections in this expression in general represent transitions between degenerate sublevels, they are not the physically observed quantities. The latter, which we will call σ (P.E.) and $\sigma(R)$ can be obtained by averaging and summing the above relation over initial and final states. That is to say, by definition we have

$$\sigma\,(\text{P.E.}) \equiv \frac{1}{g_{i'}} \sum_{\alpha(i')} \sum_{\beta(i')} \sigma_{\alpha\beta}\,(\text{P.E.}),$$

and (3.4.24)

$$\sigma(R) \equiv \frac{1}{g_{j'}} \sum_{\alpha(i')} \sum_{\beta(i')} \sigma_{\alpha\beta}(R) \cdot$$

where we have used i' to denote the collection of degenerate states of the bound system (ion plus active bound electron) *and* the photon. The collection of degenerate states of the residual ion (minus a bound electron) *and* the free electron has been denoted by j'. The appropriate statistical weights are, of course, $g(i')$ and $g(j')$, respectively. If we now sum eqn. (3.4.23) according to eqn. (3.4.24), we obtain

$$\sigma\,(\text{P.E.}) = \frac{g_{j'}}{g_{i'}} \frac{m^2 v^2 c^2}{\hbar^2 \omega^2}\, \sigma(R). \tag{3.4.25}$$

We have primed the state index in order to be able to conveniently differentiate between the composite systems of atom/ion plus photon/electron, and the bound systems—atom or ion alone. Thus, g_j and g_i will be used to represent the statistical weights of the ion and atom alone. This is usually called the "Milne Relation" after E. A. Milne (1924) who first derived it (from more general considerations than we have used here). To return to our original argument, eqn. (3.4.22) can be written, by use of eqn. (3.4.8), as

$$\frac{m^2 v^2 c^2}{\hbar^2 \omega^2} \frac{d\sigma_{\alpha\beta}(R)}{d\Omega_\gamma} = \frac{d\sigma_{\beta\alpha}\,(\text{P.E.})}{d\Omega_e} = \frac{m^2 e^2 \omega_{\alpha\beta}}{2\pi\hbar^2 c}\, vL^3\, |\boldsymbol{r}_{\alpha\beta} \cdot \hat{\varepsilon}|^2. \tag{3.4.26}$$

To obtain the physically observed photoelectric cross section in terms of the matrix element we must now average eqn. (3.4.26) over initial states and sum it over final states. We can explicitly perform the sum over the two independent polarization directions as in eqn. (3.4.9) but without the factor $\frac{1}{2}$ since we now want a sum, not an average). We obtain the result

$$\frac{m^2 v^2 c^2}{\hbar^2 \omega^2} \frac{d\sigma(R)}{d\Omega_\gamma} = \frac{m^2 e^2 \omega v L^3}{2\pi\hbar^3 c g_{j'}} \sum_{\alpha(i)} \sum_{\beta(j')} |\boldsymbol{r}_{\alpha\beta}|^2 \sin^2\theta. \tag{3.4.27}$$

(The prime on the i has been dropped, since we have now explicitly included the sum over polarization directions.)

For a fixed direction $\boldsymbol{r}_{\alpha\beta}$, this result can be immediately integrated over all photon propagation directions $\boldsymbol{k}_\gamma$ (viz. over the angle θ) by use of

$$\int \sin^2\theta\, d\Omega = 8\pi/3.$$

Calling

$$\int \frac{d\sigma(R)}{d\Omega_\gamma}\, d\Omega_\gamma \equiv \sigma_R,$$

and setting $g_{j'} = 2g_j$ to account explicitly for the sum over initial electron spins (j now refers to the degenerate states of the ion only), we obtain

$$\sigma_R = \frac{2e^2\omega^3L^3}{3\hbar c^3 v g_j} \sum_{\alpha(i)} \sum_{\beta(j)} |\boldsymbol{r}_{\alpha\beta}|^2 \tag{3.4.28}$$

for the recombination cross section.

For the photoelectric cross section, this yields by use of eqn. (3.4.25), the result

$$\sigma\,(\text{P.E.}) = \frac{2m^2e^2\omega vL^3}{3\hbar^3 c g_i} \sum_{\alpha(i)} \sum_{\beta(j)} |\boldsymbol{r}_{\alpha\beta}|^2 \tag{3.4.29a}$$

where the asymptotic electronic wave function is normalized according to $e^{i\boldsymbol{k}_e\cdot\boldsymbol{r}}L^{-3/2}$. Since the L-factors cancel in the above equation we can drop them altogether and refer to the electronic wave function $e^{i\boldsymbol{k}_e\cdot\boldsymbol{r}}$ as normalized to unit amplitude asymptotically. With this proviso, eqn. (3.4.29a) can be written as

$$\sigma\,(\text{P.E.}) = \frac{32\pi^4e^2m^2\nu v}{3h^3 c g_i} \sum_{\alpha} \sum_{\beta} |\boldsymbol{r}_{\alpha\beta}|^2 \tag{3.4.29b}$$

in agreement with Bates (1946).

Analogous to the electron-spin-weight case, we must set $g_i = 2g_{i'}$, since we are explicitly accounting for the two polarization directions as is customarily done; g_i is then, as stated previously, the weight of the atomic bound state. It is now evident that $|\boldsymbol{r}_{\alpha\beta}|^2$ cannot actually depend on the direction of the photoejected electron.

In the recombination process, the direction of the incoming electron is, for the differential cross section, a preferred direction as is the direction of the outgoing photon. However, once one integrates over all outgoing photon directions to obtain the total cross section, and sums over the (degenerate) m-sublevels of the electron in the final bound state, the "preferred" nature of the incoming electron direction disappears. This is so, since there are no longer any other preferred directions to relate it to, and a change in the incoming electron direction would merely correspond to rotating the laboratory apparatus. This is the reason one can take as the wave function for the total cross section for recombination, a free-electron function of the form

$$\psi(\boldsymbol{r}) = \sum_{l=1}^{\infty} (2l + i)\, i^l e^{-i\delta_l} R_{El}(r)\, P_l(\cos\theta')$$

(cf. eqn. (3.4.30) which follows), even though this form implies that the z-direction coincides with the beam direction. Then our result above, based on detailed balance, implies that this same form is valid for the photoelectric process.

Another way to view this is as follows. Let us pick some "absolute" coordinate system unconnected with the atom and unconnected with the photoelectron direction, $\boldsymbol{p}$. Then our photoelectric or recombination process determines four vectors: $\boldsymbol{k}_\gamma$, $\boldsymbol{p}$, $\boldsymbol{r}_{\alpha\beta}$ and $\boldsymbol{a}$, where we denote by $\boldsymbol{a}$ the atom orientation, and where $\boldsymbol{a}$ and $\boldsymbol{p}$ completely determine $\boldsymbol{r}_{\alpha\beta}$. We can

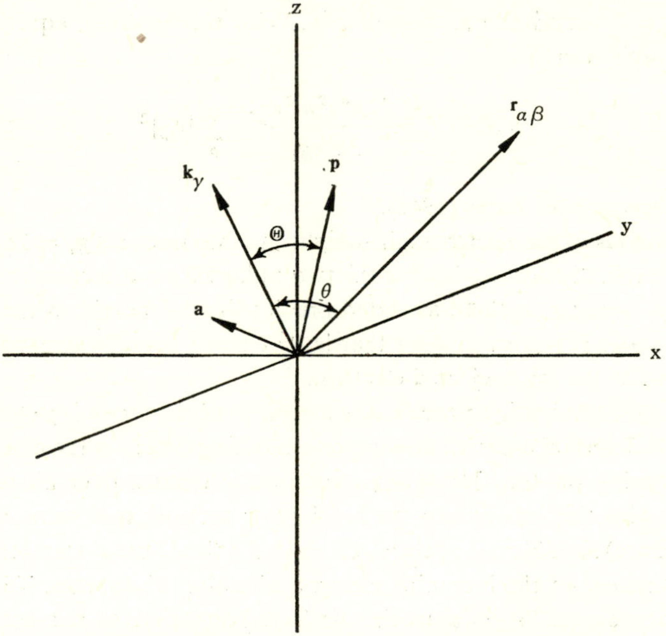

FIG. 3.3. Diagram indicating the vectors involved in the photoelectric or recombination process, and the angles between them.

draw a diagram as shown in Fig. 3.3. Now $|\boldsymbol{r}_{\alpha\beta}|^2$ cannot depend on the *absolute* orientation of any of these vectors (i.e. their orientation relative to the arbitrary system we have chosen), because this can be changed at will by merely rotating the axis. Furthermore, $\boldsymbol{r}_{\alpha\beta}$ does not depend on $\boldsymbol{k}_\gamma$ at all (recall that the use of $\boldsymbol{r}_{\alpha\beta}$ means that we are taking the dipole approximation) so that the angle between $\boldsymbol{p}$ and $\boldsymbol{r}_{\alpha\beta}$ does not depend on $\boldsymbol{k}_\gamma$, either, nor on the coordinate system chosen. It is fixed once and for all by the magnitude of $\boldsymbol{p}$ and the nature of the interaction between the electron and the ion. Because of this fixed relationship between $\boldsymbol{p}$ and $\boldsymbol{r}_{\alpha\beta}$ one can obtain the correct total cross section from eqn. (3.4.9) by replacing $d\Omega_e$ by $d\Omega_r$, the element of solid angle about $\boldsymbol{r}_{\alpha\beta}$ and integrating over θ, the angle between $\boldsymbol{k}_\gamma$ and $\boldsymbol{r}_{\alpha\beta}$. Once the integration is carried out over all angles θ the result is equivalent to having integrated over Θ (the angle between $\boldsymbol{k}_\gamma$ and $\boldsymbol{p}$) in spite of the fact that the *differential cross section* expressed in terms of θ is not a physically appropriate one.

In order to obtain the differential cross section accurately, considerable pains must be taken in the specification of the final state wave function. This is not a trivial problem and we do not wish to discuss it here, since our only interest is in the total cross section. For a discussion of this problem, and references to the literature, see Schiff (1968, sect. 46) and Bethe and Salpeter (1957, sects. 69 and 70). Bethe and Salpeter also give a discussion of the angular distribution of photoejected electrons in sect. 72.

The quantity $\boldsymbol{r}_{\alpha\beta}$ in the present case, wherein one of the states α, β is asymptotically a plane wave, viz. asymptotically the momentum has a definite direction, is an essentially different quantity than in the case of two bound states. In the case of two bound states, we could consider $\boldsymbol{r}_{\alpha\beta}$ connected in a one-to-one correspondence with the orientation of the atom (viz. a system of coordinates fixed in the atom in some prescribed fashion). In the present case, however, the free wave function makes $\boldsymbol{r}_{\alpha\beta}$ depend on a quantity—the free electron momentum direction—which is disconnected from the atom, and which asymptotically has a definite directional dependence. If we were considering an angular momentum eigenstate for the free electron, the situation would remain analogous to the bound case. But we would then pass directly to a total cross section rather than through the intermediary of a differential cross section as in the present case. We will consider this situation in the next section.

Thus, we cannot now strictly average over orientations of the atom, or employ the integral over the three spherical harmonics inside the matrix element integration in just the same way as we did in eqns. (3.3.19) to (3.3.27). However, the angle between $\boldsymbol{r}_{\alpha\beta}$ and $\boldsymbol{k}_\gamma$ as we have seen is in a one-to-one correspondence with the angle between the free electron momentum $\boldsymbol{p}$ and the photon vector $\boldsymbol{k}_\gamma$. Therefore, we can also average $d\sigma_{\alpha\beta}/d\Omega_e$ as given in eqn. (3.4.9) over the angle θ between $\boldsymbol{r}_{\alpha\beta}$ and $\boldsymbol{k}_\gamma$, in analogy to the "average over atom orientations" carried out in the discrete transition case. The resulting *average* differential cross section is then independent of Θ as well as θ and equals $d\sigma_{\alpha\beta}/d\Omega_e$ averaged over outgoing momentum directions. Although this isotropic average is no longer of any value as a differential cross section, the total cross section follows from it by simply multiplying by 4π. The result, of course, is the same as the results obtained by the other methods described above. We append one further observation to this discussion.

In a plane wave approximation for the final state (which is not very accurate at low energies) the angle between $\boldsymbol{r}_{\alpha\beta}$ and $\boldsymbol{k}_\gamma$ is *equal* to the angle between $\boldsymbol{p}$ and $\boldsymbol{k}_\gamma$. We can see this as follows. If either state α or β in eqn. (3.3.7) is a plane wave (momentum eigenfunction) the momentum operator eigenvalue can be taken from under the integral sign. If α, say, is the state to be approximated by a plane wave, we can write eqn. (3.3.7) as

$$\boldsymbol{p}_\alpha \cong \{(im\omega_{\alpha\beta})/\int \psi_\beta^* \psi_\alpha \, dV\} \, \boldsymbol{r}_{\alpha\beta}.$$

Since the quantities in the brackets are scalars, this result shows that the momentum vector $\boldsymbol{p}_\alpha$ in the final state lies in the same direction as $\boldsymbol{r}_{\alpha\beta}$. Hence the angle θ in eqn. (3.4.9) is approximately equal to the angle between the momentum vector of the ejected photoelectron and the photon propagation vector $\boldsymbol{k}_\gamma$. Also, in this approximation, it is clear that $|\boldsymbol{r}_{\alpha\beta}|^2$ is proportional to p_α^2 and does not depend on the direction of $\boldsymbol{p}_\alpha$.

The result we have obtained so far [eqn. (3.4.29)] for the total photoelectric cross section is valid for a linear-momentum-eigenfunction final state (viz. a final state which is *asymptotically* a linear momentum eigenfunction). Since this is not the most generally useful form, we will illustrate the passage from the linear momentum form to a form with angular momentum eigenfunctions in the final state.

The general solution for a positive-energy electron in any central potential that has the proper asymptotic behavior appropriate to the definition of the differential cross section can be written (Mott and Massey, 1965, p. 24) as

$$u(r, \theta', \varphi') = \frac{1}{L^{3/2}} \sum_{l=0}^{\infty} \sqrt{4\pi}\,(2l+1)^{1/2}\, i^l\, e^{-i\delta_l}\, R_{El}(r)\, Y_{l0}(\theta', 0). \tag{3.4.30}$$

This is an expansion in angular momentum eigenfunctions of a wave which asymptotically has a definite direction (the z-axis). In the case of a Coulomb potential Ze/r, the phase shift δ_l (Coul.) is given by†

$$\delta_l\,(\text{Coul.}) = \arg \Gamma(l + 1 + iZ/k_e a_0),$$

where a_0 is the Bohr radius, and $R_{El}(r)$ is the radial wave function computed in the appropriate potential and has the asymptotic behavior

$$R_{El}(r) \sim \frac{\sin(k_e r - \pi l/2 - \delta_l)}{k_e r}. \tag{3.4.31}$$

For phase shifts $\delta_l = 0$ this becomes a plane wave in the z-direction normalized to a volume L^3, and $R_{El}(r)$ becomes the spherical Bessel function $j_{l+1/2}(k_e r)$. The distorting effects of a non-vanishing potential enter asymptotically through the phase shifts δ_l. For the above wave function $u(r, \theta', \varphi')$, eqn. (3.4.29a) for the photoelectric cross section becomes

$$\sigma\,(\text{P.E.}) = \frac{2m^2e^2\omega v}{3\hbar^3 c g_i} \sum_{m_\alpha} \Big| \sum_l \int R_\alpha(r)\, Y_{l_\alpha m_\alpha}(\theta', \varphi')\, r \sqrt{4\pi}\,(2l+1)^{1/2}\, i^l e^{-i\delta_l} \times$$

$$\times\, R_{El}(r)\, Y_{l0}(\theta', 0)\, r^2\, dr\, d\Omega' \Big|^2. \tag{3.4.32}$$

We have dropped the final state summation over β indicated previously as the final state represented by $u(r, \theta)$ has no further degenerate sublevels.

We can carry out the angular integrations required in eqn. (3.4.32) by means of our previous eqn. (3.3.33). If we set $l_\alpha = l$, then non-vanishing

† The Coulomb phase shift will be denoted by σ_1 hereafter.

terms occur only for $l_\beta = l \pm 1$. Equation (3.4.32) becomes

$$\begin{aligned}\sigma\,(\text{P.E.}) = \frac{2m^2e^2\omega v}{3\hbar^3 c g_i} \sum_{m_\alpha} 2\pi\,(2l+1)\,\langle\{&\mathscr{R}\,(l-1,l)\,C\,(l,1,l-1;000)\\ &\times [C\,(l,1,l-1;m_\alpha,-1,0) - C\,(l,1,l-1;m_\alpha,1,0)]\\ &+ \mathscr{R}\,(l+1,l)\,C\,(l,1,l+1;000)\,[C\,(l,1,l+1;m_\alpha,-1,0)\\ &- C\,(l,1,l+1;m_\alpha,1,0)]\}^2\\ &+ \{\mathscr{R}\,(l-1,l)\,C\,(l,1,l-1;000)\,[C\,(l,1,l-1;m_\alpha,-1,0)\\ &+ C\,(l,1,l-1;m_\alpha,1,0)] + \mathscr{R}\,(l+1,l)\,C\,(l,1,l+1;000)\\ &\times [C\,(l,1,l+1,m_\alpha,-1,0) + C\,(l,1,l+1;m_\alpha,1,0)]\}^2\\ &+ 2\,\{\mathscr{R}\,(l-1,l)\,C\,(l,1,l-1;000)\,C\,(l,1,l-1;m_\alpha,0,0)\\ &+ \mathscr{R}(l+1,l)C(l,1,l+1;000)C(l,1,l+1;m_\alpha,0,0)\}^2\rangle \qquad (3.4.33)\end{aligned}$$

where $\mathscr{R}(l \pm 1, l) \equiv \int R_l(r)\,R_{E,\,l\pm 1}(r)\,r^3\,dr$, as defined in eqn. (3.2.5).

We now note that in the sum over m_α only one component contributes for a given projection quantum number of the free electron in the final state. Since we have taken the free electron m-value to be zero, the particular value of m_α selected out is determined by $m_\alpha = 0 - m_2$ where m_2 is 0 or ± 1. For reasons to be discussed later, we do not ascribe this numerical value to m_α, but we designate the one m_α value which yields a non-vanishing term in the sum over m_α as m_α, m'_α, or m''_α, depending on m_2, and drop the $\sum_{m_\alpha}$. If one now expands the brackets in this expression, some of the terms cancel. One then obtains by collecting the remaining terms the result

$$\begin{aligned}\sigma\,(\text{P.E.}) = \frac{2m^2e^2\omega v}{3\hbar^3 c g_i}\,(2l+1)\,4\pi\,\{&\mathscr{R}^2\,(l-1,l)\,C^2\,(l,1,l-1;000)\\ &\times [C^2\,(l,1,l-1;m_\alpha,-1,0) + C^2\,(l,1,l-1;m'_\alpha,1,0)\\ &+ C^2\,(l,1,l-1;m''_\alpha,0,0)]\\ &+ \mathscr{R}^2\,(l+1,l)\,C^2\,(l,1,l+1;000)\,[C^2\,(l,1,l+1;m_\alpha,-1,0)\\ &+ C^2\,(l,1,l+1;m'_\alpha,1,0) + C^2\,(l,1,l+1;m''_\alpha,0,0)]\\ &+ 2\mathscr{R}(l-1,l)\,\mathscr{R}(l+1,l)\,C(l,1,l-1;000)\,C(l,1,l+1;000)\\ &\times [C\,(l,1,l-1;m_\alpha,-1,0)\,C\,(l,1,l+1;m_\alpha,-1,0)\\ &+ C\,(l,1,l-1;m'_\alpha,1,0)\,C\,(l,1,l+1;m'_\alpha,1,0)\\ &+ C\,(l,1,l-1;m''_\alpha,0,0)\,C\,(l,1,l+1;m''_\alpha,0,0)]\}. \qquad (3.4.34)\end{aligned}$$

The orthogonality relation eqn. (3.3.24), along with the symmetry relation eqn. (3.3.34), can be employed to simplify this complicated expression. The transposition of the first two indices by use of the symmetry relation shows that the first two terms in square brackets are again orthonormal sums over the first m-index and, hence, equal unity. The third expression in square brackets corresponds to the orthonormality relation above for $j \neq j'$ and, hence, vanishes. Since $g_i = 2l + 1$, we obtain from eqn. (3.4.34),

$$\sigma\,(\text{P.E.}) = \frac{2m^2e^2\omega v}{3\hbar^3 c}\,\{4\pi C^2\,(l, 1, l-1; 000)\,\mathscr{R}^2\,(l-1, l)$$

$$+\, 4\pi C^2\,(l, 1, l+1; 000)\,\mathscr{R}^2\,(l+1, l)\}. \qquad (3.4.35)$$

This agrees with the result first given in this form by Bates (1946), when account is taken [cf. eqn. (3.3.28)] of the values of the C^2-factors, and the energy quantum numbers which we have suppressed in our definition of $\mathscr{R}\,(l \pm 1, l)$. Because of our transposition of the first and second projection quantum numbers in the C-coefficients in the square brackets of eqn. (3.4.34), the foregoing derivation does not actually depend on the m-value of the free electron which we took to be zero (the derivation would have been a little simpler had we allowed it to so depend). Because of this, our result is independent of which direction we choose as our z-axis which is the implied momentum direction for the wave function of eqn. (3.4.30). We will not demonstrate this in the detail of the preceding derivation, but by noting the following situation. Our result, eqn. (3.4.35), shows that we could have taken the l-sum in eqn. (3.4.32) out of the absolute value-squared brackets, as all the cross terms vanish. The reason must lie in the fact that eqn. (3.4.30) represents a superposition of degenerate l-sublevels, and we could have considered transitions into them individually and summed the resulting individual transition probabilities in the usual fashion after the rule of "averaging over initial states and summing over final states". Had we done this, we would not have had to examine the cross terms at all. Now from the spherical harmonics addition theorem cited earlier, eqn. (3.3.19), we can decouple the electron coordinate-vector direction from the z-axis, or momentum direction. The relative angle between these two directions was taken as θ' in eqn. (3.4.30).

We now refer each of these directions to an independent fixed axis and label the electron coordinate vector orientation relative to this new axis as θ_e, φ_e, and the electron momentum orientation relative to this axis as θ_p, φ_p. To consider a transition into a single l, m sublevel we would use one term of eqn. (3.4.30) in the formula for the cross section with $Y_{l0}(\theta', 0)$ replaced by eqn. (3.3.19) with $\Theta = \theta_p$ and $\theta_1 = \theta_e$, say. Summing over all sublevels as final degenerate states, we would have

$$\sigma\,(\text{P.E.}) = \frac{2m^2e^2\omega v}{3\hbar^3 c}\Bigg\{4\pi\mathscr{R}^2\,(l-1, l)\,C^2\,(l, 1, l-1; 000) \times \left[\sum_M \frac{4\pi}{2l-1}\,|Y_{l-1,M}\,(\theta_p, \varphi_p)|^2\right] + 4\pi\mathscr{R}^2\,(l+1, l) \times C^2\,(l, 1, l+1; 000)\left[\sum_M \frac{4\pi}{2l+3}\,|Y_{l+1,M}\,(\theta_p, \varphi_p)|^2\right]\Bigg\} \tag{3.4.36}$$

since the formulas are just the same as we used before except for the additional factor

$$Y^*_{lm}\,(\theta_p, \varphi_p) \times \left(\frac{4\pi}{2l+1}\right)^{1/2}$$

and the non-zero value of M. If we now make use of the formula for the sum over the squares of the spherical harmonics, eqn. (3.3.25), the equation above reduces back to the result eqn. (3.4.35), and we see explicitly that our result is independent of the direction that the wave function eqn. (3.4.30) implies for the asymptotic momentum direction. If this were not so, we would, of course, be in trouble, for we have already integrated over the outgoing momentum directions! The foregoing considerations, however, afford an explicit substantiation of our statement, made previously (see Fig. 3.3, for example) that $|\boldsymbol{r}_{\alpha\beta}|^2$ is independent of the direction of the electron momentum.†

† When β designates the asymptotic linear momentum eigenstate, not one of the sublevels M defined above. When these are considered, they must be summed over as above to get all the components of $x_{\alpha\beta}$, $y_{\alpha\beta}$, etc., in the "rotated" coordinate system.

TOTAL PHOTOELECTRIC CROSS SECTION—ANGULAR MOMENTUM FINAL EIGENSTATE

It is interesting now to rederive the photoelectric cross section making direct use of an angular momentum eigenfunction for the free electron state, and of spherical-box normalization for the radial function. The wave function for a free particle with a specific angular momentum l and confined to a sphere of radius R is (Goldberger and Watson, 1964, pp. 18–20)

$$\psi_{k_e lm}(\boldsymbol{r}) = \left(\frac{2}{R}\right)^{1/2} k_e j_l(k_e r)\, Y_{lm}(\theta, \varphi). \tag{3.4.37}$$

The normalization factor $(2/R)^{1/2} k_e$ can be obtained from the properties of the spherical Bessel functions j_l as given, for example, by Schiff (1968) or, more simply, from the asymptotic sinusoidal form of $j_l(kr)$. This asymptotic form of the spherical Bessel functions is:

$$j_l(kr) \sim \frac{\sin(kr - l\pi/2)}{kr}. \tag{3.4.38}$$

If a short-range potential exists near the origin, the radial wave function has the asymptotic form above except that the phase shift is no longer $l\pi/2$. In general, we can write the asymptotic form of the radial wave function as in eqn. (3.4.31), recalling that in the case of a Coulomb potential contribution, the phase shift δ_l will be a slowly varying function of r.

For spherical-box normalization, we want the wave function to vanish at the radius R. In order for this to occur, we must have

$$k_e R - \pi l/2 - \delta_l = n\pi, \tag{3.4.39}$$

in view of the asymptotic behavior, eqn. (3.4.31), of the radial wave function. Differentiating this relation with respect to n we find, denoting the value of k_e which satisfies eqn. (3.4.39) as k_n (Blatt and Weisskopf, 1952),

$$dn = \frac{R}{\pi}\, dk_n. \tag{3.4.40}$$

Since $k_e = (2mE/\hbar^2)^{1/2}$, we can write this as

$$\frac{dn}{dE_n} = \frac{R}{\pi}\left(\frac{m}{\hbar^2 k_n}\right); \tag{3.4.41}$$

this gives us the number density of radial eigenstates over the wave-number range dk_n or the energy range dE_n. The sum over states in the present case can be indicated by

$$\sum_{n=0}^{\infty} \sum_{l=0}^{\infty} \sum_{m=-l}^{m=+l} .$$

The asymptotic form of the radial wave function eqn. (3.4.31) can now be used along with eqn. (3.4.37) (which provides the normalization) to obtain the result

$$|\psi_{klm}(r)|^2 \frac{dn}{dE} \sim \frac{2m}{\pi\hbar^2 k_e} \frac{\sin^2(k_e r - \pi l/2 - \delta_l)}{r^2} |Y_{lm}|^2 \quad (3.4.42)$$

which is independent of R, just as the corresponding quantity is independent of L^3 in the rectangular-box normalization. For this reason, we can assume that the limit $R \to \infty$ will ultimately be taken so the index n on k_e and E need not be explicitly acknowledged. The matrix element for photoelectric absorption is again as given by eqn. (3.4.2), and as usual we convert from $\langle\alpha|\boldsymbol{p}|\beta\rangle$ to $\langle\alpha|\boldsymbol{r}|\beta\rangle$ by means of eqn. (3.3.7). For simplicity we take n_ω in the matrix element equal to 1, which implies that the cross section is given by [cf. eqn. (3.4.7)]

$$\sigma_{\alpha\beta} = \frac{w_{\alpha\beta} V}{c} \quad (3.4.43)$$

where V is the photon normalization volume. Combining eqns. (3.2.1), (3.3.7), (3.4.41) and (3.4.43), we obtain

$$\sigma_{\alpha\beta} = \frac{4\pi m e^2 \omega_{\alpha\beta} R}{\hbar^2 k_e c} |\boldsymbol{r}_{\beta\alpha} \cdot \hat{\varepsilon}|^2. \quad (3.4.44)$$

In this result, the dipole moment matrix element vector $\boldsymbol{r}_{\beta\alpha}$ is defined as

$$\boldsymbol{r}_{\beta\alpha} \equiv \int \psi_{nlm}(\boldsymbol{r})\, \boldsymbol{r} \psi^*_{k_e l' m'}(\boldsymbol{r})\, dV$$

in agreement with the earlier definition given following eqn. (3.3.10). Using eqn. (3.4.37) we see that this is approximately equal to

$$\boldsymbol{r}_{\beta\alpha} \cong \left(\frac{2}{R}\right)^{1/2} k_e \int j_l(k_e r)\, Y^*_{l'm'} \boldsymbol{r} \psi_{nlm}\, dV. \quad (3.4.45)$$

Since $j_l\,(k_e r)$ has the asymptotic behavior

$$\frac{\sin k_e r}{k_e r},$$

we can use this result to pass to other equations for σ involving different asymptotic normalization conventions. For example, for normalization on the energy scale, we would have

$$\boldsymbol{r}_{\beta\alpha} = \left(\frac{\pi k_e \hbar^2}{Rm}\right)^{1/2} \int u_w^* \boldsymbol{r} \psi_b \, dV \tag{3.4.46}$$

where ψ_b is a bound state and u_w has the asymptotic behavior

$$u_w \sim \sqrt{\left(\frac{2m}{\pi \hbar^2 k_e}\right)} \left(\frac{\sin k_e r}{r}\right) Y_{l'm'}. \tag{3.4.47}$$

If we insert eqn. (3.4.46) into eqn. (3.4.44) we obtain

$$\sigma_{\alpha\beta}^{(w)} = \frac{4\pi^2 e^2 \omega}{c} \left| \int u_w^* \boldsymbol{r} \cdot \hat{\varepsilon} \psi_b \, dV \right|^2. \tag{3.4.48}$$

This is the formula given by Bethe and Salpeter (1957) in their eqn. (71.1). The normalization of u_w has been chosen so that

$$\int u_w^*(r)\, u_{w'}(r)\, r^2 \, dr = \delta\,(E - E') \quad \text{where} \quad E = \frac{\hbar^2}{2m} k_e,$$

as will be discussed later. With the usual average†

$$\overline{|\boldsymbol{r}_{\alpha\beta} \cdot \hat{\varepsilon}|^2} = |\boldsymbol{r}_{\alpha\beta}|^2 \, \overline{\cos^2 \theta} = \tfrac{1}{3} \, |\boldsymbol{r}_{\alpha\beta}|$$

over atom orientations and the formal designation of the sum and average over initial and final states

$$\frac{1}{g_i} \sum_{\alpha(i)} \sum_{\beta(j)}$$

† It is fair to do this again as in the discrete case because of our choice of final state eigenfunction.

eqn. (3.4.48) can also be written as

$$\sigma = \frac{8\pi^3 e^2 \nu}{3cg_i} \sum \sum |\int u_w^* r \psi_b \, dV|^2 \tag{3.4.49}$$

in agreement with eqn. (1) of Burgess and Seaton (1960).

Let us now define a matrix element $\boldsymbol{r}_{nlm}^{El'm'}$ by the relation

$$\boldsymbol{r}_{nlm}^{El'm'} \equiv \int [R_{nlm}(r) \, Y_{lm}(\theta, \varphi)]^* \, \boldsymbol{r} \, [R_{El'}(r) \, Y_{l'm'}(\theta, \varphi)] \, dV \tag{3.4.50}$$

where $R_{El'}(r)$ is normalized according to eqn. (3.4.31). By comparing this form with eqn. (3.4.45), it is easy to see that eqn. (3.4.44) for the cross section can now be written as

$$\sigma_{\alpha\beta} = \frac{8\pi^2 m^2 e^2 \omega v}{3\hbar^3 c} |\boldsymbol{r}_{nlm}^{El'm'}|^2 \tag{3.4.51}$$

where the factor $\frac{1}{3}$ appears again if we average over atom orientations, and where k_e has been replaced according to $k_e = mv/\hbar$. If we now insert the explicit sum and average over final and initial states,

$$\frac{1}{2l+1} \sum_{ml} \sum_{m'l'}$$

we obtain from eqn. (3.4.51)

$$\sigma_{l,l+1} = \frac{2m^2 e^2 \omega v}{3\hbar^3 c} \cdot \frac{4\pi}{2l+1} \sum_m \sum_{m'} \{|\boldsymbol{r}_{nlm}^{E,l-1,m'}|^2 + |\boldsymbol{r}_{nlm}^{E,l+1,m'}|^2\}. \tag{3.4.52}$$

We now make use of the relation [Bethe and Salpeter, 1957, eqns. (60.12) and (60.13)]

$$\sum_{m'} |\boldsymbol{r}_{nlm}^{n',l\pm 1,m'}|^2 = \frac{(l+\frac{1}{2}) \pm \frac{1}{2}}{2l+1} (R_{nl}^{n',l+1})^2. \tag{3.4.53}$$

The notation for matrix elements and radial integrals used in eqns. (3.4.50) to (3.4.53) is that of Bethe and Salpeter. The factors $R_{nl}^{n',l'}$ are those we have designated as $\mathscr{R}(l, l')$. Although this relation was derived for bound-state radial matrix elements, we can make use of it in the bound-free case. If we note that the sum over m yields simply a factor $2l + 1$ after carrying out the sum over m' above, we obtain

$$\sigma_{l,l+1} = \frac{2me^2\omega v}{3\hbar^3 c}\left\{\frac{4\pi l}{2l+1}\mathscr{R}^2(l-1,l) + \frac{4\pi(l+1)}{2l+1}\mathscr{R}^2(l+1,l)\right\}$$

in agreement with eqn. (3.4.35). Note that in obtaining this result using an angular momentum final state function we did not need to pass through the intermediate step of computing a differential cross section. The specification of the final state as an angular momentum eigenstate allows us to proceed directly to a total absorption cross section.

Normalization of Free-electron Radial Wave Functions

Since the transition probability depends on the square of the final state wave function multiplied by the density of final states dn/dE, we can inquire as to what normalization of the radial wave function will permit us to set $dn/dE = 1$. In the case of angular momentum final eigenstates, since the angular states are discrete they do not contribute to dn/dE. Therefore, we need only to incorporate the factor $\sqrt{(dn/dE)}$ into $R_{El}(r)$ in the radial eigenfunction.† From eqn. (3.4.42) we see that the radial eigenfunction normalized in this fashion will have the asymptotic form

$$\sqrt{\left(\frac{dn}{dE}\right)}R_{El}(r) \sim \left(\frac{2}{\pi}\right)^{1/2}\left(\frac{m}{\hbar^2}\right)^{1/2} \cdot \frac{\sin(kr+\delta_l)}{\sqrt{kr}}. \quad (3.4.54)$$

One further question is of interest: the above normalization permits us to take a unit density of final states, but what does it yield for the quadratic integral of the wave function itself? In other words, what is the relationship of the asymptotic normalization of the free-state wave function, and the square integral normalization of the same function? To obtain this relationship, we can compute the value of the integral

$$\int\left[\left(\frac{dn}{dE}\right)^{1/2}R_{E,l}(r)\right]^*\left[\left(\frac{dn}{dE'}\right)^{1/2}R_{E',l}(r)\right]r^2\,dr \equiv I_l(E,E'). \quad (3.4.55)$$

In order to do this, we consider the *radial* Schrödinger equation obeyed by

$$\frac{f(E,l;r)}{r} \equiv R_{E,l}(r)\left(\frac{dn}{dE}\right)^{1/2}. \quad (3.4.56)$$

† For the free-state eigenfunction this is just $R_{El}(r) = (2/R)^{1/2}\,k_e j_l(k_e r)$, of course, as in eqn. (3.4.37). In the interests of simplicity we now drop the subscript e on the electron propagation number for the remainder of this section.

This equation is

$$\left[\frac{d^2}{dr^2} - \frac{l(l+1)}{r^2} - \frac{2m}{\hbar^2} V(r) + k^2\right] f(E, l; r) = 0. \quad (3.4.57)$$

If we consider this equation for two values of E, E and E', say, multiply it in each case by the function belonging to the other eigenvalue, integrate both resulting equations from zero to R and subtract them, we obtain

$$\int_0^R dr\,[fD^2f' - f'D^2f] + (k'^2 - k^2) \int ff'\,dr = 0. \quad (3.4.58)$$

The terms in $V(r)$ and l cancel, and we have set $f(E, l; r) \equiv f$, $f(E', l; r) \equiv f'$, and $d/dr \equiv D$. Integration by parts twice shows that

$$\int_0^R dr\,(fD^2f') = (fDf' - f'Df)_R + \int_0^R dr\, f'D^2f. \quad (3.4.59)$$

With this result, we obtain from eqn. (3.4.58) the result

$$\int_0^\infty ff'\,dr = \frac{1}{k^2 - (k')^2} \lim_{R\to\infty} [fDf' - f'Df]_R. \quad (3.4.60)$$

$I_l(E, E')$ as defined by eqn. (3.4.55) can now be evaluated. That is to say, we can now write

$$I_l(E, E') = \frac{1}{k^2 - (k')^2} \lim_{R\to\infty} [fDf' - f'Df]_R \quad (3.4.61)$$

where the asymptotic form of f is given by eqns. (3.4.54) and (3.4.56):

$$f \sim \left(\frac{2m}{\pi\hbar^2}\right)^{1/2} \frac{\sin(kr + \delta_l)}{\sqrt{k}}. \quad (3.4.62)$$

Inserting eqn. (3.4.62) into eqn. (3.4.61), we obtain

$$\left(\frac{\pi\hbar^2}{2m}\right) I_l(E, E') = \lim_{R\to\infty} \frac{1}{k^2 - (k')^2}$$

$$\times \left[\frac{\sin kR\,(k')^{1/2} \cos k'R}{k^{1/2}} - \frac{\sin k'R\,(k)^{1/2} \cos kR}{(k')^{1/2}}\right] \quad (3.4.63)$$

where δ_l has been neglected for simplicity. Using the identity $2 \sin \alpha \cos \beta = \sin(\alpha + \beta) + \sin(\alpha - \beta)$, this becomes

$$\left(\frac{\pi\hbar^2}{2m}\right) I(E, E') = \lim_{R\to\infty} \frac{1}{2\sqrt{(kk')}} \left[-\frac{\sin(k + k')R}{(k + k')} + \frac{\sin(k - k')R}{(k - k')} \right] \tag{3.4.64}$$

where we can now disregard the l-dependence of the integral.

Now the δ-function can be represented by (Schiff, 1968)

$$\pi\delta(k) = \lim_{R\to\infty} \frac{\sin kR}{k}. \tag{3.4.65}$$

With this representation, we obtain from eqn. (3.4.64),

$$I(E, E') = \left(\frac{2m}{\pi\hbar^2}\right) \times \frac{\pi}{2k} \delta(k - k') \tag{3.4.66}$$

since the first term in eqn. (3.4.64) does not contribute if we limit ourselves to $k, k' > 0$. Since $\delta(k - k') = 2k\delta(k^2 - (k')^2)$ this result can also be written as

$$I(E, E') = \left(\frac{2m}{\hbar^2}\right) \delta(k^2 - (k')^2) \tag{3.4.67}$$

or, since

$$\delta(k^2) = \delta(E) \frac{dE}{d(k^2)} = \delta(E) \left(\frac{\hbar^2}{2m}\right), \tag{3.4.68}$$

we can finally write our result as

$$I(E, E') \equiv \int \left[\left(\frac{dn}{dE}\right)^{1/2} R_{E,l}(r)\right]^* \left[\left(\frac{dn}{dE}\right)^{1/2} R_{E',l}(r)\right] r^2\, dr = \delta(E - E'). \tag{3.4.69}$$

The asymptotic behavior of the wave function so normalized is given by eqn. (3.4.54). Therefore, as stated previously, with the use of a wave function having this asymptotic form we can set $dn/dE = 1$ in the equation for the transition probability [eqn. (3.2.1)], as the final state density will have been included in the matrix element. We can now correlate our results with a formula given by Bethe and Salpeter (1957). The matrix element for photoelectric absorption, eqn. (3.4.2), can be written for $n_\omega = 1$ as

$$\langle ml|\, H' \,|m'l'\rangle = \frac{1}{V^{1/2}} \frac{e\hbar}{m} \left(\frac{2\pi\hbar}{\omega}\right)^{1/2} \boldsymbol{D}|^{(e)}_{ml;\,m'l'}|^2 \tag{3.4.70}$$

where

$$\boldsymbol{D}^{(e)}_{ml;\,m'l'} \equiv \int \psi^*_{ml}(r)\, \nabla_e \psi_{m'l'}(r)\, dV \tag{3.4.71}$$

and the subscript e on the ∇ operator has been used to denote the polarization direction. If now, we agree to normalize the radial factor of $\psi_{ml}(\boldsymbol{r})$ according to eqn. (3.4.54) and consider light of a particular polarization only, the formula for the photoelectric cross section becomes

$$\sigma_{ml-m'l'} = \frac{2\pi V}{\hbar c} |\langle ml|\, H' \,|m'l'\rangle|^2. \tag{3.4.72}$$

Inserting eqn. (3.4.70) into eqn. (3.4.72) yields the result

$$\sigma_{ml-m'l'} = \frac{2\pi e^2 \hbar^2}{m^2 c \nu} |\boldsymbol{D}^{(e)}_{ml,\,m'l'}|^2 \tag{3.4.73}$$

for a transition between particular angular momentum sublevels m, l and $m'l'$. Equation (3.4.73) is the result given by Bethe and Salpeter (1957) in their eqn. (69.2).

3.5. Reduction of the Angular Dependence of Many-Electron Matrix Elements to Analytic Formulas

Before we can apply the preceding formulas to real physical problems in general, we must account, in some approximate fashion, for the possession of more than one electron per atom. In so doing, we must indicate how to average and sum degenerate many-electron transition probabilities over the appropriate initial and final states. For angular momentum eigenstates, this sum and average is dependent on the angular quantum numbers which uniquely specify the angular wave functions that appear in the matrix element. For practical problems it is usually necessary to approximate "true" atomic wave functions by separable product wave functions wherein the coordinates of a given electron appear in only one factor of the product. We will follow this procedure, and, in addition, will neglect configuration interaction. For a discussion of this topic, the reader may consult Hartree (1957), pp. 17 and 159, and Dalgarno (1969).

Our formulas, therefore, are applicable to Hartree–Fock wave functions, for example, or hydrogenic product wave functions, but are not appropriate for non-separable wave functions such as the Hylleraas variational functions often used for helium. In the separable approximation to bound-state wave functions, the angular dependence reduces to known spherical harmonics so that all the angular integrations involved in the matrix elements with which we are concerned can be performed (Gaunt, 1928). This is by no means a trivial task, especially for equivalent electrons, so we will not go through all the details. Rather, we confine ourselves to indicating the procedures and giving some results. By the methods of Racah algebra (Racah, 1942a, b, 1943, 1949), developed in the early 1940s some time after the initial development period of quantum mechanics, and by subsequent developments thereof (e.g. Kelly, 1959; Rohrlich, 1959; Judd, 1963; Levinson and Nikitin, 1965; Shore and Menzel, 1968; Wybourne, 1970; Mizushima, 1970) these angular integrations can be reduced to analytic formulas involving functions of the angular momentum quantum numbers. The earlier methods of Condon and Shortley (1935) can also be used for this purpose, but are more tedious.

We will not discuss the angular factor reduction for the photoelectric cross sections *per se*. For more specific application to this case, the reader is referred to Burgess and Seaton (1960), and to Armstrong *et al.* (1967).

We begin this procedure with the total cross section for absorption as given by eqn. (3.3.49) which we can now write formally as

$$\sigma(\omega) = \frac{4\pi^2 e^2 \omega}{3\hbar c} \frac{b(\omega)}{g_i} \sum_{\alpha} \sum_{\beta} |\boldsymbol{r}_{\alpha\beta}| \tag{3.5.1}$$

to indicate the required summation over final states and average over initial states.

The number of degenerate initial states, or the statistical weight, of the initial level or term is designated g_i and is equal to $2J + 1$ for an atomic level or $(2L + 1)(2S + 1)$ for a term (see below). Referring to eqn. (3.5.1), it is convenient to define and consider the quantity called the absolute *line strength*:

$$S_{ij} \equiv \sum_{\alpha(i)} \sum_{\beta(j)} |\boldsymbol{r}_{\alpha\beta}|^2 \tag{3.5.2}$$

as this quantity is symmetric in the initial and final states and more or less independent of the remaining factors in the formula for the cross section σ. We follow the usual notation and employ capital letters J, L and S for the total angular momentum, total orbital angular momentum, and total spin quantum numbers, respectively. For the total magnetic quantum numbers, or z-components of J, L and S, we employ M_J, M_L and M_S, respectively. The corresponding lower-case letters are employed for the individual electron quantum numbers. For the details of spectroscopic notation the reader is referred to Allen (1963, chap. 4). The basic degeneracy of an atomic state when there are no external fields present (as in the cases we are considering) is the degeneracy of the $2J + 1$ sublevels, or states, pertaining to the possible values of M_J. We will limit ourselves to LS coupling (Condon and Shortley, 1935) which is usually sufficient for light atoms.

With these limitations in mind, the angular integrations and sums indicated in eqn. (3.5.1) can be carried out. This analysis has been performed explicitly by Rohrlich (1959) for all cases of astrophysical interest. These cases fall into four categories (see also Levinson and Nikitin, 1965, p. 69):

$$\begin{aligned} &\text{(a)} \quad l^n \cdot l' \leftrightarrow l^n \cdot l'', \\ &\text{(b)} \quad l^n \leftrightarrow l^{n-1} l', \\ &\text{(c)} \quad l^n \cdot l' \leftrightarrow l^{n-1}(l')^2, \\ &\text{(d)} \quad l^n l' \cdot l'' \leftrightarrow l^n (l')^2, \end{aligned} \tag{3.5.3}$$

where n, or the numerical superscript, indicates the number of electrons having the l-value to which the superscript is attached. The first two categories normally dominate problems of practical interest involving large numbers of transitions. As examples of these four categories of interest as applied to heated-air problems, we cite the configurational transitions:

(a) $(1s)^2\,(2s)^2\,2p \leftrightarrow (1s)^2\,(2s)^2\,3s$ (N III, O IV)

$l = 0,\ l' = 1,\ l'' = 0,\ n = 2$

(b) $(1s)^2\,(2s)^2\,(2p)^4 \leftrightarrow (1s)^2\,(2s)^2\,(2p)^3\,3d$ (O I)

$$l = 1,\ l' = 2,\ n = 4$$

(c) $(1s)^2\,(2s)^2\,(2p)^3 \leftrightarrow (1s)^2\,2s\,(2p)^4$ (O II, N I)

$$l = 1,\ l' = 0,\ n = 4$$

(d) $(1s)^2\,(2s)\,(2p) \leftrightarrow (1s)^2\,(2p)^2$ (O V, N IV)

$$l = 0,\ l' = 1,\ l'' = 0,\ n = 2$$

It is important to note that in case (a) there is a well-defined core ($1s^2 2s^2$) to which the outer electron ($2p$ or $3s$) couples. Thus, the coupling of the core electrons to each other must be specified in addition to the final total coupling of the outer electron to this core. In the other cases, various *parent* couplings occur so that one has, effectively, an outer electron coupling to a linear combination of cores. Let us take the specific example under case (a)

$$(1s)^2\,(2s)^2\,{}^1S\,2p\,{}^2P \leftrightarrow (1s)^2\,(2s)^2\,{}^1S\,3s\,{}^2S.$$

In each term the core state is a ${}^1S$ and the outer ($2p$, $3s$) electron couples to this to yield a ${}^2P$, ${}^2S$ *term* respectively. As another example, in case (b), the wave function of the $(2p)^4$ combination can be expanded in a fractional parentage coefficient expansion (Racah, 1943) as† (for the specific case of the $1s^2\,2s^2\,2p^4\,{}^3P$ ground term of O I):

$$1s^2\,2s^2\,2p^4\,{}^3P = 1s^2\,2s^2\,2p^3\,\{+\sqrt{\tfrac{1}{3}}\cdot{}^4S - \sqrt{\tfrac{5}{12}}\,{}^2D + \sqrt{\tfrac{1}{4}}\cdot{}^2P\}\,p\,{}^3P. \tag{3.5.4}$$

We will use the subscript p as in L_p to denote that the quantum number to which it is appended belongs to a parent configuration (a core is also a parent—with a fractional parentage coefficient of unity—so the same designation will be used for this case). For the formal definition of the fractional parentage coefficients, see eqn. (3.5.29).

After summation over all magnetic quantum numbers (viz. all strictly

† The p^4 fractional parentage coefficients can be derived from eqns. (19) and (65) of Racah (1943), using the p^3 fractional parentage coefficients given in table I of that paper. Table 3.3, given later in this section, collects all the coefficients for the p-shell.

degenerate sublevels) the line strength S_{ij} can be written as (Rohrlich, 1959; Bates and Damgaard, 1949):

$$S_{ij} = \mathscr{S}(\mathscr{L}_i, \mathscr{L}_j)\, \mathscr{S}(\mathscr{M}_i, \mathscr{M}_j)\, \sigma_{ij}^2 \tag{3.5.5}$$

where $\mathscr{L}_i$ symbolizes the triplet of quantum numbers $S^{(i)}\, L^{(i)}\, J^{(i)}$ belonging to state i, and $\mathscr{L}_j$ the set belonging to state j. $\mathscr{M}_i$ and $\mathscr{M}_j$ designate for the states i and j the set of all pertinent quantum numbers except J, e.g. for case (a), we have

$$\mathscr{M}_i \equiv \{n, l, l', S_pL_p, L^{(i)}, S^{(i)}\},$$

$$\mathscr{M}_j \equiv \{n, l, l'', S_pL_p, L^{(j)}, S^{(j)}\}.$$

Hence, $\mathscr{S}(\mathscr{L}_i\mathscr{L}_j)$ depends only on the quantum numbers that specify the line, and $\mathscr{S}(\mathscr{M}_i,\mathscr{M}_j)$ depends only on the quantum numbers that specify the particular multiplet of the transition array (Allen, 1963). In this definition, we have taken the left-hand side of eqn. (3.5.3) as the "i" state and the right-hand side as the "j" state. The remaining factor, σ_{ij}^2, called the *radial dipole integral*, is defined as in eqn. (3.2.6),

$$\sigma_{ij}^2 = \frac{1}{4l_>^2 - 1}\left[\int_0^\infty R_i^*(r)\, R_j(r)\, r^3\, dr\right]^2, \tag{3.5.6}$$

where $R_i(r)$ and $R_j(r)$ are the radial wave functions for the two states i and j, normalized so that

$$\int_0^\infty R^2(r)\, r^2\, dr = 1 \tag{3.5.7}$$

and $l_>$ is the greater of the two orbital angular momentum values which the "jumping", or "active", electron has in states i and j.

As stated above, eqn. (3.5.5) shows that the line strength splits into two factors, one of which depends on the line quantum numbers (and not on the parent) and the other depends only on the multiplet to which the line belongs. The definition of $\mathscr{S}(\mathscr{L}_i, \mathscr{L}_j)$ is constructed so that if one sums it over all the lines of a multiplet, the result is unity:

$$\sum_{i,j} \mathscr{S}(\mathscr{L}_i, \mathscr{L}_j) = 1. \tag{3.5.8}$$

Thus $\mathcal{S}(\mathcal{L}_i, \mathcal{L}_j)$ is called the *relative line strength.* The strength of a multiplet then is given by

$$\sum_{\substack{\text{all lines of } i \text{ and } j\\ \text{for a given multiplet}}} S_{ij} = [\sum_{ij} \mathcal{S}(\mathcal{L}_i, \mathcal{L}_j)]\, \mathcal{S}(\mathcal{M}_i, \mathcal{M}_j)\, \sigma_{ij}^2 = \mathcal{S}(\mathcal{M}_i, \mathcal{M}_j)\, \sigma_{ij}^2 \tag{3.5.9}$$

where $\mathcal{S}(\mathcal{M}_i, \mathcal{M}_j)$ is called the relative multiplet strength.

The sum indicated in eqn. (3.5.9) can only be performed if σ_{ij}^2 is the same for all lines in a multiplet. This implies that the radial wave functions are the same for all states in a term. While, of course, this is not strictly true, it is a satisfactory practical approximation for the circumstances with which we are concerned.

Case (a)

For this simplest case, Rohrlich (1959) gives the value of $\mathcal{S}(\mathcal{L}_i, \mathcal{L}_j)$ as

$$\mathcal{S}(\mathcal{L}_i, \mathcal{L}_j) = (2J^{(i)} + 1)(2J^{(j)} + 1)\, W^2(L^{(i)}J^{(i)}L^{(j)}J^{(j)}; S1)/(2S + 1) \tag{3.5.10}$$

where $W(abcd; ef)$ is the Racah coefficient [Racah, 1942b, eqn. (36); also Simon, Van der Sluis, and Biedenharn, 1954, tables. See also the discussion following eqn. (3.5.33)].

We have set $S^{(i)} = S^{(j)} = S$ since here we are interested only in transitions which obey the selection rule for ordinary electric dipole radiation that the spin does not change. Condon and Shortley give the line strengths S_{ij} in terms of a factor $f(SL^{(i)}J^{(i)}, SL^{(j)}J^{(j)})$ such that

$$S_{ij} = f(SL^{(i)}J^{(i)}, SL^{(j)}J^{(j)})\, |\gamma^{(i)}L^{(i)}|\, P\, |\gamma^{(j)}L^{(j)}|^2 \tag{3.5.11}$$

where the reduced matrix element† $(\gamma^{(i)}L^{(i)}\vdots P \vdots \gamma^{(j)}L^{(j)})$ does not depend on the line quantum number J. This equation factorizes the line strengths into line and multiplet factors as does eqn. (3.5.5), but f is not normalized to unity over all lines. Since Condon and Shortley give the values of the sum of the strengths over all lines, we can easily relate f to our $\mathcal{S}(\mathcal{L}_i, \mathcal{L}_j)$.

† γ represents the remaining quantum numbers, of the complete set of computing observables, which do not need to be considered explicitly, but which exist in principle. As an example, the energy E could belong to the set γ.

The sums needed are:

transition $\sum_{J^{(i)},J^{(j)}} S_{ij}$,

$$\begin{aligned} &L \to L+1 \quad (2S+1)\, L_>\, (4L_>^2 - 1)\, |(\gamma^{(i)}L \,\vdots P \vdots\, \gamma^{(j)}L+1)|^2, \\ &L \to L \qquad\;\; (2S+1)(2L+1)\, L(L+1)\, |(\gamma^{(i)}L \,\vdots P \vdots\, \gamma^{(j)}L|^2, \\ &L \to L-1 \quad (2S+1)\, L_>\, (4L_>^2 - 1)\, |(\gamma^{(i)}L \,\vdots P \vdots\, \gamma^{(j)}L-1)|^2, \end{aligned} \tag{3.5.12}$$

where the symbol $L_>$ has been used for the greater of $L^{(i)}$ and $L^{(j)}$. Thus we see that

$$\begin{aligned} \mathscr{S}(\mathscr{L}_i, \mathscr{L}_j) &= \frac{f(S, L^{(i)}, J^{(i)}; S, L^{(j)}, J^{(j)})}{(2S+1)\, L_>\, (4L_>^2 - 1)}; \quad L \to L \pm 1 \\ &= \frac{f(S, L^{(i)}, J^{(i)}; S, L^{(j)}, J^{(j)})}{(2S+1)\, L\, (L+1)(2L+1)}; \quad L \to L. \end{aligned} \tag{3.5.13}$$

The relative multiplet strengths at this point would be given by equation (3.5.12) in terms of the reduced matrix elements† $(\gamma^{(i)}, L^{(i)} \vdots P \vdots \gamma^{(j)}, L^{(j)})$. However, the multiplet strengths can be reduced further and cast into a simpler form and we shall do this below. We note here that by comparing eqns. (3.5.13) and (3.5.10) an explicit expression for $f(SL^{(i)}, J^{(i)}; SL^{(j)}, J^{(j)})$ can be obtained for the case $L \to L \pm 1$:

$$\begin{aligned} &f(SL^{(i)}, J^{(i)}; SL^{(j)}, J^{(j)}) \\ &\quad = L_>\, (4L_>^2 - 1)(2J^{(i)}+1)(2J^{(j)}+1)\, W^2(L^{(i)}, J^{(i)}, L^{(j)}, J^{(j)}, S1) \end{aligned} \tag{3.5.14}$$

where again $L_>$ is the greater of $L^{(i)}$ and $L^{(j)}$. We shall not concern ourselves with applying eqns. (3.5.10) to (3.5.14) to spectrum lines since it has so far proved impractical to incorporate the details of actual individual lines into a large-scale absorption coefficient calculation. We will assume

† An expression for the reduced matrix elements $|(\gamma^{(i)}, L^{(i)} \vdots P \vdots \gamma^{(j)}, L^{(j)})|^2$ can be obtained from the preceding formulas. We have not done this herein as we wish to avoid explicit use of this quantity. We use $\mathscr{S}(\mathscr{M}_i, \mathscr{M}_j)$ and σ_{ij}^2 instead, and have only related Condon and Shortley's f to $\mathscr{S}(\mathscr{L}_i, \mathscr{L}_j)$ so that the reader who wishes to can use their tables of the f-factors (table 1[9], p. 241, Condon and Shortley, 1935).

in this section that all the atomic states belonging to a term are degenerate and deal with entire multiplets (or in some circumstances even broader groups of lines) instead of the individual lines themselves. However, the foregoing formulas for the relative line strengths are still of interest because they are closely related to the formulas for the relative multiplet strengths. In fact in some cases the formulas are the same and one needs only to interchange the set of quantum numbers to obtain a multiplet strength from a line strength, as will be seen below.

Recalling that we are still dealing with case (a) of (3.5.3), the formula for $\mathscr{S}\,(m_i, m_j)$ is (Rohrlich, 1959)

$$\begin{aligned}\mathscr{S}\,(\mathscr{M}_i, \mathscr{M}_j) &\equiv \mathscr{S}\,(l^n l' S_p L_p S L^{(i)}, l^n l'' S_p L_p S L^{(j)}) \\ &= (2S+1)(2L^{(i)}+1)(2L^{(j)}+1)\, l_> (4l_>^2 - 1) \\ &\quad \times W^2 (l' L^{(i)} l'' L^{(j)}; L_p 1) \qquad (3.5.15)\end{aligned}$$

where $l_>$ is the greater of l' and l''. If we compare this equation to equation (3.5.10), we see that the correspondence $S \to L_p$, $L^{(i)} \to l'$, $J^{(i)} \to L^{(i)}$, $L^{(j)} \to l''$, $J^{(j)} \to L^{(j)}$ between the line quantum numbers (given to the left of the arrows) and the multiplet quantum numbers (given to the right of the arrows) carries the formula for $\mathscr{S}\,(\mathscr{L}_i, \mathscr{L}_j)$ into the formula for $\mathscr{S}\,(\mathscr{M}_i, \mathscr{M}_j)$ except for factors which are constant within a transition array. These latter factors (l', l'', and S) do not affect the relative multiplet strengths within a transition array so that the relative multiplet strengths can be obtained from the relative line strengths by the above correspondence. In terms of the Condon and Shortley f-factors, eqn. (3.5.15) becomes

$$\mathscr{S}\,(\mathscr{M}_i, \mathscr{M}_j) = (2S + 1)\, f(L_p l' L^{(i)}; L_p l'' L^{(j)}). \qquad (3.5.16)$$

The reason that the f-function of Condon and Shortley yields both $\mathscr{S}\,(\mathscr{L}_i, \mathscr{L}_j)$ and the case (a) $\mathscr{S}\,(\mathscr{M}_i, \mathscr{M}_j)$ values is that in both cases one seeks to reduce the matrix element of an operator that depends on only one of two coupled angular momenta. In determining $\mathscr{S}\,(\mathscr{L}_i, \mathscr{L}_j)$ one has S and $L^{(i)}$ coupled to $J^{(i)}$ (as well as S and $L^{(j)}$ coupled to $J^{(j)}$) with the matrix element of the dipole interaction operator being independent of spin coordinates. Thus, the matrix element is reducible to a function of $L^{(i)}$ and $L^{(j)}$ alone. In the case of $\mathscr{S}\,(\mathscr{M}_i, \mathscr{M}_j)$, L_P and l' are coupled to $L^{(i)}$

(and L_p and l'' are coupled to $L^{(j)}$). The matrix element in this case can be reduced to a function of l' and l'' alone since only this element is non-vanishing. The situation is covered by eqn. (44) given by Racah (1942b). Upon inserting the appropriate quantum numbers LSJ or $L_p lL$, and summing over magnetic quantum numbers, it is possible to form two functions which in combination yield the product

$$\mathcal{S}(\mathcal{L}_i, \mathcal{L}_j)\, \mathcal{S}(\mathcal{M}_i, \mathcal{M}_j)\, \sigma_{ij}^2$$

for case (a). The functions are

$$|\langle J^{(i)}|\, P_1\, |J^{(j)}\rangle|^2$$
$$= (2J^{(i)} + 1)(2J^{(j)} + 1)\, W^2\, (L^{(i)}J^{(i)}L^{(j)}J^{(j)}; S1)\, |\langle L^{(i)}|\, |P_1|\, |L^{(j)}\rangle|^2$$

and

$$|\langle L^{(i)}|\, |P_1|\, |L^{(j)}\rangle|^2$$
$$= (2L^{(i)} + 1)(2L^{(j)} + 1)\, W^2\, (l'L^{(i)}l''L^{(j)}; L_p 1)\, l_>\, (4l_>^2 - 1)\, \sigma_{ij}^2.$$

Going back to eqn. (3.5.1) now, bearing in mind eqns. (3.5.5), (3.5.6) and (3.5.9), we obtain the total cross section for absorption of a photon by an atom which makes a transition from one term to another:

$$\sigma(\omega) = \frac{4\pi^2 e^2 \omega}{3\hbar c}\, \frac{b(\omega)}{g_i}\, \mathcal{S}(\mathcal{M}_i, \mathcal{M}_j)\, \sigma_{ij}^2. \qquad (3.5.17)$$

This is an important practical formula for large-scale calculations of absorption in heated gases. Many values of the relative multiplet strengths $\mathcal{S}(\mathcal{M}_i, \mathcal{M}_j)$ are tabulated in Allen's (1963) book, in addition to the Condon and Shortley tabulation of f-factors from which the $\mathcal{S}(\mathcal{M}_i, \mathcal{M}_j)$ can be obtained through eqn. (3.5.16). In order to better illustrate the use of the relative multiplet factors, let us consider some examples in detail. Our original example for case (a) was the non-equivalent electron transition

$$1s^2\, 2s^2\, 2p\, {}^2P \rightarrow 1s^2\, 2s^2\, 3s\, {}^2S$$

for N III, say. Equation (3.5.16) yields

$$\mathcal{S}(\mathcal{M}_i, \mathcal{M}_j) = 2 \cdot 3 \cdot 1 \cdot 1(1)(3)\, W^2\, (1100; 01) = 6$$

since Simon, Van der Sluis and Biedenharn (1954) give the value $\frac{1}{3}$ for the square of the Racah coefficient. This value agrees with that listed by Allen (1963) on p. 62 in the upper left-hand corner of the page.

The arguments of the W coefficients can be permuted in a variety of ways without changing their values (Edmonds, 1957, p. 94†).

For example,

$$W(l'L^{(i)}l''L^{(j)}; L_p 1) = W(l'l''L^{(i)}L^{(j)}; 1L_p). \qquad (3.5.18)$$

The order of the arguments on the right-hand side of eqn. (3.5.18) is that which is employed by Burgess and Seaton (1960) and for convenience in comparing our analysis with their paper we shall employ it in the sequel. For electric-dipole transitions, in which $L^{(j)} = L^{(i)} \pm 0, 1$, the Racah coefficient reduces to a simple analytic form (see, for example, Condon and Shortley, p. 238, for the analytic form of the f-functions). In Table 3.2 we give the formulas for $\mathcal{S}(\mathcal{M}_i, \mathcal{M}_j)$ for case (a) based on this analytic form. It is worthy of note that the $l \to l+1$ and $l \to l-1$ cases are not independent, but are derivable, one from the other, by an interchange of arguments based on the invariance

$$W^2(l'l''L^{(i)}L^{(j)}; 1L_p) = W^2(l''l'L^{(j)}L^{(i)}; 1L_p). \qquad (3.5.19)$$

The formulas of Table 3.2 may be summed over $L^{(j)}$ to obtain the total strength of the lines of the array which originate in the term $\gamma_p L_p S_p nlSL^{(i)}$. The result is (Condon and Shortley, 1935, p. 248):

$$l \to l-1 \quad \sum_{L(j)} \mathcal{S}(\mathcal{M}_i, \mathcal{M}_j) = (2S+1)(2L^{(i)}+1)\, l(2l-1) \qquad (3.5.20a)$$

$$l \to l+1 \quad \sum_{L(j)} \mathcal{S}(\mathcal{M}_i, \mathcal{M}_j) = (2S+1)(2L^{(i)}+1)(l+1)(2l+3). \qquad (3.5.20b)$$

We return now to our examples.

First let us compute the line strength factor $f(SLJ, SL'J')$ for

$$SLJ = 323,$$

$$S'L'J' = 323.$$

† The relationship between the 6-j coefficients in terms of which Edmonds gives these permutations and the Racah coefficients is given by Edmonds on p.97 and later in this section by eqn. (3.5.34a).

TABLE 3.2. Relative multiplet strengths $\mathscr{S}(\mathscr{M}_i, \mathscr{M}_j)$ for non-equivalent electrons

$$\mathscr{S}(\mathscr{M}_i, \mathscr{M}_j) = (2S+1)(2L^{(i)}+1)(2L^{(j)}+1)\, l_>\, (4l_>^2 - 1)\, W^2(l'l''L^{(i)}L^{(j)}; 1L_p)$$
$$= (2S+1)\, f(L_p l' L^{(i)}; L_p l'' L^{(j)})$$

(A) $l \to l-1$

	$\mathscr{S}(\mathscr{M}_i, \mathscr{M}_j)$
$L \to L+1$	$\dfrac{(2S+1)(L_p+L-l+1)(L_p-L+l)(L_p+L-l+2)(L_p-L+l-1)}{4(L+1)}$
$L \to L$	$\dfrac{(2S+1)(2L+1)(-L_p+L+l)(L_p+L-l+1)(L_p+L+l+1)(L_p-L+l)}{4L(L+1)}$
$L \to L-1$	$\dfrac{(2S+1)(L-L_p+l-1)(L-L_p+l)(L+L_p+l+1)(L+L_p+l)}{4L}$

(B) $l \to l+1$

	$\mathscr{S}(\mathscr{M}_i, \mathscr{M}_j)$
$L+1 \to L$	$\dfrac{(2S+1)(L_p+L-l)(L_p-L+l+1)(L_p+L-l+1)(L_p-L+l)}{4(L+1)}$
$L \to L$	$\dfrac{(2S+1)(2L+1)(-L_p+L+l+1)(L_p+L-l)(L_p+L+l+2)(L_p-L+l+1)}{4L(L+1)}$
$L-1 \to L$	$\dfrac{(2S+1)(L-L_p+l)(L-L_p+l+1)(L+L_p+l+2)(L+L_p+l+1)}{4L}$

The formula, from Condon and Shortley (1935, eqn. 2^9 2b), is

$$f = (2J+1)\frac{[J(J+1) - S(S+1) + L(L+1)]^2}{4J(J+1)}. \quad (3.5.21)$$

Inserting the above values into the formula, one obtains $f = \frac{21}{4}$. This is the correct absolute value.

To find Condon and Shortley's tabular entry in table 1^9 this value must be normalized so that the maximum f-value is 100. That is

$$f_{\text{tab}} = \tfrac{21}{4}(a) \quad (3.5.22)$$

where a is determined by

$$af_{\max} = 100. \quad (3.5.23)$$

Now, according to table 1[9], f_{max} occurs for the $S = 3, L = 2$ to $S = 3, L = 2$ transition for $J = 5, J' = 5$, viz. the 325 to 325 transition. We use eqn. (3.5.21) again and obtain $f_{max} = (11 \times 24)/5$. Again, this is the correct absolute value. We determine $\boldsymbol{a}$ from eqns. (3.5.22) and (3.5.23):

$$a = \frac{100}{f_{max}} = \frac{500}{11 \times 24}$$

Inserting this value in eqn. (3.5.22) we obtain

$$f_{tab} = 9.94.$$

The tabular entry is 9.9. As noted on p. 240 of Condon and Shortley, the entries have been sharply rounded off. The line strength $\mathcal{S}(\mathcal{L}_i, \mathcal{L}_j)$ is given by eqn. (3.5.13) in terms of f, and from this relation we obtain†

$$\mathcal{S}(\mathcal{L}_i, \mathcal{L}_j) = \frac{f}{(2S+1)L(L+1)(2L+1)} = \frac{\frac{21}{4}}{7 \cdot 2 \cdot 3 \cdot 5} = \frac{1}{40}.$$

This agrees with the result obtained from eqn. (3.5.10):

$$\mathcal{S}(\mathcal{L}_i, \mathcal{L}_j) = \frac{7 \cdot 7}{7} W^2(2323; 31) = 7 \cdot \frac{1}{5 \cdot 7 \cdot 8} = \frac{1}{40}.$$

Next, let us consider a multiplet. We take the example quoted by Condon and Shortley, p. 245:

$$TiId^2\, 4s(^4F)\, 5s\; ^5F \rightarrow d^2\, 4s(^4F)\, 4p\; ^5DFG$$

for which:

$$L_p = 3 \qquad S = 2$$
$$l' = 0 \qquad l'' = 1$$
$$L^{(i)} = 3 \quad L^{(j)} = 2, 3, 4.$$

† This example can be obtained as follows from Allen's (1963) tables on p. 57, which are normalized still a different way. The maximum J-value, J_m, for this line is $L + S = 5$. Hence, the small box on p. 56 for DD multiplets such as this $(L, L' = 2)$ gives x_3 as Allen's designation for the line strength for $J_m - 2$. Returning to the table on p. 57 one finds for x_3 in the DD table the value 4.37 in the column under multiplicity $2S + 1 = 7$. The total line strength given at the head of this column is 175. Dividing these two values yields 0.025, the result above. To obtain this value from Levinson and Nikitin (1965), we refer to the table for septet transitions ($S = 3$) on p. 173. In the box for DD transitions, we consult the row and column for $J = J' = 3$. The entry is 2.5, which is the line strength $\mathcal{S}$ normalized to 100.

Relative values of $\mathscr{S}(\mathscr{M}_i, \mathscr{M}_j)$ can again be obtained from table 1[9] of Condon and Shortley by the correspondence $SLJ \rightarrow L_p l' L^{(i)}$, and $SL'J' \rightarrow L_p l'' L^{(j)}$ which carries the line-strength $f(SLJ;\ SL'J')$ into the multiplet strength $f(L_p L' L^{(i)};\ L_p l'' L^{(j)})$. Thus, the effective $SLJ \rightarrow SL'J'$ values are $303 \rightarrow 312; 3; 4$. The three entries for $L^{(j)}$ (or J') $= 2, 3, 4$ are listed in the box above the "spin = 3" diagonal which appears as:

	2	3	4
0	55.6	77.8	100

In their example on p. 245, Condon and Shortley quote these numbers as the ratios 50 : 70 : 90, which can be obtained from the tabular entries by multiplying the latter by 0.90. The values of this configuration are not listed as such† by Allen (1963), but do appear as the first entries (45, 35 and 25) under the transition $sd^3 \rightarrow pd^3$ on p. 64. This is so because the only parent quantum number that appears in eqn. (3.5.15) is L_p, and therefore, in this transition all ^4F and ^6F parent states are equivalent for calculating $\mathscr{S}(\mathscr{M})$. To obtain values from our formulas one may use eqn. (3.5.15) or Table 3.2. From Table 3.2, for $l \rightarrow l + 1$ we find

(a) $L + 1 \rightarrow L \qquad (F \rightarrow D,\ L = 2,\ l = 0)$

$$\mathscr{S}(\mathscr{M}_i, \mathscr{M}_j) = 5\,\frac{(3+2-0)(3-2+0+1)(3+2-0+1)(3-2+0)}{4(3)}$$

$$= 25.$$

(b) $L \rightarrow L \qquad (F \rightarrow F,\ L = 3,\ l = 0)$

$$\mathscr{S}(\mathscr{M}_i, \mathscr{M}_j) =$$

$$5\,\frac{(7)(-3+3+0+1)(3+3-0)(3+3+0+2)(3-3+0+1)}{4 \cdot 3 \cdot 4}$$

$$= 35.$$

† As with the Condon and Shortley table, relative values (unnormalized) can be obtained from Allen's (1963) line strength tables by the line-to-multiplet strength correspondence listed above. The maximum J-value is $J_m = 4$, so from the SP multiplet box on p. 56 of Allen's book one finds z, y and x, as his designations for the $J' = 2, 3$ and 4 strengths, respectively. The box at the top of p. 57 for SP transitions then yields the relative values 5.0, 7.0 and 9.0 for these strengths under the column for multiplicity $2S + 1 = 7$.

(c) $L-1 \to L \qquad (F \to G,\ L=4,\ l=0)$

$$\mathscr{S}(\mathscr{M}_i, \mathscr{M}_j) = 5\,\frac{(4-3+0)(4-3+0+1)(4+3+0+2)(4+3+0+1)}{16}$$

$$= 45.$$

These display the ratio 50:70:90 quoted by Condon and Shortley.

To evaluate this example from the tables of Levinson and Nikitin (1965), we consult their table on p. 188 for *s–p* transitions outside an S_0L_0 core. Since $L_0 = L$ for an *s* electron outside the core, these transitions are specified by the final *L*-values of (in this example) 2, 3 and 4. Thus, we read the entries for the columns D, F and G at the row for multiplicity 5. These entries are 4.17, 5.83 and 7.50, which in their terminology are "reduced multiplet strengths". To convert these to absolute values, according to p. 80 of their book, they must be multiplied by the "one-electron multiplet strength" factor

$$s(nl, n'l') \equiv 2l_> (4l_>^2 - 1)\, \sigma^2_{nl,n'l'}$$

where σ is the radial dipole integral (see eqns. (6.3) and (6.4) on p. 73 or eqn. (8) on p. 181, *op cit.*). This means that we must multiply the above entries by $2l_> (4l_>^2 - 1)$ to convert them to values of our $\mathscr{S}(\mathscr{M}_i, \mathscr{M}_j)$. For $l_> = 1$, this factor equals 6 which converts the above entries to 25.0, 35.0 and 45.0.

In the case of hydrogen, our formula eqn. (3.5.17) must, of course, reduce back to the same result as is obtainable by elementary means. To see that this occurs, we note that

$$L_p = 0,$$

$$L^{(i)} = l',$$

$$L^{(j)} = l'',$$

where, by eqn. (3.5.15) we obtain

$$\mathscr{S}(\mathscr{M}_i, \mathscr{M}_j) = (2S+1)(2l'+1)(2l''+1)\, l_> (4l_>^2 - 1)\, W^2(l'l''l'l''; 10). \tag{3.5.24}$$

From the formula for W for this special case, given by Edmonds [1957, p. 98, eqn. (6.3.2)] we have

$$W^2 (ll''l'l''; 10) = [(2l' + 1)(2l'' + 1)]^{-1}. \tag{3.5.25}$$

Using this result and eqn. (3.5.24) in eqn. (3.5.17) we obtain for the hydrogen cross section:

$$\sigma(\omega) = \frac{4\pi^2 e^2 \omega}{3\hbar c} b(\omega) \frac{l_> (4l_>^2 - 1)}{2l' + 1} \sigma_{l',l''}^2. \tag{3.5.26}$$

We have used the correspondence $L^{(i)} \to l'$, which implies that $g_i = (2S + 1)(2l' + 1)$. An average f-number is often defined by [Bethe and Salpeter, 1597, eqn. (61.2); our eqns. (3.3.58) and (3.1.33)].

$$\bar{f}_{n'l',n''l''} \equiv \frac{l_>}{3(2l' + 1)} \left(\frac{\hbar\omega_{n'l',n''l''}}{\text{Ryd}} \right) \left[\int_0^\infty R_{n'l'}(r) R_{n''l''}(r) r^3 dr^2 \right] \Big/ a_0^2 \tag{3.5.27}$$

where Ryd $= e^2/2a_0$ is the Rydberg energy unit. In terms of this average f-number, our cross section as given by eqn. (3.5.26) can be written

$$\sigma(\omega) = \frac{2\pi^2 e^2}{mc} \bar{f}_{n'l',n''l''} b(\omega) \tag{3.5.28}$$

in agreement with eqn. (3.3.59) (the averaging does not alter the constants in the formula).

Coefficients of Fractional Parentage

Rohrlich's case (a), which we have been discussing up to this point, does not materially involve equivalent electrons. The core may, of course, involve them, but case (a) assumes that the core electron quantum numbers remain unchanged. To proceed to the remaining cases which do involve equivalent electrons, we must define the *coefficients of fractional parentage*, or fpc, previously alluded to. These arise in the problem of factoring a one-electron wave function out of an antisymmetrized product wave function in such a way that the overall wave function remains antisymmetric. They can be defined by the equation

$$\psi(l^n SL) = \sum_{S_p L_p} F(l^n SL; S_p L_p) \psi([l^{n-1} S_p L_p] lSL) \tag{3.5.29}$$

where $\psi([l^{n-1}S_pL_p]\,lSL)$ denotes the wave function for a state in which $n-1$ electrons couple together into a parent term S_pL_p, and the remaining electron couples to this parent to yield the overall coupling SL. The Racah notation for the fpc which we have denoted by $F(l^nSL, S_pL_p)$ in eqn. (3.5.29) is

$$F(l^nSL; S_pL_p) \equiv (l^nSL\,\{|l^{n-1}(S_pL_p)\,lSL). \tag{3.5.30}$$

The values of these coefficients for the p-shell ($l = 1$) are given in Table 3.2. Equation (3.5.29) may be lengthened to include explicitly the angular momentum coupling of the substates distinguished by different magnetic quantum numbers. The more explicit statement is:

$$\begin{aligned}\psi(l^nSLM_SM_L) = \sum_{S_pL_p} F(l^nSL, S_pL_p) \sum_{M(L_p),m(l)} \; \sum_{M(S_p),m(s)} & C(L_plL; M(L_p), m(l), M(L)) \\ & \times C(S_p\tfrac{1}{2}S; M(S_p), m(s), M(S)) \\ & \times \psi(l^{n-1}S_pL_p\,M(S_p)\,M(L_p))\,\Phi(l\,m(l)\,m(s)).\end{aligned} \tag{3.5.31}$$

The coefficients $C(j_1, j_2, j_3; m_1, m_2, m_3)$ are Clebsch–Gordan coefficients used earlier [cf. eqn. (3.3.20)]. They are defined (Condon and Shortley, 1935, chap. 3) to give the correct fraction of each substate when coupling two angular momenta (j_1 and j_2) to get a third (j_3). They are also known as Wigner, or vector-coupling, coefficients and are referred to by Condon and Shortley as transformation amplitudes for vector addition. They are closed related to the Wigner 3-j symbols (Edmonds, 1957, p. 46):

$$C(l_1l_2l_3; m_1m_2m_3) = (-1)^{l_1-l_2+m_3}(2l_3+1)^{1/2}\begin{pmatrix} l_1 l_2 l_3 \\ m_1 m_2, -m_3\end{pmatrix}.$$

Magnetic quantum numbers are not usually included explicitly in matrix element formulas because the results do not depend on them unless the operator whose matrix element is under consideration is itself not symmetric with respect to the magnetic angle φ. This occurs, for example, in the case of an atom perturbed by an external magnetic field. In the absence of such asymmetric perturbations, the factors of the wave func-

TABLE 3.3. Fractional parentage coefficients for the p-shell

$$F_p\,(nS'L';\,SL) = [p^{n-1}\,(S'L')\,p\,|\}\,p^n\,(SL)]$$

(The square root sign has been omitted from all magnitudes, and the squares of the fpc are normalized to n, the number of electrons in the shell.)

$F_p\,(0\;\;S'L';\;SL) = 0$
$F_p\,(1\;\;0\;\;0;\;\tfrac{1}{2}\;\;1) = 1$
$F_p\,(2\;\;\tfrac{1}{2}\;\;1;\;0\;\;0) = 2$
$\quad 1\;\;1) = 2$
$\quad 0\;\;2) = 2$

$F_p\,(3\;\;S'L';\;SL)$:

$S'L'$ / SL	0 0	1 1	0 2
$\tfrac{3}{2}$ 0	0	3	0
$\tfrac{1}{2}$ 1	$\tfrac{2}{3}$	$-\tfrac{3}{2}$	$-\tfrac{5}{6}$
$\tfrac{1}{2}$ 2	0	$\tfrac{3}{2}$	$-\tfrac{3}{2}$

$F_p\,(4\;\;S'L';\;SL)$:

$S'L'$ / SL	$\tfrac{3}{2}$ 0	$\tfrac{1}{2}$ 1	$\tfrac{1}{2}$ 2
0 0	0	4	0
1 1	$\tfrac{4}{3}$	1	$-\tfrac{5}{3}$
0 2	0	1	3

$F_p\,(5\;\;0\;\;0;\;\tfrac{1}{2}\;\;1) = \tfrac{1}{3}$
$\quad 1\;\;1;\quad\;) = 3$
$\quad 0\;\;2;\quad\;) = \tfrac{5}{3}$
$F_p\,(6\;\;\tfrac{1}{2}\;\;1;\;0\;\;0) = 6$

tions which depend on m and φ integrate to one or zero, leaving only sums of products of Clebsch–Gordan coefficients, which also reduce to unity because of the sum rule, eqn. (3.3.24), obeyed by these coefficients, and in view of the fact that $C\,(l_1l_2l_3;\,m_1m_2m_3) = 0$ unless $m_1 + m_2 = m_3$. With the foregoing definition of the fpc, the relative multiplet strength for Rohrlich's case (b) ($l^n \to l^{n-1}l'$) may be written as

$$\begin{aligned}\mathcal{S}\,(\mathcal{M}_i,\,\mathcal{M}_j) &\equiv \mathcal{S}\,(l^nSL^{(i)};\,l^{n-1}S_pL_pl'SL^{(j)})\\ &= nF^2\,(l^nSL^{(i)};\,S_pL_p)\,\mathcal{S}\,(l^{n-1}lS_pL_pSL^{(i)},\,l^{n-1}l'S_pL_pSL^{(j)}).\end{aligned} \tag{3.5.32}$$

To obtain a formula for case (c), $l^nl' \to l^{n-1}(l')^2$, we must expand both l^n and $(l')^2$ with fpc in order to isolate the active electron. However, the fpc for $(l')^2$ are all either zero or one, since there is only one possible parent state. In addition, we must recouple the expanded function $l^{n-1}ll'$ in the order $l^{n-1}l'l$ to cause the active electron (viz. the electron for which $l \to l'$) to be the last to be coupled on in both states. This is accomplished by the use of the Racah (1943) eqn. (6), which in our case has the form

$$\psi(l^{n-1}S_pL_p, lS'L', l'SL) = \sum_{S''L''} U(\tfrac{1}{2}S_pS\tfrac{1}{2}; S'S'')\, U(lL_pLl'; L'L'') \times \psi(l^{n-1}S_pL_p, l'S''L'', lSL) \times (\text{phase factor}). \quad (3.5.33)$$

In this formula, the Jahn coefficients U are a particular form of the Racah coefficients defined by a sum of products of four Clebsch–Gordan coefficients. Their virtue is to again relieve us of any need to consider Clebsch–Gordan coefficients explicitly.

Tables of the Racah coefficients W and the closely related 6-j coefficients are now generally available (see Simon, Van der Sluis and Biedenharn, 1954; Howell, 1959; Rotenberg, 1959). The relationship of the three symbols to each other is given by (see Edmonds, 1957, chap. 6):

$$U(abcd; ef) = \sqrt{(2e+1)}\sqrt{(2f+1)}\, W(abcd; ef) = \sqrt{(2e+1)}\sqrt{(2f+1)} \begin{Bmatrix} a & b & e \\ d & c & f \end{Bmatrix} (-1)^{a+b+c+d} \quad (3.5.34a)$$

and their relationship to the Clebsch–Gordan coefficients is given by

$$U(abcd; ef) = \sum_{m_a m_b m_d m_e m_f} C(edc; m_e m_d m_c)\, C(abe; m_a m_b m_e) \times C(bdf; m_b m_d m_f)\, C(afc; m_a m_f m_c). \quad (3.5.34b)$$

Using eqn. (3.5.33) and the expansion (3.5.29), we can now write the angular factor for Rohrlich's case (c) as

$$\mathcal{S}(\mathcal{M}_i, \mathcal{M}_j) = \sum_{S''L''} U^2(\tfrac{1}{2}S_pS\tfrac{1}{2}; S'S'') \cdot U^2(lL_pL^{(i)}l'; L'L'') \cdot 2n \times F^2(l^nS'L', S_pL_p) \cdot \mathcal{S}(l^{n-1}l'lS''L''SL^{(i)}, l^{n-1}l'l'S''L''SL^{(j)}). \quad (3.5.35)$$

We note that S_p, L_p, S', L', S and L all have fixed values determined by the chosen initial and final states, while $S''L''$ takes on all the permitted values for the configuration $l^{n-1}S_pL_p, l'S''L''$.

To illustrate the use of eqn. (3.5.35), consider our example of case (c):

$$1s^2\,2s\,2p^4\,(^4P) \to 1s^2\,2s^2\,2p^3\,(^4S).$$

The fact that this is really $l'l^n \to l'^2l^{n-1}$ can introduce no more than a phase change into the wave function and, therefore, has no effect upon $\mathscr{S}(\mathscr{M}_i, \mathscr{M}_j)$.

We find

$$L_p = 0 \quad S_p = \tfrac{3}{2}$$

$$L' = 1 \quad S' = 1 \quad S'' = 2 \text{ or } 1$$

$$L^{(i)} = 1 \qquad L'' = 0$$

$$S = \tfrac{3}{2}$$

$$L^{(j)} = 0$$

$$l = 1 \quad l' = 0$$

$$nF^2\,(p^4\,{}^3P, {}^4S) = \tfrac{4}{3}.$$

From the tables:

$$U^2\,(\tfrac{1}{2}\tfrac{3}{2}\tfrac{3}{2}\tfrac{1}{2}; 1S'') = \tfrac{1}{16} \quad \text{if} \quad S'' = 1$$

$$= \tfrac{15}{16} \quad \text{if} \quad S'' = 2,$$

and $U^2\,(10L1; 10) = 1$.

Consequently, we find for the multiplet strength,

$$\mathscr{S}(\mathscr{M}_i, \mathscr{M}_j) = \tfrac{1}{16}\cdot 2\cdot 4\cdot \tfrac{1}{3}\,\mathscr{S}(p^3sp\,{}^3S\,{}^4P, p^3ss{}^3S\,{}^4S)$$

$$+ \tfrac{15}{16}\cdot 2\cdot 4\cdot \tfrac{1}{3}\,\mathscr{S}(p^3sp\,{}^5S\,{}^4P, p^3ss{}^5S\,{}^4S)$$

$$= \tfrac{1}{6}\cdot 9 + \tfrac{5}{2}\cdot 15 = 39,$$

where the values of $\mathscr{S}$ for the non-equivalent electron transitions are obtained from eqn. (3.5.15) as before.

Turning to case (d), $l^nl'l'' - l^nl'^2$ may be computed without use of fpc since it is not necessary to alter any shell having no more than two equivalent electrons. The non-equivalent electron formulas of case (a) may be employed as before, after $l^n l'^2$ has been recoupled to $(l^n l')\,l'$. This is accomplished by using the Racah (1943), eqn. (6), in the form

$$\psi\,(l^n S_l L_l, l'^2 S'L', SL^{(j)}) = \sum_{S''L''} U\,(\tfrac{1}{2}\tfrac{1}{2}SS_l;\, S'S'')\, U\,(l'l'L^{(j)}L_l;\, L'L'')$$

$$\times\, \psi\,(l^n S_l L_l l' S'' L'', l', SL^{(j)}) \times \text{(phase factor)}. \quad (3.5.36)$$

Non-zero matrix elements can occur only if S'' and L'' have values equal to those specified for $l^n l'$ in the state $l^n S_l L_l l' S_p L_p l'' SL$. Therefore, we keep only that term in our expansion for which $S'' = S_p$ and $L'' = L_p$.

Then we have

$$\mathcal{S}\,(\mathcal{M}_i, \mathcal{M}_j) = U^2\,(\tfrac{1}{2}\tfrac{1}{2}SS_l;\, S'S_p)\, U^2\,(l'l'L^{(j)}L_l;\, L'L_p)$$

$$\times\, 2\mathcal{S}\,(l^n l' l'' S_p L_p SL^{(i)}, l^n l' l' S_p L_p SL^{(j)}). \quad (3.5.37)$$

We illustrate the use of eqn. (3.5.37) with the case (d) example, $1s^2 2s 2p\,{}^3P - 1s^2 2p^2\,{}^3P$. Since $1s^2$ is a closed shell the U-coefficients are both equal to one. On this account the transition may also be seen as a simple example of case (b), with fpc equal to one.

Using eqn. (3.5.15) in (3.5.37), we find

$$\mathcal{S}\,(\mathcal{M}_i, \mathcal{M}_j) = 1 \cdot 1 \cdot 2\mathcal{S}\,(1s^2\, 2p\, 2s\, {}^2P\, {}^3P, 1s^2\, 2p\, 2p\, {}^2P\, {}^3P)$$

$$= 2 \cdot 3 \cdot 3 \cdot 3 \cdot 1 \cdot 3 \cdot W^2(1101;\, 11) = 18, \quad \text{since} \quad W^2 = \tfrac{1}{9}.$$

This result agrees with Allen's tables.

A more complex example will pair the ground state of titanium, which we write (ignoring the closed inner shells) as $3d^2 4s^2\,({}^3F)$, with the excited state $3d^2\,({}^3F)\,4s\,4p\,({}^3G)$. This excited state is actually degenerate, consisting of the two states $3d^2\,({}^3F)\,4s\,({}^4F)\,4p\,({}^3G)$ and $3d^2\,({}^3F)\,4s\,({}^2F)\,4p\,({}^3G)$. We shall compute the transition to these two states separately.

The Racah coefficients are

$$U^2\,(\tfrac{1}{2}\tfrac{1}{2}11;\, 0\tfrac{3}{2}) = \tfrac{2}{3} \text{ for the first transition}$$

$$U^2\,(\tfrac{1}{2}\tfrac{1}{2}11;\, 0\tfrac{1}{2}) = \tfrac{1}{3} \text{ for the second,}$$

and

$$U^2\,(0033;\, 03) = 1 \text{ for both.}$$

Again using eqns. (3.5.15) and (3.5.37) we find

$$\mathscr{S}(\mathscr{M}_i, \mathscr{M}_j) = \tfrac{2}{3} \cdot 1 \cdot 2 \cdot 3 \cdot 7 \cdot 9 \cdot 1 \cdot 3 \cdot W^2(0314; 31)$$

$$= 36 \quad \text{since} \quad W^2 = \tfrac{1}{21}.$$

$\mathscr{S}(\mathscr{M}_i, \mathscr{M}_j)$ for the second transition differs only in the Racah coefficient, since S_p does not appear explicitly in eqn. (3.5.15). Therefore, its value is 18. The sum of these numbers, $36 + 18 = 54$, is the result given by Allen who lumps degeneracies of this type. A very general formula for the relative multiplet strength can be given for the case of many shells of electrons coupled to a particular state $\mathcal{L}\mathcal{S}$ and making a dipole transition to a state $\mathcal{L}'$G (Kelly, 1964a). In the formula which we give below, α and β designate the shells occupied by the active electron in states i and j. Primed quantum numbers refer to the upper state (j) and unprimed numbers to the lower state (i). While L_α, S_α stand for the resultant orbital momentum and spin of all electrons in shell α, $\mathcal{L}_\alpha$,$\mathcal{S}_\alpha$ refer to the total resultants for all shells out to and including shell α. It is assumed that α is more tightly bound than β and that the order of angular momentum coupling is from the most bound shell to the least bound one (which is denoted as shell t). The order of coupling is significant only when we consider degenerate states (same $\mathcal{L}$ and $\mathcal{S}$) of a configuration having three or more open shells. This general formula is

$$\begin{aligned}
\mathscr{S} = {} & l_> (4l_>^2 - 1)\,[L'_\alpha, l_\alpha|\}\, L_\alpha]^2\, [L_\beta, l_\beta|\}\, L'_\beta]^2 \\
& \times \prod_{p=\beta}^{t-1} U^2(\mathcal{L}_{p+1} L_{p+1} 1 \mathcal{L}'_p; \mathcal{L}_p \mathcal{L}'_{p+1}) \\
& \times (2\mathcal{L}_{\beta-1} + 1)(2L'_\beta + 1)(2\mathcal{L}'_\beta + 1)(2\mathcal{L} + 1) \begin{Bmatrix} 1 l_\beta l_\alpha \\ \mathcal{L}_\beta L_\beta \mathcal{L}_{\beta-1} \\ \mathcal{L}'_\beta L'_\beta \mathcal{L}'_{\beta-1} \end{Bmatrix}^2 \\
& \times \delta(\mathcal{S}'_\beta \mathcal{S}_\beta)(2\mathcal{S} + 1)\, U^2(\tfrac{1}{2} \mathcal{S}'_{\beta-1} S_\beta \mathcal{S}_\beta; \mathcal{S}_{\beta-1} S'_\beta) \\
& \times \prod_{q=\alpha}^{\beta-2} U^2(l_\alpha \mathcal{L}'_q \mathcal{L}_{q+1} L_{q+1}; \mathcal{L}_q \mathcal{L}'_{q+1})\, U^2(\tfrac{1}{2} \mathcal{S}'_q \mathcal{S}_{q+1} S_{q+1}; \mathcal{S}_q \mathcal{S}'_{q+1}) \\
& \times U^2(l_\alpha L'_\alpha \mathcal{L}_\alpha \mathcal{L}_{\alpha-1}; L_\alpha \mathcal{L}'_\alpha)\, U^2(\tfrac{1}{2} S'_\alpha \mathcal{S}_\alpha \mathcal{S}_{\alpha-1}; S_\alpha \mathcal{S}'_\alpha).
\end{aligned} \tag{3.5.38}$$

The symbol $\begin{Bmatrix} a & b & c \\ d & e & f \\ g & h & i \end{Bmatrix}$ denotes a 9-j coefficient (see Edmonds, 1957, chap. 6).

This formula is strictly correct only if there is no change in the inactive orbitals during a transition. Such changes are normally quite small, leading to alterations in f numbers of a few percent at most.

Shore and Menzel (1965, 1968) have followed the Racah algebra approach to transition-probability calculations in their compilation of extensive tables of line-strength factors. Their tables cover a large number of the possible transitions involving s-, p- and d-shells of electrons. Their line strength factor differs from that of Rohrlich only by the factor $2S+1$. All values of J have been covered for which $L < 8$ and $S < 4$. The corresponding multiplet strength factors are presented as a table of formulas for transitions among large classes of open-shell configurations. These multiplet factors are functions of three quantities which are also tabulated. These three quantities are: (1) multiplet factors for transitions involving non-equivalent electrons only (corresponding to our eqn. (3.5.15) and Table 3.1, but omitting the factors $(2S+1)\, l_> \,(4l_>^2 - 1)$ found in eqn. (3.5.15)); (2) fractional parentage coefficients—a complete tabulation for the p- and d-shells, including phases; (3) recoupling coefficients required when the order of angular momentum coupling of a state must be changed before (1) can be applied. The multiplet factor table does not cover such cases as pp', dd' (two open p- or d-shells), $d^n p^n$ (except for $p^n d$ and $d^n p$), and $f^n (n > 1)$.

An excellent compendium of classic papers on angular momentum coupling is now available, including work of Wigner, Pauli, Racah, and many later contributions (Biedenharn and Van Dam, 1965).

3.6. Free–Free Radiative Transitions†

Although a free electron cannot absorb a photon and simultaneously conserve both energy and momentum, a nearby third particle may accept the necessary recoil momentum and thereby permit the photon absorption. Such a process is the inverse of bremsstrahlung and is called free–

† This section follows closely the article by Johnston (1967).

free absorption. It is generally the dominant continuous radiation absorption effect for photons of energy less than the important photoionization thresholds. The absorption and emission processes can, of course, be related by detailed balance arguments as in the photoelectric case.

For densities that are not too high, one can associate the photon-absorbing electron with a particular momentum-absorbing ion or atom (hereafter called ion), and the free–free absorption process may be treated as a radiative absorption transition between two continuum electron states of the ion. Corrections due to momentum absorption by pairs of ions may become important at very high ion densities, and Debye shielding effects and photon absorption by electron pairs may become important at high electron densities. Such effects are ignored here; only the Debye shielding effect will be mentioned briefly.

The classical calculation of the radiation emitted by an electron moving in a Coulomb field was first carried out by Kramers (1923). Gaunt (1930) provided a definitive quantum-theory calculation and applied his results to certain important astrophysical questions. Gaunt's calculations were done explicitly for the absorption case. Somewhat later, Menzel and Pekeris (1935) gave more detailed results and a number of analytic approximations for hydrogen. Further papers appeared giving increasing detail primarily for hydrogen, e.g. Berger (1956; 1957), Grant (1958), Karzas and Latter (1961), culminating in the broad review by Brussaard and Van de Hulst (1962). These, along with some other significant references, are collected in Table 3.4. Essentially non-hydrogenic results began to appear in the early 1960s (Breene and Nardone, 1960, 1961, 1963), and the quantum defect method of Burgess and Seaton was generalized to the free–free case by Peach (1965) and applied (Peach, 1967a) to a wide variety of atoms. Extensive computations of bremsstrahlung emission have also been made for themonuclear plasma physics applications (see, for example, Bekefi, 1966), although these have been largely of a hydrogenic nature. An interesting impact parameter treatment of the classical limit has recently been given by Gould (1970), who also obtains quantum mechanical corrections to the Born approximation. The case of bremsstrahlung from highly relativistic electrons has also recently been reviewed by Blumenthal and Gould (1970).

Since the physical system involved in free–free absorption is more complicated than that involved in either discrete or bound–free absorption, we might expect the mathematical description to be more difficult. This turns out to be the case, particularly as regards normalization and the counting of states. For this reason, we will now undertake a more general approach to these problems based on the elementary theory of transition probabilities and wave-function normalization of Sections 3.2–3.4, and on the results of formal scattering theory (Goldberger and Watson, 1964; Shore, 1967).

TABLE 3.4. Selected references on free–free transition cross sections

Kramers (1923)	Greene (1959)
Gaunt (1930)	Karzas and Latter (1961)
Menzel and Pekeris (1935)	Breene, Jr., and Nardone (1960, 1961, 1963)
Sommerfeld (1939)	Brussaard and Van de Hulst (1962)
Wheeler and Wildt (1942)	Peach (1965)
Chandrasekhar and Breen (1946)	De Vore (1964)
Mayer (1947)	John (1966)
Berger (1956; 1957)	Ashkin (1966)
Bethe and Salpeter (1957)	Gould (1970)
Grant (1958)	

Our basic quantum states χ_α, χ_β, say, are such that, if we expand an arbitrary state Ψ in terms of the χ_α, e.g.

$$\Psi = \sum_\alpha c(\alpha)\, \chi_\alpha, \tag{3.6.1}$$

the probability of obtaining the result "α" when measuring the quantity α in the system whose wave function is Ψ will be $|c(\alpha)|^2$. Since

$$\langle \Psi | \Psi \rangle = 1 = \sum_{\alpha,\alpha'} c^*(\alpha)\, c(\alpha')\, \langle \chi_\alpha | \chi_{\alpha'} \rangle \tag{3.6.2}$$

we must have

$$\langle \chi_{\alpha'} | \chi_\alpha \rangle = \langle \alpha' | \alpha \rangle = \delta(\alpha' - \alpha) \tag{3.6.3}$$

and our basic states $|\alpha\rangle$ are said to be normalized on the "α-scale". Then the dimension of $|\alpha\rangle$ is $[|\alpha\rangle] = [\alpha]^{-1/2}$. We now recall the discussions of the Fermi "Golden Rule", eqn. (3.2.1) or (3.2.3) and the requirement of a continuous range of states in the initial or final state of the system. This latter requirement permits one to write the sum over continuous states which leads to the Rule as an integral over $|\langle\alpha| H' |\beta\rangle|^2 \, \delta(E_\alpha - E_\beta) \, dn$ where dn is the number of states centered about α or β, whichever is continuous. If now we *postpone* the sum over continuous states, and do not perform it yet, we can write the basic transition probability formula as

$$\frac{\delta w_{\beta\alpha}}{\delta\alpha} = \delta\beta \, \frac{2\pi}{\hbar} \, \delta(E_\beta - E_\alpha) \, |\langle\beta|H'|\alpha\rangle|^2 \qquad (3.6.4)$$

where $\delta\alpha$ is the number of states at α so that $\delta w_{\alpha\beta}/\delta\alpha$ is the transition probability per initial state and $\delta\beta$ is the number of states at β. Usually we are concerned with transitions in which one state (say α) is discrete and labelled with a dimensionless quantum number n, say. For such a state u_n we have

$$\int u_n^* \, u_{n'} \, dv = \delta_{nn'} \qquad (3.6.5)$$

and $\delta\alpha = 1$ in the above formula.

As an example of this approach we will repeat one of our earlier calculations of a radiative transition from a discrete atomic state $|i\rangle$, energy E_i, to a continuum of states $|f\rangle$ in δf [see eqn. (3.4.15)]. The interaction with the radiation field H' is $-(e/mc)\,\boldsymbol{p}\cdot\boldsymbol{A}$ as given by eqn. (3.3.1), and the intensity I_0 is given by $\omega^2/2\pi c \, |A_0|^2$ where $A = A_0 \exp(i\boldsymbol{k}\cdot\boldsymbol{r})$. Then the number of transitions per unit time from an initial discrete state $|i\rangle$ (c-number normed) into a final state $|f\rangle$ in df at f is, in the dipole approximation

$$\delta w_{fi} = \delta f \, \frac{2\pi}{\hbar} \, \delta(E_f - E_i) | \langle f| \left(-\frac{e}{mc} A_0\right) \hat{\boldsymbol{\varepsilon}}\cdot\boldsymbol{p}\,|i\rangle|^2$$

$$= \delta f \, \frac{2\pi}{\hbar} \, \delta(E_f - E_i) \, \frac{e^2}{m^2c^2} \, \frac{2\pi c}{\omega^2} \, I_0 \, |\langle f| \, \hat{\boldsymbol{\varepsilon}}\cdot\boldsymbol{p}\,|i\rangle|^2. \qquad (3.6.6)$$

$|f\rangle$ and $|i\rangle$ are energy eigenstates, so we can use eqn. (3.3.7)

$$\langle f| \, \hat{\boldsymbol{\varepsilon}}\cdot\boldsymbol{p}\,|i\rangle = -im\omega_{fi} \, \langle f| \, \hat{\boldsymbol{\varepsilon}}\cdot\boldsymbol{r}\,|i\rangle$$

and find

$$\delta w_{fi} = \delta f \, \frac{4\pi^2 c^2}{\hbar c} \, I_0 \, \delta(E_f - E_i) \, | \, \hat{\boldsymbol{\varepsilon}}\cdot\langle f| \, \boldsymbol{r}\,|i\rangle|^2. \qquad (3.6.7)$$

Now $I_0/\hbar\omega$ is the photon flux, and the cross section as defined earlier is $\delta\sigma = \delta w \hbar\omega / I_0$. Therefore, the cross section for transition into any of the δf states centered at f is given by integration of eqn. (3.6.7) over df:

$$\sigma_{fi} = \int df \left(\frac{4\pi^2 e^2}{\hbar c}\right)(\hbar\omega)\, \delta(E_f - E_i) \,|\, \hat{\varepsilon} \cdot \langle f|\, \boldsymbol{r} \,|i\rangle|^2. \qquad (3.6.8)$$

If the wave function in the final state is normalized on the energy scale, $df = dE_f$, and if we average the above result over polarizations (which produces the factor $\frac{1}{3}$), eqn. (3.6.8) becomes

$$\sigma_{fi} = \frac{8\pi^3 e^2 \nu}{3c} |\boldsymbol{r}_{fi}|^2 \qquad (3.6.9)$$

in agreement with eqn. (3.4.49). If the wave function of the free electron in the final state is normalized on the wave number scale,

$$\Psi_k(\boldsymbol{r}) = (2\pi)^{-3/2} \exp(i\boldsymbol{k} \cdot \boldsymbol{r}), \qquad (3.6.10)$$

we can write the number of states df as

$$df = \frac{dn}{dE_f} dE_f = \frac{4\pi p^2 (dp/dE_f)}{(2\pi\hbar)^3} dE_f = \frac{m^2 v}{2\pi^2 \hbar^3} dE_f. \qquad (3.6.11)$$

Using the result in eqn. (3.6.8) and performing the usual polarization average, we obtain

$$\sigma_{fi} = \frac{2e^2 \omega m^2 v}{3\hbar^3 c} |\boldsymbol{r}_{fi}|^2$$

in agreement with eqn. (3.4.29b).

The Fundamental Cross Section

Both the initial and final states of the absorbing electron are in the continuum spectrum of the atom (ion + electron), so the choice of wave-function normalization must be consistent with the density of states factor in the Fermi "Golden Rule" [eqn. (3.2.1)]. This is made evident by writing the basic expression for the transition probability as

$$dw_{\alpha\beta} = \frac{2\pi}{\hbar} |\langle\beta|\, H' \,|\alpha\rangle|^2\, \delta(E_i - E_f) \left[\frac{d^3 k_\gamma}{(2\pi)^3}\right] d\alpha\, d\beta. \qquad (3.6.12)$$

The number of initial states is the product of the number of states $d\alpha$ of the atomic system and the number of photon states

$$dn_R = \frac{\omega^2 \, d\Omega}{(2\pi c)^3 \hbar} \, dE = \frac{d^3 k_\gamma}{(2\pi)^3} \tag{3.6.13}$$

[cf. Heitler, 1954, p. 244, eqn. (11)]. This formula gives the number of transitions per second from an atomic (ion + electron) state $|\alpha\rangle$ in $d\alpha$ to a state $|\beta\rangle$ in $d\beta$ due to the interaction H'. Here E_i and E_f are the total energy of the system in the initial and final states, respectively, and the integration over the energy-conserving delta-function has not yet been carried out. The photon density of states $d^3k_\gamma/(2\pi)^3$ is consistent with the photon amplitude normalization implicit in H' presented previously [eqn. (3.3.3)],

$$H' = -\frac{e}{m}\left(\frac{2\pi\hbar}{\omega} n_\omega\right)^{1/2} \boldsymbol{p} \cdot \hat{\varepsilon} \, e^{-i\boldsymbol{k}_\gamma \cdot \boldsymbol{r}}, \tag{3.6.14}$$

which corresponds to an initial state of n_ω photons in a volume $(2\pi)^3$ with polarization $\hat{\varepsilon}$ and wave number $\boldsymbol{k}_\gamma$ in $d^3\boldsymbol{k}_\gamma$. That is, the incident photon flux is given by:

$$\text{photon flux} = n_\omega \left[\frac{d^3 k_\gamma}{(2\pi)^3}\right] c \; (\text{cm}^{-2}\,\text{sec}^{-1}). \tag{3.6.15}$$

The electron momentum operator $\boldsymbol{p}$ here and in the following is to be understood as the sum

$$\sum_{j=1}^{J} \boldsymbol{p}_j$$

over the J electrons of the system. The state $|\alpha\rangle$ is normalized on the "α-scale",

$$\langle\alpha|\alpha'\rangle = \int \langle\alpha|\boldsymbol{r}\rangle \, d^3\boldsymbol{r} \, \langle\boldsymbol{r}|\alpha'\rangle = \delta(\alpha - \alpha'), \tag{3.6.16}$$

corresponding to one state per interval $d\alpha$, as is evident from the closure relation:

$$\sum_\alpha |\alpha\rangle\langle\alpha| = 1 \rightarrow \int \langle\boldsymbol{r}|\alpha\rangle \, d\alpha \, \langle\alpha|\boldsymbol{r}'\rangle = \delta(\boldsymbol{r} - \boldsymbol{r}'). \tag{3.6.17}$$

Clearly the dimensions of $\langle \boldsymbol{r} | \alpha \rangle^2 \, d\alpha$ are (length)$^{-3}$ and the matrix element of any operator O between such states has the dimensions of O. Thus, one may symbolically indicate that $(|\alpha\rangle)^2 \, d\alpha$ is dimensionless.

Dividing the transition probability eqn. (3.6.12) by the incident flux eqn. (3.6.15) and using the matrix element of eqn. (3.6.14) one obtains, in the usual electric dipole approximation, the result

$$d\sigma_{\alpha\beta} = \frac{4\pi^2 e^2}{m^2 \omega c} |\langle \beta | \boldsymbol{p} \cdot \hat{\varepsilon} | \alpha \rangle|^2 \, \delta(E_f - E_i) \, d\alpha \, d\beta \quad (\text{cm}^2). \qquad (3.6.18)$$

This is the cross section per incident photon with energy in $d(\hbar\omega)$ at $\hbar\omega$ for a transition from a state $|\alpha\rangle$ in $d\alpha$ to state $|\beta\rangle$ in $d\beta$, and the indicated dimensions follow from the discussion above.

Alternate forms of the matrix elements equivalent to eqn. (3.6.18), the so-called dipole velocity form, for *exact* wave functions are easily obtained from the commutation relations of the operators $\boldsymbol{p}$ and $\boldsymbol{r}$ with the Hamiltonian. These are [see eqn. (3.3.7) and Bethe and Salpeter, 1957, § 59β]:

$$\langle \beta | \boldsymbol{p} \cdot \hat{\varepsilon} | \alpha \rangle = im\omega_{\alpha\beta} \langle \beta | \boldsymbol{r} \cdot \hat{\varepsilon} | \alpha \rangle \qquad (3.6.19a)$$

$$= \frac{i}{\omega_{\alpha\beta}} \langle \beta | \hat{\varepsilon} \cdot \nabla V | \alpha \rangle = i \frac{Ze^2}{\omega_{\alpha\beta}} \left\langle \beta \right| \frac{\hat{\varepsilon} \cdot \boldsymbol{r}}{r^3} \left| \alpha \right\rangle, \qquad (3.6.19b)$$

where V is the potential. The first relation connects the dipole velocity form with the dipole length form which was used in Section 3.4. The last of these is the so-called Coulomb "dipole acceleration" form and is most frequently used in free–free calculations. As usual $\hbar\omega_{\alpha\beta} = E_\beta - E_\alpha$ and Z is the nuclear charge. For approximate wave functions the validity of these alternative forms should be considered from the standpoint of the underlying commutation relations. Thus, the dipole acceleration form is obtained from matrix elements of

$$[\boldsymbol{p}, H] = \boldsymbol{p}H - H\boldsymbol{p} = \frac{\hbar}{i} \nabla V, \qquad (3.6.20)$$

in terms of the exact Hamiltonian H, written $H = \sum (p^2/2m) + V$. For exact states $|\alpha\rangle$ and $|\beta\rangle$, V is the sum of the Coulomb interactions and eqn. (3.6.19b) follows. For states $|\alpha\rangle$ and $|\beta\rangle$ which are two eigenstates of some approximate Hamiltonian, use of the associated approximate energies and eqn. (3.6.20) with the approximate potential V provides an alternate form of the dipole-acceleration matrix element which is useful for non-Coulombic cases (Bethe and Salpeter, 1957, p. 252). Finally, if the states $|\alpha\rangle$ and $|\beta\rangle$ are eigenstates of different approximate Hamiltonians, even the validity of eqn. (3.6.20) must be reconsidered, as discussed by Mjolsness and Ruppel (1967) in connection with the use of Hartree–Fock wave functions.

Averaging over the polarization directions of the incident photon as before [eqn. (3.3.11)],

$$\overline{|\langle\beta|\boldsymbol{p}\cdot\hat{\boldsymbol{\varepsilon}}|\alpha\rangle|^2} = \tfrac{1}{3}\sum_{\mu}|\langle\beta|p_\mu|\alpha\rangle|^2 \equiv \tfrac{1}{3}|\langle\beta|\boldsymbol{p}|\alpha\rangle|^2, \qquad (3.6.21)$$

we find that the cross section (3.6.18) may be written in the form

$$d\sigma_{\alpha\beta} = \frac{4\pi^2e^2}{3m^2\omega c}|\langle\beta|\boldsymbol{p}|\alpha\rangle|^2\,\delta(E_i - E_f)\,d\alpha\,d\beta. \qquad (3.6.22)$$

To proceed further, the states $|\alpha\rangle$ and $|\beta\rangle$ and their normalization should be prescribed in greater detail. This is most simply done in terms of their asymptotic forms. Neglecting for the present effects due to electron exchange, the initial state $|\alpha\rangle$ asymptotically describes an ion in some state i—with wave function, say, $\varphi_i(\boldsymbol{R})$ and energy W_i—and a free electron in a state described by the wave function $\chi_{k_i}^{(+)}(\boldsymbol{r})$ with energy $\varepsilon_i = \hbar^2k_i^2/2m$. That is,

$$\langle\boldsymbol{r},\boldsymbol{R}|\alpha\rangle \xrightarrow[|r|\to\infty]{} \varphi_i(\boldsymbol{R})\,\chi_{k_i}^{(+)}(\boldsymbol{r}), \qquad (3.6.23)$$

and $E_i = W_i + \varepsilon_i + \hbar\omega$. $\boldsymbol{R}$ is the ion coordinate vector.

If one assumes the long-range Coulomb potential to be shielded at large distances, the state $\chi_{k_i}^{(+)}$ asymptotically represents an incident plane wave and outgoing scattered waves:

$$\chi_{k_i}^{(+)}(\boldsymbol{r}) \xrightarrow[|r|\to\infty]{} A\,[e^{i\boldsymbol{k}_i\cdot\boldsymbol{r}} + f^{(+)}(\varepsilon_i, \hat{\boldsymbol{k}}_i\cdot\hat{\boldsymbol{r}})\,e^{ik_ir}/r]. \qquad (3.6.24)$$

Here $f^{(+)}(\varepsilon_i, \hat{\boldsymbol{k}}_i\cdot\hat{\boldsymbol{r}})$ is the usual elastic scattering amplitude at energy ε_i in terms of directions measured from that of the incident beam. Normalizing on the $\boldsymbol{k}$-scale determines $A = (2\pi)^{-3/2}$ and $d\alpha = d^3k_i$, as may be seen from the closure relation, eqn. (3.6.17):

$$\int \langle \boldsymbol{r}' | \alpha \rangle \, d\alpha \, \langle \alpha | \boldsymbol{r} \rangle \xrightarrow[|r| \to \infty]{} \int \frac{e^{i\boldsymbol{k}\cdot\boldsymbol{r}'}}{(2\pi)^{3/2}} \, d^3\boldsymbol{k} \, \frac{e^{-i\boldsymbol{k}\cdot\boldsymbol{r}}}{(2\pi)^{3/2}} = \delta(\boldsymbol{r} - \boldsymbol{r}').$$

The final state $|\beta\rangle$ similarly represents asymptotically the ion state i and a free electron in a final plane wave state with *ingoing* spherical waves,

$$\chi_{k_f}^{(-)}(\boldsymbol{r}) \xrightarrow[|r| \to \infty]{} (2\pi)^{-3/2} \left[e^{i\boldsymbol{k}_f\cdot\boldsymbol{r}} + f^{(-)}(\varepsilon_f, \hat{\boldsymbol{k}}_f \cdot \hat{\boldsymbol{r}}) \, e^{-ik_f r}/r \right], \qquad (3.6.25)$$

where again the normalization is on the $\boldsymbol{k}$-scale and $d\beta = d^3k_f$. The states $\chi_k^{(-)}$ are simply related to $\chi_k^{(+)}$ in terms of the Wigner time-reversal operation (Goldberger and Watson, 1964, p. 170),

$$\chi_k^{(-)} = T_0 \chi_k^{(+)}, \qquad (3.6.26)$$

where T_0 is simply the operator of complex conjugation with neglect of spin. Thus, we have

$$f^{(-)}(\varepsilon, \hat{\boldsymbol{k}} \cdot \hat{\boldsymbol{r}}) = [f^{(+)}(\varepsilon, -\hat{\boldsymbol{k}} \cdot \hat{\boldsymbol{r}})]^*.$$

Finally, $E_f = W_i + \varepsilon_f$, and

$$\langle \boldsymbol{r}, \boldsymbol{R} | \beta \rangle \xrightarrow[|r| \to \infty]{} \varphi_i(\boldsymbol{R}) \, \chi_{k_f}^{(-)}(\boldsymbol{r}). \qquad (3.6.27)$$

With this choice of continuum states, the basic cross section eqn. (3.6.22) becomes, after summing over the final electron states,

$$d\sigma_{\alpha\beta} = \frac{4\pi^2 e^2 k_f}{3m\hbar^2 \omega c} \{ \textstyle\int d\Omega_f \, |\langle \beta | \boldsymbol{p} | \alpha \rangle|^2 \} \, d^3k_i \quad (\text{cm}^2), \qquad (3.6.28)$$

where we have used $d^3k_f = (mk_f/\hbar^2) \, d\varepsilon_f \, d\Omega_f$, and k_f is to be determined from energy conservation, $k_f^2 = k_i^2 + (2m/\hbar^2) \, \hbar\omega$.

The physically significant quantity is the radiation absorption coefficient $N_v \sigma(\omega)$, which is the probability per unit length that a photon of angular frequency ω will be absorbed in the medium [see eqn. (3.3.51)]. For bound–bound and bound–free processes the radiation-absorbing electron is uniquely associated with a momentum-absorbing ion, so it is convenient to write

$$\mu(\omega) = N_v \sigma(\omega) \quad (\text{cm}^{-1})$$

where N_v is the number density, and the cross section $\sigma(\omega)$ has the dimensions (length)^2 = area. For free–free transitions, however, the electrons are no longer uniquely associated with any particular ion, so the electron density is in general independent of the ion density unless thermodynamic equilibrium prevails. Thus, it is convenient to take the free–free absorption coefficient $\mu^{FF}(\omega)$ proportional to both the electron and the ion densities,

$$\mu^{FF}(\omega) = N_v N_e \sigma^{FF}(\omega), \tag{3.6.29}$$

and the coefficient of proportionality, $\sigma^{FF}(\omega)$, is frequently called the "free–free cross section" in spite of the fact that it has dimensions (length)^5.

The quantity $d\sigma_{\alpha\beta}$ of eqn. (3.6.28) is the cross section per ion for absorption of a photon of angular frequency ω by a free electron of density $(2\pi)^{-3}$ per unit volume with wave number $\boldsymbol{k}_i$ in d^3k_i, as is evident from the chosen asymptotic forms [eqns. (3.6.24) and (3.6.25)]. Then, if $F_e(\boldsymbol{k})\, d^3k$ is the fractional number of electrons per unit volume with $\boldsymbol{k}$ in d^3k—normalized to unity—one obtains

$$\mu^{FF}(\omega) = N_v N_e \int F_e(\boldsymbol{k}_i)\, [(2\pi)^3\, d\sigma_{\alpha\beta}].$$

Writing the free–free cross section as

$$\sigma^{FF}(\omega) = \int F_e(\boldsymbol{k}_i)\, \sigma^{FF}(\boldsymbol{k}_i, \omega)\, d^3k_i, \tag{3.6.30}$$

one evidently obtains from eqn. (3.6.28) the result

$$\sigma^{FF}(\boldsymbol{k}_i, \omega) = \frac{(2\pi)^5 e^2 k_f}{3m\hbar^2\omega c} \int d\Omega_f\, |\langle\beta|\, \boldsymbol{p}\, |\alpha\rangle|^2 \quad (\text{cm}^5). \tag{3.6.31}$$

The One-electron Approximation and Radial Decomposition of the Matrix Elements

In most applications a one-electron approximation is assumed, according to which the photon-absorbing electron is assumed to move in a static potential $V(\boldsymbol{r})$ determined by the average charge distribution of the ion and possibly including some semi-empirical corrections for the effects of ion polarizability and electron exchange (Kivel, 1961; DeVore, 1964). As mentioned above, the choice of asymptotic forms eqns. (3.6.23) and (3.6.27) precludes a correct treatment of electron exchange. Accordingly, in this section and in most of the following, electron exchange will be ignored, except as it can be approximately included in the potential $V(\boldsymbol{r})$, and will be discussed only at the conclusion of the next section.

Thus, in the one-electron approximation one assumes equalities for all $\boldsymbol{r}$ in eqns. (3.6.23) and (3.6.27), with the states $\chi_k^{(+)}(\boldsymbol{r})$ taken to be continuum solutions of the single-particle Schrödinger equation

$$\left[-\frac{\hbar^2}{2m}\nabla^2 + V(\boldsymbol{r})\right]\chi_k^{(+)}(\boldsymbol{r}) = \left(\frac{\hbar^2k^2}{2m}\right)\chi_k^{(+)}(\boldsymbol{r}),$$

satisfying the boundary conditions eqns. (3.6.24) and (3.6.25). In this approximation the necessary matrix elements become one-electron quantities and are most frequently expressed in the dipole-length or dipole-acceleration forms eqns. (3.6.19a) and (3.6.19b):

$$\begin{aligned}\langle\beta|\,\boldsymbol{p}\,|\alpha\rangle &= im\omega \int d^3r\,[\chi_{k_f}^{(-)}(\boldsymbol{r})]^*\,\boldsymbol{r}\chi_{k_i}^{(+)}(\boldsymbol{r}) \\ &= (i/\omega)\, d^3r\,[\chi_{k_f}^{(-)}(\boldsymbol{r})]^*\,(\nabla V)\,\chi_{k_i}^{(+)}(\boldsymbol{r}).\end{aligned} \tag{3.6.32}$$

If the effective single-particle potential is taken to be spherically symmetric a significant simplification results from an angular momentum decomposition of the one-electron states. Let

$$\chi_{k_i}^{(+)}(\boldsymbol{r}) = \sum_{l=0}^{\infty} R_l^{(+)}(k_i, r)\sum_m Y_{lm}^*(\hat{\boldsymbol{k}}_i)\,Y_{lm}(\hat{\boldsymbol{r}}), \tag{3.6.33}$$

where the normalized spherical harmonics Y_{lm} are from Edmonds (1957). Similarly decomposing the asymptotic form [eqn. (3.6.24)] with $A = (2\pi)^{-3/2}$, gives

$$\chi_{k_i}^{(+)}(\boldsymbol{r}) \xrightarrow[r\to\infty]{} \sum_{l=0}^{\infty}\left[\sqrt{\frac{2}{\pi}}\, i^l\, e^{i\delta_l}\,\frac{\sin(k_i r - l\pi/2 + \delta_l)}{k_i r}\right]\sum_m Y_{lm}^*(\hat{\boldsymbol{k}}_i)\,Y_{lm}(\hat{\boldsymbol{r}}),$$

where the scattering amplitude $f^{(+)}(\varepsilon_i, \hat{\boldsymbol{k}}_i\cdot\hat{\boldsymbol{r}})$ has been expressed in terms of the elastic scattering phase shifts $\delta_l(\varepsilon_i)$ in the usual way:

$$f^{(+)}(\varepsilon, \hat{\boldsymbol{k}}\cdot\hat{\boldsymbol{r}}) = \frac{4\pi}{k}\sum_{lm} e^{i\delta_l}\sin\delta_l\, Y_{lm}^*(\hat{\boldsymbol{k}})\,Y_{lm}(\hat{\boldsymbol{r}}). \tag{3.6.34}$$

Therefore, if we write

$$R_l^{(+)}(k, r) = i^l e^{i\delta_l(\varepsilon)} \frac{w_l(k, r)}{kr}. \tag{3.6.35}$$

the function $w_l(k, r)$ is the solution of

$$\frac{d^2}{dr^2} w_l(k, r) - \left[\frac{l(l+1)}{r^2} + \frac{2m}{\hbar^2} V(r) - k^2\right] w_l(k, r) = 0. \tag{3.6.36}$$

It satisfies the boundary conditions

$$w_l(k, 0) = 0$$

$$w_l(k, r) \xrightarrow[r\to\infty]{} \sqrt{\frac{2}{\pi}} \sin(kr - l\pi/2 + \delta_l) \tag{3.6.37}$$

and is normalized according to

$$\int_0^\infty w_l(k, r)\, w_l(k', r)\, dr = \delta(k - k'). \tag{3.6.38}$$

The final state wave function may be similarly expressed:

$$\chi_{k_f}^{(-)}(\boldsymbol{r}) = \sum_{l=0}^{\infty} R_l^{(-)}(k_f, r) \sum_m Y_{lm}^*(\hat{\boldsymbol{k}}_f)\, Y_{lm}(\hat{\boldsymbol{r}}). \tag{3.6.39}$$

Using eqns. (3.6.26), (3.6.33), (3.6.35) and (3.6.39) one easily finds the result

$$R_l^{(-)} = (-1)^l T_0 R_l^{(+)} = i^l e^{-i\delta_l(\varepsilon)} \frac{w_l(k, r)}{kr} = e^{-2i\delta_l(\varepsilon)} R_l^{(+)}. \tag{3.6.40}$$

The one-electron matrix elements (3.6.32) then become

$$\begin{aligned}|\langle\beta|\boldsymbol{p}|\alpha\rangle|^2 = (4\pi)^{-2} \sum_{l_i l_i' l_f l_f'} &\mathscr{R}(l_f l_i)\, \mathscr{R}^*(l_f', l_i')\,(2l_i+1)(2l_i'+1)(2l_f+1)(2l_f'+1)\\ &\times \begin{pmatrix} l_i l_f 1 \\ 0\,0\,0 \end{pmatrix} \begin{pmatrix} l_i' l_f' 1 \\ 0\,0\,0 \end{pmatrix} \sum_l (-1)^{l+1} \begin{pmatrix} l_i l_i' l \\ 0\,0\,0 \end{pmatrix} \begin{pmatrix} l_f l_f' l \\ 0\,0\,0 \end{pmatrix}\\ &\times \begin{Bmatrix} l_i l_i' l \\ l_f' l_f 1 \end{Bmatrix} (2l+1)\, P_l(\hat{k}_f \cdot \hat{k}_i)\end{aligned} \tag{3.6.41}$$

in terms of radial integrals defined by

$$\mathscr{R}(l_i, l_f) = im\omega \int_0^\infty r^2\, dr\, R_{l_f}^{(-)}(k_f, r)^*\, r R_{l_i}^{(+)}(k_i, r) \quad (3.6.42a)$$

$$= \frac{i}{\omega} \int_0^\infty r^2\, dr\, R_{l_f}^{(-)}(k_f, r)^* \frac{\partial V}{\partial r} R_{l_i}^{(+)}(k_i, r) \quad (3.6.42b)$$

and in terms of the Wigner 3-*j* and 6-*j* symbols. In obtaining this result the integral over three spherical harmonics was used,

$$\int d\Omega\, Y_{l_1 m_1} Y_{1\mu} Y_{l_2 m_2} = \left[\frac{3}{4\pi}(2l_1+1)(2l_2+1)\right]^{1/2} \begin{pmatrix} l_1 l_2 1 \\ 0\,0\,0 \end{pmatrix} \begin{pmatrix} l_1 l_2 1 \\ m_1 m_2 \mu \end{pmatrix}, \quad (3.6.43)$$

which can be obtained from eqn. (3.3.20) and the relation between the Clebsch–Gordan and 3-*j* coefficients [see eqn. (3.5.31) et seq.]. In addition, the following vector-coupling relations are needed:

$$Y_{lm}(\hat{k})\, Y_{l'm'}(\hat{k}) = \sum_{LM} \left[\frac{(2l+1)(2l'+1)(2L+1)}{4\pi}\right]^{1/2} \times \begin{pmatrix} l l' L \\ 0\,0\,0 \end{pmatrix} \begin{pmatrix} l l' L \\ m m' M \end{pmatrix} Y_{LM}^*(\hat{k}) \quad (3.6.44)$$

and

$$\sum_{\mu_1\mu_2\mu_3} (-1)^{l_1+l_2+l_3+\mu_1+\mu_2+\mu_3} \begin{pmatrix} j_1 l_2 l_3 \\ m_1 \mu_2 -\mu_3 \end{pmatrix} \begin{pmatrix} l_1 j_2 l_3 \\ -\mu_1 m_2 \mu_3 \end{pmatrix} \begin{pmatrix} l_1 l_2 j_3 \\ \mu_1 -\mu_2 m_3 \end{pmatrix} = \begin{pmatrix} j_1 j_2 j_3 \\ m_1 m_2 m_3 \end{pmatrix} \begin{Bmatrix} j_1 j_2 j_3 \\ l_1 l_2 l_3 \end{Bmatrix}. \quad (3.6.45)$$

The 6-*j* symbol has been denoted, as is customary, by brackets to distinguish it from the 3-*j* symbol in the large parentheses. The relationships between the 6-*j* symbol, the Racah coefficient *W*, and the Jahn coefficient *U* have been given in eqn. (3.5.34a). Equations (3.6.44) and (3.6.45) are eqns. (4.6.5), and (6.2.8), respectively, from Edmonds (1957). An integration over the direction of the outgoing electron and an average over the direction of the incident electron of the free–free cross section [eqn. (3.6.31)] yields, finally,

$$\sigma^{FF}(\boldsymbol{k}_i, \omega) = \frac{8\pi^4}{3}\left(\frac{e^2}{\hbar c}\right)\frac{k_f}{m\hbar\omega}\sum_{l_i=0}^{\infty}(l_i+1)\{|\mathscr{R}(l_i, l_i+1)|^2 + |\mathscr{R}(l_i+1, l_i)|^2\} \tag{3.6.46}$$

in terms of the radial integrals eqns. (3.6.42a) and (3.6.42b). We can compare this result to that of Gaunt (1930) by noting that his radial wave function has the asymptotic form

$$\left(\frac{8\pi m}{hk}\right)^{1/2}\frac{\sin kr}{r}$$

(from eqn. 5.28ff. of Gaunt, 1930) compared to

$$\left(\frac{2}{\pi k^2}\right)^{1/2}\frac{\sin kr}{r}$$

here. If we set $\partial V/\partial r = Ze^2/r^2$ in eqn. (3.6.42b) we can write, in view of the difference in wave function normalization,

$$|\mathscr{R}(l_f, l_i)|^2 = \frac{Z^2e^4h^2}{16\pi^4m^2\omega^2k_ik_f}|\mathscr{R}^G(l_f, l_i)|^2 \tag{3.6.47}$$

where the radial integral labelled $\mathscr{R}^G$ is now equivalent to Gaunt's radial integral. If we substitute this result into eqn. (3.6.46) we obtain

$$\sigma^{FF} = \frac{Z^2e^6h}{24\pi^2m^4cv_iv^3}\sum_{l=0}(l+1)\{|\mathscr{R}^G(l+1, l)|^2 + |\mathscr{R}^G(l, l+1)|^2\} \tag{3.6.48}$$

which agrees with eqn. (5.24) of Gaunt (1930).

Relation to Elastic Scattering

For low-energy photons the free–free absorption cross section may be approximately expressed in terms of elastic scattering amplitudes. This relation is discussed by Hundley (1962) and Low (1958) and utilized by many authors (for example, Ohmura and Ohmura, 1960; Firsov and Chibisov, 1961; and Ashkin, 1966).

The relation may be most easily derived in the one-electron approximation with neglect of exchange, although the result is of more general validity (John, 1966). In this approximation the basic matrix element [eqn. (3.6.32)] may be written†

$$\langle\beta|\,\boldsymbol{p}\,|\alpha\rangle = (\chi_{k_f}^{(-)}, \boldsymbol{p}\chi_{k_i}^{(\pm)}),$$

where $\chi_k^{(+)}$ satisfies a one-electron Schrödinger equation with Hamiltonian $H = H_0 + V$. This equation may be written in the Lippmann–Schwinger form (Goldberger and Watson, 1964, sect. 5.3)

$$\chi_k^{(\pm)} = \varphi_k + \frac{1}{\varepsilon_k \pm i\eta - H_0} V\chi_k^{(\pm)},$$

where η is a small parameter used to select the proper contour around the singularity and is ultimately set to zero, and φ_k is a plane-wave eigenstate of H_0 with energy ε_k. Then, using eqn.(3.6.20), we can construct the relation

$$\begin{aligned}\langle\beta|\,\boldsymbol{p}\,|\alpha\rangle &= \frac{1}{\varepsilon_f - \varepsilon_i}(\chi_{k_f}^{(-)}, (H\boldsymbol{p} - \boldsymbol{p}H)\,\chi_{k_i}^{(+)})\\ &= \frac{1}{\hbar\omega}\left\{\left(\chi_{k_f}^{(-)}, V\boldsymbol{p}\left(\varphi_{k_i} + \frac{1}{\varepsilon_{k_i} + i\eta - H_0}V\right)\chi_{k_i}^{(+)}\right)\right.\\ &\quad \left. - \left(\left(\varphi_{k_f} + \frac{1}{\varepsilon_{k_f} - i\eta - H_0}V\right)\chi_{k_f}^{(-)}, \boldsymbol{p}V\chi_{k_i}^{(+)}\right)\right\}.\end{aligned}$$

The φ_k are momentum eigenstates, so we have

$$\begin{aligned}\langle\beta|\,\boldsymbol{p}\,|\alpha\rangle = \frac{1}{\hbar\omega}\Bigg\{&\hbar\boldsymbol{k}_i\,(\chi_{k_f}^{(-)}, V\varphi_{k_i}) - \hbar\boldsymbol{k}_f\,(\varphi_{k_f}, V\chi_{k_i}^{(+)})\\ &+ \left(\chi_{k_f}^{(-)}, V\boldsymbol{p}\left(\frac{1}{\varepsilon_i + i\eta - H_0} - \frac{1}{\varepsilon_f + i\eta - H_0}\right)V\chi_{k_i}^{(+)}\right)\Bigg\}.\end{aligned} \tag{3.6.49}$$

† This equation is actually an identity. In this section, we use the scalar product notation $(\Psi', \boldsymbol{p}\Psi)$ rather than the Dirac notation when we wish to call explicit attention to the wave function. The reader is cautioned to remember that this scalar product notation still permits the operation toward the left-hand side by means of the Hermitian conjugate. See Goldberger and Watson (1964), sect. 1.1.

The scattering amplitudes for elastic electron scattering may be written (*op. cit.*, sect. 6.2)

$$f(\varepsilon, \hat{\boldsymbol{k}}_i \cdot \hat{\boldsymbol{k}}_f) = -(2\pi)^2 \frac{m}{\hbar^2} (\varphi_{k_f}, V\chi_{k_i}^{(+)}) = -(2\pi)^2 \frac{m}{\hbar^2} (\chi_{k_f}^{(-)}, V\varphi_{k_i})$$

when $\varepsilon_f = \varepsilon_i$. Thus, in the limit $\omega \to 0$, eqn. (3.6.49) may be written

$$\langle\beta|\, \boldsymbol{p}\, |\alpha\rangle \xrightarrow[\omega\to 0]{} \frac{(\hat{\boldsymbol{k}}_i - \hat{\boldsymbol{k}}_f)}{\omega} \left[\frac{-\hbar^2}{(2\pi)^2 m} f(\varepsilon_i, \hat{\boldsymbol{k}}_i \cdot \hat{\boldsymbol{k}}_f)\right]. \quad (3.6.50)$$

By expanding the energy-dependent quantities in eqn. (3.6.49) about the mean energy $\bar{\varepsilon} \equiv \frac{1}{2}(\varepsilon_i + \varepsilon_f)$, Low (1958, eqn. 1.7 N.R.) has shown

$$\langle\beta|\, \boldsymbol{p}\, |\alpha\rangle = \frac{-\hbar^2}{(2\pi)^2 m} \left\{ \frac{(\boldsymbol{k}_i - \boldsymbol{k}_f)}{\omega} f(\bar{\varepsilon}, \hat{\boldsymbol{k}}_i \cdot \hat{\boldsymbol{k}}_f) - \left(\frac{\boldsymbol{k}_i + \boldsymbol{k}_f}{2}\right) \frac{\partial f(\bar{\varepsilon}, \hat{\boldsymbol{k}}_i \cdot \hat{\boldsymbol{k}}_f)}{\partial \bar{\varepsilon}} + O\left(\frac{\hbar\omega}{\bar{\varepsilon}}\right) \right\}. \quad (3.6.51)$$

Because the cross-term in the square of this matrix element contains the factor $(k_f^2 - k_i^2)$ which is proportional to ω, retention of the first term alone should lead only to relative errors of order $(\hbar\omega/\bar{\varepsilon})^2$ in the cross section, aside from possible elastic scattering resonances near the energy $\bar{\varepsilon}$.

Thus, an approximation to the free–free absorption coefficient useful at low photon energies is given by

$$\sigma^{\mathrm{FF}}(k_i, \omega) = \frac{8\pi}{3} \frac{e^2 k_f \bar{\varepsilon}}{m^2 c \omega^3} Q_d(\bar{\varepsilon}) + O\left(\frac{\hbar\omega}{\bar{\varepsilon}}\right)^2$$

$$\cong \frac{8\pi}{3} \alpha a_0^3 \left\{\frac{k_f \bar{\varepsilon}}{E_\omega^3}\right\}_{\mathrm{a.u.}} Q_d(\bar{\varepsilon}) \ (\mathrm{cm}^5) \quad (3.6.52)$$

in terms of the momentum transfer (or "diffusion") cross section Q_d (in cm^2) which is defined as (Massey and Burhop, 1969, p. 47)

$$Q_d(\bar{\varepsilon}) = \int d\Omega_f\, (1 - \cos\theta)\, |f(\bar{\varepsilon}, \cos\theta)|^2 \quad (\mathrm{cm}^2)$$

where θ is the electron deflection angle between $\hat{\boldsymbol{k}}_i$ and $\hat{\boldsymbol{k}}_f$.

The bracketed quantity in eqn. (3.6.52) is in atomic units (a.u.), E_ω is the photon energy, α the fine-structure constant, and a_0 is the Bohr radius. Expressing the scattering amplitude in terms of phase shifts, eqn. (3.6.34), one immediately obtains

$$Q_d(\bar{\varepsilon}) = \frac{2\pi a_0^2}{\bar{\varepsilon}_{\text{a.u.}}} \sum_{l=0} \{(2l+1)\sin^2\delta_l - 2(l+1)\sin\delta_l \sin\delta_{l+1}\cos(\delta_l - \delta_{l+1})\}$$

$$= \frac{2\pi a_0^2}{\bar{\varepsilon}_{\text{a.u.}}} \sum_{l=0} (l+1)\sin^2[\delta_l(\bar{\varepsilon}) - \delta_{l+1}(\bar{\varepsilon})].$$

The energy dependence of the phase shifts has been explicitly displayed in the last equation above. This result yields

$$\sigma^{\text{FF}}(k_i, \omega) \cong \frac{16\pi^2}{3}\alpha a_0^5 \left\{\frac{k_f}{E_\omega^3}\right\}_{\text{a.u.}} \sum_{l=0} (l+1)\sin^2[\delta_l(\bar{\varepsilon}) - \delta_{l+1}(\bar{\varepsilon})] \tag{3.6.53}$$

when inserted into eqn. (3.6.52).

The zero-frequency relation can be derived directly from the asymptotic forms of the wave functions, eqns. (3.6.24) and (3.6.25), by using the asymptotic form over all space instead of just at large r. This concurs with the familar arguments that small momentum transfers preferentially sample large radii, and, hence, at low energy the cross section is not sensitive to the value of the wave function at small r. Results equivalent to eqn. (3.6.53) have been obtained in this fashion by several authors and differ only in the electron energy at which the phase shifts δ_l are evaluated. Consideration of the validity criteria contained in eqn. (3.6.51) will resolve such ambiguity in favor of $\bar{\varepsilon}$, to order $(\hbar\omega/\bar{\varepsilon})^2$ with neglect of resonances, and is substantiated by direct comparison of eqn. (3.6.53) with numerical results by Ashkin (1966).

Modifications to eqn. (3.6.53) due to effects of electron exchange have been studied by John (1966), for the particular case of free–free absorption on the neutral hydrogen atom. The cross section is obtained from consideration of the asymptotic forms alone, with due regard for exchange, for low photon energies and for electron energies below the first excitation threshold. The result is the same as eqn. (3.6.52), with $Q_d(\bar{\varepsilon})$ modified to the form

$$Q_d(\bar{\varepsilon}) = \frac{2\pi a_0^2}{\bar{\varepsilon}_{\text{a.u.}}} \sum_s \sum_l \omega_s (l + 1) \sin^2 [\delta_l (\bar{\varepsilon}, s) - \delta_{l+1} (\bar{\varepsilon}, s)]. \qquad (3.6.54)$$

The phase shifts are now spin dependent and there appears a spin statistical factor

$$\omega_s = \frac{(2s + 1)}{\sum_s (2s + 1)}$$

where s is the total spin of the system (electron + atom). The phase shifts $\delta_l (\varepsilon, s)$ are those obtained from a correct treatment of electron exchange. The resultant symmetry dependence is manifested in their dependence on s. As discussed by John, for photon energies such that eqn. (3.6.54) is no longer valid, an adequate treatment of exchange effects requires a detailed many-electron calculation, such as, for example, the close-coupling computations of Burke and Smith (1962), or the treatment of Mjolness and Ruppel (1967).

Free–Free Absorption by Hydrogenic Ions

The long-range of the Coulomb potential guarantees that for reasonable positive ion densities, free–free transitions in the Coulomb field will dominate the low-frequency radiation absorption mechanisms. Most effects peculiar to such Coulomb transitions are present in the simplest case of free–free absorption in the field of hydrogen nuclei, and accordingly have been most extensively studied. A summary of the hydrogenic results is included below.

Radiative transitions in the field of a *point* charge Ze are called hydrogenic transitions. For such transitions the relevant wave functions and matrix elements may be directly obtained to any desired degree of accuracy and so have been extensively studied, both as a precise test of physical theory and as an initial approximation to transitions in the fields of ions other than hydrogen.

An extensive treatment of continuum Coulomb wave functions and transitions as well as useful approximations thereto is provided by the review article of Alder *et al.* (1956). Specific application to free–free absorption and detailed numerical results are presented by Grant (1958) and by Karzas and Latter (1961). Only results relevant to the foregoing general theory will be presented here; for details, the above references should be consulted.

As is well known, the long range of the Coulomb potential modifies the asymptotic forms, eqns. (3.6.24) and (3.6.25), so these become (Schiff, 1968, sect. 21):

$$\chi_k^{(\pm)}(\boldsymbol{r}) \xrightarrow[|r|\to\infty]{} (2\pi)^{-3/2} \left\{ e^{i[\boldsymbol{k}\cdot\boldsymbol{r}+\eta \ln (kr-\boldsymbol{k}\cdot\boldsymbol{r})]} + f_c^{(\pm)}(\hat{\boldsymbol{k}}\cdot\hat{\boldsymbol{r}}) \frac{1}{r} e^{[\pm i(kr-\eta \ln 2kr)]} \right\}.$$

$f_c^{(\pm)}(\hat{\boldsymbol{k}}\cdot\hat{\boldsymbol{r}})$ is the Coulomb scattering amplitude, and the dimensionless parameter η essentially characterizes the interaction of the electron with the field:

$$\eta = -\frac{Ze^2}{\hbar v} = -\frac{Z}{ka_0} = -Z\alpha\left(\frac{c}{v}\right). \qquad (3.6.55)$$

Here v and k are the electron velocity and wave number, respectively, a_0 is the Bohr radius, and α is the fine structure constant. Classical theory applies for $|\eta| \gg 1$ or $(v/c) \ll Z/137$. Further, for $|\eta| \ll 1$ the Coulomb field produces only a small perturbation and the Born approximation should be valid. The critical value $|\eta| = 1$ corresponds to an electron kinetic energy of Z^2 Rydbergs $= 13.6Z^2$ eV.

For a pure Coulomb field the one-electron states are well known. The complete wave functions with the prescribed normalization and asymptotic form are (Alder *et al.*, 1956, eqn. II-B40, 41)

$$\langle \boldsymbol{r}|\alpha\rangle = (2\pi)^{-3/2} e^{-\eta_i\pi/2}\,\Gamma(1+i\eta_i)\, e^{i\boldsymbol{k}_i\cdot\boldsymbol{r}}\, {}_1F_1\,[-i\eta_i; 1; i(k_ir-\boldsymbol{k}_i\cdot\boldsymbol{r})], \qquad (3.6.56)$$

$$\langle \boldsymbol{r}|\beta\rangle = (2\pi)^{-3/2} e^{-\eta_f\pi/2}\,\Gamma(1-i\eta_f)\, e^{i\boldsymbol{k}_f\cdot\boldsymbol{r}}\, {}_1F_1\,[i\eta_f; 1; -i(k_fr+\boldsymbol{k}_f\cdot\boldsymbol{r})]$$

in terms of the confluent hypergeometric function. The basic matrix elements are given by

$$|\langle\beta|\,\boldsymbol{p}\,|\alpha\rangle|^2 = \frac{9Z^2e^4}{2\,(2\pi)^5\omega^2k_ik_f}\;\frac{df(\eta_i,\eta_f)}{d\Omega_f}$$

in terms of the dimensionless function (Alder *et al.*, 1956, eqn. II-E.62)

$$\frac{df(\eta_i, \eta_f)}{d\Omega_f} = \frac{32\pi^3\eta_i\eta_f}{9\xi^2} \frac{e^{2\pi\eta_i}}{(e^{2\pi\eta_i} - 1)(e^{2\pi\eta_f} - 1)} \times \frac{d}{dx}\left\{-x \frac{d}{dx}|F(-i\eta_i, -i\eta_f; 1; x)|\right\}.$$

where the parameters x and ξ are defined by

$$x = -\frac{4\eta_i\eta_f}{\xi^2}\sin^2\theta/2, \quad \xi = \eta_f - \eta_i$$

where θ is again the electron deflection angle. Substituting these equations into eqn. (3.6.31) the hydrogenic free–free cross section is obtained:

$$\sigma^{FF}(k_i, \omega) = \frac{3}{2}\alpha \frac{Z^2e^4}{m\hbar\omega^3k_i} f(\eta_i, \eta_f) \quad (\text{cm}^5). \qquad (3.6.57)$$

As mentioned earlier, the semiclassical calculation of the radiation emitted by an electron moving in a Coulomb field was first carried out by Kramers (1923). Expressed in terms of absorption, the resulting "Kramers cross section" is given by

$$\sigma_K^{FF}(k_i, \omega) = \frac{16\pi^3}{3\sqrt{3}}\alpha \frac{Z^2e^4}{m\hbar\omega^3k_i} \quad (\text{cm}^5) \qquad (3.6.58)$$

(see also Landau and Lifshitz, 1962).

Some forms of this equation useful in applications are:

$$\sigma_K^{FF}(k_i, \omega) = \frac{16\pi^3}{3\sqrt{3}}\alpha Z^2 (E_\omega^3 k_i)^{-1}_{\text{a.u.}} a_0^5 \quad (\text{cm}^5)$$

$$= \frac{4}{3\sqrt{3}}\alpha^3\lambda^3 \left(\frac{v_i}{c}\right)\left(\frac{Z^2\,\text{Ryd}}{\varepsilon_i}\right) a_0^2 \quad (\text{cm}^5) \qquad (3.6.59)$$

in terms of the fine structure constant α, the Bohr radius a_0, the photon energy E_ω (a.u.), and wavelength λ (cm). The quantum-theory calculation was studied very early by Gaunt (1930), and it has since been customary to write the result [eqn. (3.6.57)] in terms of the Kramers cross section:

$$\sigma^{FF}(k_i, \omega) = g_{FF}(k_i, k_f)\,\sigma_K^{FF}(k_i, \omega). \qquad (3.6.60)$$

This expression defines the "Gaunt factor" g_{FF} as the quantum theory correction to the semiclassical cross section. By means of eqn. (3.6.57) it can be expressed in terms of the dimensionless Coulomb function $f(\eta_i, \eta_f)$. The relation is

$$g_{FF}(k_i, k_f) = \frac{9\sqrt{3}}{32\pi^3} f(\eta_i, \eta_f) = \pi\sqrt{3} \cdot e^{2\pi\eta_f}(e^{2\pi\eta_i} - 1)^{-1}(e^{2\pi\eta_f} - 1)^{-1}$$

$$\times x_0 \frac{d}{dx_0} |F(-i\eta_i, -i\eta_f; 1; x_0)| \qquad (3.6.61)$$

where

$$x_0 = -4\eta_i\eta_f/\xi^2.$$

This result was first obtained by Sommerfeld (1939) and numerical values have been extensively tabulated (Karzas and Latter, 1961; Grant, 1958).

A spherical harmonic decomposition of the Coulomb states [eqn. (3.6.56)] may be similarly performed (Alder *et al.*, 1956, II-B.3). The radial functions $w_l(k, r)$ are given in terms of the regular Coulomb functions $F_l(kr)$

$$w_l(k, r) = \sqrt{\frac{2}{\pi}} F_l(kr) \qquad (3.6.62)$$

and the phase shifts $\delta_l(\varepsilon)$ become the Coulomb phase shifts

$$\sigma_l(\eta) = \arg\Gamma(l + 1 + i\eta).$$

Evaluation of the radial matrix elements [eqn. (3.6.42)] is discussed in detail by Alder *et al.* (1956), and Biedenharn (1956) has shown the equivalence of the resulting equation [eqn. (3.6.46)] with eqn. (3.6.57). As discussed by Alder *et al.* (1956, p. 452) the main contribution to the l-sum in eqn. (3.6.46) is from values $l \sim \eta_i$, although the convergence is slow, particularly for low photon energies. It is of interest to study the dependence of this dominant $l\ (\equiv l_d)$ on the photon energy $\hbar\omega$. The semiclassical, most-probable electron deflection angles are tabulated by Alder *et al.* (1956) in their table II.7 in terms of $\xi = (\eta_f - \eta_i)$. Interpreting the deflection angles in terms of angular momenta, the rough correspondence shown in Table 3.5 is obtained.

Approximating $l_d/\eta \sim 1/\xi$, one obtains

$$l_d \sim (\varepsilon_i/\hbar\omega).$$

A similar result is obtained from the expansions of Burgess (1958).

TABLE 3.5. Dominant angular momentum contribution to the Coulomb Gaunt factor as a function of η and ξ—from the semiclassical treatment of Alder *et al.* (1956)

ξ	$l_d/\eta \sim$	ξ	$l_d/\eta \sim$
0.1	12.0	0.6	2.5
0.2	5.7	0.8	1.8
0.4	3.2	1.0	1.5

The complexity of the expression [eqn. (3.6.61)] for the Gaunt factor has stimulated considerable effort to obtain approximations which are reliable in the various domains of interest. Critical summaries of the approximations are provided by Grant (1958) and by Brussaard and Van de Hulst (1962).

From an asymptotic evaluation of the hypergeometric function in eqn. (3.6.61), Menzel and Pekeris (1935) have obtained an approximation suitable for $\varepsilon_i \ll 1$ and $\varepsilon_i \ll \varepsilon_f$,

$$g_{\mathrm{FF}}^{\mathrm{MF}}(k_i, k_f) \sim 1 + \frac{0.1728\,(1 + \varepsilon_i/\varepsilon_f)}{(1 - \varepsilon_i/\varepsilon_f)^{2/3}} \left(\frac{\mathrm{Ryd}}{\varepsilon_f}\right)^{1/3} - \frac{0.0496\,[1 - (484/15)(\varepsilon_i/\varepsilon_f) + (\varepsilon_i^2/\varepsilon_f^2)]}{(1 - \varepsilon_i/\varepsilon_f)^{4/3}} \left(\frac{\mathrm{Ryd}}{\varepsilon_f}\right)^{2/3} + \cdots. \tag{3.6.63}$$

A similar technique was employed by Burgess (1958) for asymptotic approximations to the Coulomb radial matrix elements [eqn. (3.6.42)] valid for $\varepsilon_i \ll 1$, $\varepsilon_i \ll \varepsilon_f$ and $l \ll \eta_i$.

As discussed at the beginning of this section, the Born approximation is expected to be valid if $|\eta| \ll 1$, or if the photon energy is small. For the Gaunt factor, the Born approximation yields

$$g_{\mathrm{FF}}^{\mathrm{B}}(k_i, k_f) = \frac{\sqrt{3}}{\pi} \ln\left(\frac{k_i + k_f}{k_f - k_i}\right). \tag{3.6.64}$$

This result displays the usual logarithmic divergence at small photon energies due to the infinite range of the pure Coulomb field. In practice, shielding effects of other electrons effectively limit the range of the field and so remove this divergence, as will be discussed below. Partial allowance for the long-range Coulomb distortions of the Born approximation plane waves by Elwert (1939) lead to a simple modification of eqn. (3.6.64):

$$g_{\mathrm{FF}}^{\mathrm{BE}}(k_i, k_f) = \left(\frac{e^{2\pi\eta_i} - 1}{e^{2\pi\eta_f} - 1}\right)\left(\frac{k_i}{k_f}\right) g_{\mathrm{FF}}^{\mathrm{B}}(k_i, k_f). \tag{3.6.65}$$

This expression is of considerably wider applicability.

In the limit $|\eta| \gg 1$ the WKB (or "semiclassical") approximation is expected to be valid. Using WKB wave functions the free–free matrix elements may be evaluated (Alder *et al.*, 1956), with the resulting Gaunt factor given by

$$g_{\mathrm{FF}}^{\mathrm{C}}(k_i, k_f) = \frac{-\sqrt{3}}{2\pi} e^{+\pi\xi} \left\{x \frac{d}{dx} [K_{i|\xi|}(x)]^2\right\}_{x=|\xi|} \tag{3.6.66}$$

in terms of the modified Bessel function $K_{i|\xi|}$ [cf. eqn. (2.6.35)].

A useful approximation to $g_{\mathrm{FF}}^{\mathrm{C}}$ for $|\xi| > 1$ is

$$g_{\mathrm{FF}}^{\mathrm{C}}(k_i, k_f) \sim 1 + 0.21775\,|\xi|^{-2/3} - 0.01312\,|\xi|^{-4/3},$$

a result which more closely approximates eqn. (3.6.61) than does the Menzel–Pekeris expansion, eqn. (3.6.63), according to Grant (1958). From this, the Kramers result is seen to be valid in the limit $\xi \gg 1$.

Grant has compared the above approximation with detailed numerical evaluation of eqn. (3.6.61). Better than 1 per cent accuracy is obtained by the various approximations in the following domains of validity:

$(k_f/k_i) = 1.0 - 1.025,$	$\lvert\eta_f\rvert \geq 30.0$:	semiclassical (3.6.66)
$= 1.025 - 1.30,$	$\lvert\eta_f\rvert \leq 0.1$:	Born–Elwert (3.6.65)
	$\lvert\eta_f\rvert \geq 5.0$:	semiclassical (3.6.66)
$\geq 1.30,$	$\lvert\eta_f\rvert \leq 0.1$:	Born–Elwert (3.6.65)
	$\lvert\eta_f\rvert \geq 2.5$:	semiclassical (3.6.66)

From the tables of Alder *et al.* (1956), one finds that the semiclassical Gaunt factor agrees with the quantum mechanical to within 3 per cent of all $\eta_i \geq 1.0$ and $\xi \geq 0.1$. In most applications the electron distribution function $F_e(\boldsymbol{k}_i)$ of eqn. (3.6.30) is Maxwellian. With eqn. (3.6.59) the Coulomb free–free absorption coefficient $\mu^{FF}(\omega)$ may be written in terms of the Maxwell-average Gaunt factor $\langle g_{FF}(\omega)\rangle$,

$$\mu^{FF}(\omega) = N_v N_e \frac{4\alpha^2 Z^2}{3\sqrt{(3\pi)}} \lambda^3 \left(\frac{\text{Ryd}}{kT}\right)^{1/2} a_0^2 \langle g_{FF}(\omega)\rangle \quad (\text{cm}^{-1}). \qquad (3.6.67)$$

Some values of $\langle g_{FF}(\omega)\rangle$ are given by Karzas and Latter (1961); approximations to this average are reviewed and discussed by Brussaard and Van de Hulst (1962) and by Bekefi (1966; sect. 3.4).

Free–Free Absorption by Non-hydrogenic Ion

For all photon energies a satisfactory treatment of free–free absorption in the fields of ions other than hydrogen would require extensive numerical treatments of, for example, Hartree–Fock ionic states perturbed by continuum electron states which are obtained from, say, close-coupling calculations. This type of detail is beyond the scope of our discussion. Fortunately, however, for low-energy photons the long-range Coulomb field of charge Z_I dominates the momentum-absorbing interaction and the non-Coulomb portions of the ionic field result in only small modifications of the hydrogenic results of the preceding section. On the other hand, high energy photons tend to sample the shorter range non-Coulomb field and so require the more detailed calculations.

From the discussion of Table 3.5, above, the angular momenta l_d dominating the Coulomb absorption cross section [eqn. (3.6.46)] are in the neighborhood of $l_d \sim (\varepsilon_i/\hbar\omega)$. Further, one expects departures from the pure Coulomb field to be limited to some range R, and so to contribute to the sum [eqn. (3.6.46)] only through terms $l \lesssim L = k_i R = (R/a_0)(\varepsilon_i/\text{Ryd})^{1/2}$. Thus one expects substantial departure from the Coulomb results only when the dominant terms themselves become modified; that is, when $L \sim l_d$ or $\hbar\omega \sim (a_0/R)\,\varepsilon_i$.

In a one-electron spherically symmetric approximation, the states in a pure Coulomb field are given by eqn. (3.6.62)

$$w_l^C(k, r) = \sqrt{\frac{2}{\pi}}\, F_l(kr) \xrightarrow[r\to\infty]{} \sqrt{\frac{2}{\pi}} \sin(kr - l\pi/2 - \eta \ln 2kr + \sigma_l),$$

where σ_l is the Coulomb phase shift given by

$$\sigma_l(\eta) = \arg \Gamma(l + 1 + i\eta).$$

As described by Schiff (1968, p. 144) the short-range departure from the pure Coulomb field results in an additional phase shift δ_l for $r > R$, according to

$$\left.\begin{aligned} w_l(k, r) &= \sqrt{\frac{2}{\pi}}\, e^{i\delta_l} [F_l(kr) \cos \delta_l + G_l(kr) \sin \delta_l], \quad (r \geq R) \\ &\xrightarrow[r\to\infty]{} \sqrt{\frac{2}{\pi}}\, e^{i\delta_l} \sin \left(kr - \frac{l\pi}{2} - \eta \ln 2kr + \sigma_l + \delta_l\right) \end{aligned}\right\} \tag{3.6.68}$$

where one expects $\delta_l(k) \neq 0$ only for $l \lesssim kR$. The function $G_l(kr)$ in eqn. (3.6.68) is the solution of the Coulomb equation which is irregular at the origin and asymptotically approaches $\cos(kr - l\pi/2 - \eta \ln 2kr + \sigma_l)$.

The slow convergence of the l-sum in eqn. (3.6.46) is best treated by explicitly separating the terms $l \leq L$ for which $\delta_l \neq 0$ and summing the remaining series in terms of the Coulomb result [eqn. (3.6.60)] for charge Z_I:

$$\begin{aligned} \sigma^{\mathrm{FF}}(k_i, \omega) = g_{\mathrm{FF}}(k_i, k_f)\, \sigma_{\mathrm{K}}^{\mathrm{FF}}(k_i, \omega) + \frac{8\pi^4}{3} \alpha \frac{k_f}{m\hbar\omega} \sum_{l_i=0}^{L} (l_i + 1) \\ \times \{|\mathscr{R}(l_i+1, l_i)|^2 + |\mathscr{R}(l_i, l_{i+1})|^2 - |\mathscr{R}^{\mathrm{C}}(l_{i+1}, l_i)|^2 - |\mathscr{R}^{\mathrm{C}}(l_i, l_{i+1})|^2\}. \end{aligned} \tag{3.6.69}$$

The $\mathscr{R}(l_f, l_i)$ are from eqn. (3.6.42) in terms of the actual solutions [eqn. (3.6.68)], and the $\mathscr{R}^{\mathrm{C}}(l_f, l_i)$ are the Coulomb radial matrix elements which may be expressed in terms of hypergeometric functions (Alder *et al.*, 1956; Gordon, 1929) and have been approximated by Burgess (1958).

DeVore (1964) has recently carried out a one-electron approximate calculation for the nitrogen ion N^+. The potential $V(r)$ of eqn. (3.6.36) was taken to be the unperturbed static Hartree–Fock potential of the ion, and effects of exchange and polarization were neglected. Comparison of the resulting numerical solutions of eqn. (3.6.36) with the asymptotic form eqn. (3.6.68) for large r determined the phase shifts δ_l. For electron energies $\varepsilon \leq 0.36$ Ryd, the s- and p-wave phase shifts were large, and the d-phase shifts were less than 2 per cent of the s-values. Higher angular momenta were ignored and the free–free absorption cross section was obtained from eqn. (3.6.69) with $L = 2$ and photon energies $\hbar\omega \leq 0.01$ Ryd. Comparison with hydrogen ion results indicates only small differences for $\hbar\omega \ll \varepsilon_i$—despite the large s- and p-wave phase shifts. This result substantiates the general discussion at the beginning of this section.

The predominance of the asymptotic domain in the absorption of low-energy photons led Peach (1965) to evaluate the radial matrix elements, eqn. (3.6.42) by use of the asymptotic wave functions [eqn. (3.6.68)] alone, modified at small radii to ensure convergence of integrals over the irregular solution. This approach corresponds to the familiar Coulomb approximation of Bates and Damgaard (1949) and Burgess and Seaton (1960). The Coulomb integrals are evaluated numerically and the results tabulated in convenient form. In terms of the Kramers' cross section of eqn. (3.6.59) with $Z = Z_I$, Peach's results may be expressed as a "Gaunt factor" defined by

$$\frac{\pi}{\sqrt{3}} g_{\mathrm{FF}}^{\mathrm{P}} = \frac{1}{4} \sum_{l_i l_f} l_> G^2(\varepsilon_i, l_i, \varepsilon_f, l_f) \cos^2 [\delta_{l_f}(\varepsilon_f) - \delta_{l_i}(\varepsilon_i) + \pi\chi(\varepsilon_i, l_i, \varepsilon_f, l_f)]. \tag{3.6.70}$$

Here $l_f = l_i \pm 1$ only and $l_>$ is the greater of (l_i, l_f). The functions G and χ are tabulated for l_i, l_f equal (0–3) and a range of values of ε_i and $\hbar\omega$. For low-energy electrons the phase shifts may be obtained from the quantum defect method (Burgess and Seaton, 1960).

A similar result which neglects the modification of the wave function at small r follows from eqn. (3.6.53). The corresponding "elastic scattering" Gaunt factor may be written (since the initial and final state phase shifts are evaluated at the same energy):

$$\frac{\pi}{\sqrt{3}} g_{\mathrm{FF}}^{\mathrm{el}} = \frac{1}{\eta_i \eta_f} \sum_{l=0} (l+1) \sin^2 [\delta_l + \sigma_l - \delta_{l+1} - \sigma_{l+1}] \tag{3.6.71}$$

and

$$\sigma_{l+1} = \sigma_l + \tan^{-1}\left(\frac{\eta}{2l+1}\right).$$

In the limit of zero photon energy (Peach, 1965, eq. 32) and vanishing phase shifts δ, eqns. (3.6.70) and (3.6.71) reduce to the same expression.

In view of the slow rate of convergence of the l-sum for Coulomb interactions it appears most convenient to follow eqn. (3.6.69) and sum explicitly only over those l-values for which δ_l is non-vanishing. Then, for example, eqn. (3.6.71) becomes (in terms of the hydrogen Gaunt factor $g^{\mathrm{H}}_{\mathrm{FF}}(Z_I)$ with charge Z_I)

$$g^{\mathrm{ion}}_{\mathrm{FF}} = g^{\mathrm{H}}_{\mathrm{FF}}(Z_I) + \frac{\sqrt{3}}{\pi\eta_i\eta_f} \sum_{l=0}^{L} (l+1) \times \left[\sin^2(\delta_l + \sigma_l - \delta_{l+1} - \sigma_{l+1}) - \frac{\eta^2}{\eta^2 + (l+1)^2} \right]. \tag{3.6.72}$$

This form appears to be the most useful approximation to the anticipated small modification of the Coulomb–Gaunt factor due to the non-Coulomb short-range ionic field. Inclusion of electron exchange effects follows the treatment of John (1964, 1966) and eqn. (3.6.54).

Effects of Debye Shielding

As discussed in connection with eqn. (3.6.64) the logarithmic divergence of the Coulomb–Gaunt factor for small photon energies is a manifestation of the infinite range of the pure Coulomb potential. A small-ω expansion of the exact result eqn. (3.6.61) using eqn. II-E.66b of Alder *et al.* (1956) leads directly to eqn. (3.6.64), as does the elastic scattering approximation eqn. (3.6.52) with the known Coulomb scattering amplitude. This last formulation displays most clearly the origin of the divergence in the small angle—and hence large impact parameter scattering.

As is well known the assumption of a pure Coulomb potential is valid only near the ion; the interaction is screened at large radii by other electrons. For moderate densities and high temperatures an exponentially screened Debye potential is a useful approximation to the actual potential responsible for the free–free absorption. That is, we can set approximately,

$$V(r) = -\frac{Ze^2}{r} e^{-\alpha r}, \tag{3.6.73}$$

where

$$\alpha^2 = \frac{4\pi e^2}{kT_e} \sum_i n_i Z_i (Z_i + 1) \tag{3.6.74}$$

for a gas at an electron temperature T_e with n_i ions per cm^3 of type i and charge Z_i.

As the Born approximation correctly exhibits the low photon energy divergence, a Born calculation of the scattering amplitude for potential $V(r)$ of eqn. (3.6.73) in the low photon energy approximation [eqn. (3.6.52)] may be expected to demonstrate the effects of shielding on the low-frequency Gaunt factor. The result of such a calculation is that the Coulomb–Born–Gaunt factor eqn. (3.6.64) is replaced by a "Debye" result

$$g_{\mathrm{FF}}^{\mathrm{BD}}(k_i, k_f) = \frac{\sqrt{3}}{\pi}\left(\frac{1}{2}\right)\left\{\ln\left[\frac{(k_f + k_i)^2 + \alpha^2}{(k_f - k_i)^2 + \alpha^2}\right] + \frac{\alpha^2}{(k_i + k_f)^2 + \alpha^2} + \frac{\alpha^2}{(k_i - k_f)^2 + \alpha^2}\right\} \tag{3.6.75}$$

which does not diverge as $\omega \to 0$.

A WKB-approximation of more general validity for small α and not too small ω or too large ε_i has been obtained by Green (1958),

$$g_{\mathrm{FF}}^{\mathrm{G}}(k_i, \omega) = \left[1 - 2\mu^2\left(\frac{\mathrm{Ryd}}{\hbar\omega}\right)\right] g_{\mathrm{FF}}(k_i', \omega) + O(\mu^3), \tag{3.6.76}$$

in terms of the Coulomb–Gaunt factor $g_{\mathrm{FF}}(k_i', \omega)$ at the same frequency ω but lower energy $\varepsilon_i' = k_i'^2\,\mathrm{Ryd} = \varepsilon_i - 2\mu Z^2\,\mathrm{Ryd}$. The dimensionless parameter μ is related to the screening constant α through

$$\mu = \frac{\alpha a_0}{Z}.$$

The Maxwellian velocity average of eqn. (3.6.76) has been compared with detailed machine calculations and has proved useful over an extensive temperature-density range. To first order in μ, one may use (see eqn. (3.6.67))

$$\langle g_{\mathrm{FF}}^{\mathrm{shielded}}(\omega)\rangle \simeq \langle g_{\mathrm{FF}}(\omega)\rangle - [\langle g_{\mathrm{FF}}(\omega)\rangle - g\,(k_i = 0, \omega)] \times \left(2\mu Z^2 \frac{\mathrm{Ryd}}{kT}\right). \tag{3.6.77}$$

As this expression is only accurate to first order in μ it is most useful in determining when screening effects may safely be neglected. Numerical evaluation of the velocity averages of eqn. (3.6.76) should be suitable for most cases of interest.

FREE–FREE ABSORPTION ON NEUTRAL ATOMS

The short range of the electron–atom interaction relative to the Coulomb force results in much smaller free–free radiation absorption cross sections on atoms than on ions—typically smaller by several orders of magnitude (DeVore, 1964).

A convenient though rough measure of the relative importance of atoms in free–free absorption is provided by the work of Firsov and Chibisov (1961). Assuming *s*-wave interactions only, for electron energies less than a few eV, the momentum transfer cross section in eqn. (3.6.52) was approximated by the zero-energy elastic scattering cross section $Q_{\mathrm{el}}(0)$. The Maxwellian velocity average may then be performed and the result compared with the Kramers cross section. When the relative numbers of ions and atoms are estimated by the Saha equation, the condition that the atomic and ionic frequency-integrated free–free absorption be equal yields the result

$$N_v\,(\mathrm{atom})\, Q_{\mathrm{el}}^2 = 2.2 \times 10^{26} \theta^{-5/2}\, e^{-\varepsilon_i/\theta}. \tag{3.6.78}$$

Here θ is the temperature and ε_i the ionization potential of the atom in eV, Q_{el} is the zero-energy elastic scattering cross section in $10^{-16}\ \mathrm{cm}^2$, and N_v (atom) is the atom number density (cm^{-3}). For a given temperature, ionization energy, and cross section eqn. (3.6.78) then provides a critical density above which atomic free–free absorption is expected to dominate and below which ionic absorption should dominate. In obtaining eqn. (3.6.78), use was made of the fact that the absorption per atom is much less than the absorption per ion, so equal absorption contributions necessarily implies $N_v(\text{atom}) \sim N_v$, the total particle density (atoms and ions).

Few detailed studies of the free–free absorption on neutral atoms have as yet been carried out. A general treatment is of equivalent complexity to the case of non-hydrogenic ions (discussed above), so a resort to extensive approximation is necessary.

Due both to its relative simplicity and to its particular relevance in astrophysical applications, the free–free absorption on neutral hydrogen has been most extensively studied. Early work by Wheeler and Wildt (1942) employed the acceleration radial matrix element of eqn. (3.6.42b) in the cross section given by eqn. (3.6.46), using Born approximation wave functions and the unperturbed Hartree potential of the hydrogen atom without polarization or exchange.

Chandrasekhar and Breen (1946) employed similar approximations. These differed principally in the replacement of the Born approximation s-wave by a numerical solution in the assumed potential. The result is a much larger cross section, approaching an order of magnitude greater than the result of Wheeler and Wildt at $\lambda \sim 2$ microns and $T = 6300\,°\text{K}$.

Most recent studies have employed the phase-shift approximation. As emphasized by Ohmura (1964) this formulation has the particular virtue that variationally determined phase shifts are generally more accurate than variational wave functions. Ohmura and Ohmura (1960, 1961) included s-wave phase shifts only, with allowance for polarization and exchange according to eqn. (3.6.54), and found the results of Chandrasekhar and Breen (1946) to be reduced by 40–60 per cent as a result.

Finally, a very detailed treatment of the phase-shift approximation was carried out by John (1964, 1966). Numerical solutions in the static Hartree potential of the atom were obtained (with inclusion of exchange) for angular momentum states $l = 0, 1, 2$ and electron energies less than one Rydberg. It was found that inclusion of the effects of exchange increased the importance of the higher phase shifts ($l \geq 1$), and the results obtained were 30–40 per cent less than those of Chandrasekhar and Breen (1946). Detailed comparison of the results obtained by use of the numerical wave functions in eqn. (3.6.46) with the phase shift approximation eqn. (3.6.53) for initial and final electron energies independently ranging from 0.01 to 0.1 Ryd indicate at worst 3 per cent difference in p-waves and 12 per cent in the s-waves. As the photon energy in the worst case is $\hbar\omega = 99\varepsilon_i$, the agreement is remarkable in view of the fact that eqn. (3.6.51) is an expansion in $\hbar\omega/\varepsilon$, which is therefore expected to be small.

A variational method for direct determination of the hydrogen free–free matrix element itself has been proposed by Khare and Rudge (1965) which is somewhat reminiscent of the Schwinger variational procedures of scattering theory. Employing simple trial wave functions, the authors were able to obtain results substantially in agreement with the numerical results of John.

DeVore (1964, 1965) has studied the free–free absorption by neutral nitrogen, assuming an unperturbed Hartree–Fock potential and an effective polarization potential

$$V_p(r) = \frac{-\alpha}{(r^2 + r_p^2)^2}$$

for two values of α and r_p. Numerical solutions to eqn. (3.6.36) were obtained for $l = 0, 1, 2$, and the cross section was evaluated from eqn. (3.6.46).

Kivel (1967) has explored the dependence of the approximate exchange interaction on the configuration of the free electron plus atom through a study of the corresponding Hartree–Fock equations. To obtain results consistent with the experimental work of Taylor (1963), Kivel finds it necessary to evaluate separately the free–free cross sections for each of the participating atomic configurations. A comparison of similarly obtained cross sections for argon with various phase shift approximations has recently been published by Ashkin (1966). The various approximations differ in their choice of energy-centering of the phase shifts in eqn. (3.6.53). The photon energy illustrated is $\hbar\omega \sim 1.8$ eV and the range of initial electron energies ε_i is 0.01 to 1.0 Ryd. For $\varepsilon_i \lesssim 0.05$ Ryd ($\hbar\omega/\varepsilon_i \gtrsim 2.7$) the use of the average energy $\bar{\varepsilon}$ of eqn. (3.6.52) leads to essential agreement with the numerical results. For energies 0.03 Ryd $< \varepsilon_i < 0.05$ Ryd Ashkin finds that a formula due to Holstein (1965) gives best agreement with the numerical results. Holstein retains the difference $(\boldsymbol{k}_i - \boldsymbol{k}_f)^2$ in eqn. (3.6.50) explicitly and evaluates the scattering amplitude $f(\varepsilon, \cos\theta)$ only at the mean energy $\bar{\varepsilon}$. Clearly, for $\hbar\omega/\varepsilon_i \ll 1$, all these results are equivalent to eqn. (3.6.52).

The Rand report of Hundley (1962) introduces the use of effective-range expansions for the evaluation of the low-energy phase shifts of eqn. (3.6.53). Peach (1965) has suggested estimating ion phase shifts by the "quantum defect" method, which has been demonstrated by Moiseiwitsch (1963) to correspond to an effective-range expansion. Dalgarno and Lane (1966) have employed the effective-range expansion of O'Malley *et al.* (1961) which is more correctly applicable to atoms. The authors obtain useful velocity-averaged absorption coefficients in terms of a few atomic parameters and some conveniently tabulated numerical functions. The approximations employed assume low temperatures (as atomic absorption may be expected to be dominated by ion absorption at high temperatures) and retain only the s-wave phase shift in eqn. (3.6.53). As shown by O'Malley *et al.* (1961) the zero-order elastic scattering cross section is expected to have an energy dependence of the form

$$Q_{\mathrm{el}}(\varepsilon) \cong \frac{4\pi}{k^2}\sin^2\delta_0(k)$$

$$= 4\pi\,(D_1 + D_2\varepsilon^{1/2} + D_3\varepsilon\ln\varepsilon + D_4\varepsilon + D_5\varepsilon^{3/2} + \cdots)\,a_0^2 \quad (3.6.79)$$

where D_1, D_2 and D_3 may be written in terms of the scattering length Aa_0 and the atomic polarizability αa_0^3 according to

$$D_1 = A^2, \quad D_2 = \frac{2\pi}{3}\alpha A, \quad D_3 = \frac{8}{3}\alpha^2 A.$$

If one approximates the momentum-transfer cross section of eqn. (3.6.52) by the elastic scattering cross section $Q_{\mathrm{el}}(\varepsilon)$, the explicit energy dependence of eqn. (3.6.79) allows direct evaluation of the Maxwell-averaged absorption coefficient, which is expressed in terms of the D_i's and certain tabulated functions of temperature alone. The approach of Dalgarno and Lane appears eminently suited for most applications of astrophysical interest, particularly in view of the large numbers of atomic species and states commonly present in the gas.

3.7. Line Profiles: the Voigt Function

Reference has been made a number of times in Chapter 2 and in previous sections of this chapter to the finite width of spectral lines and to their frequency-dependent profiles $b(\nu)$ which have to be taken quantitatively into account in opacity calculations on hot gases. The study of line shapes and line-broadening influences is an active and difficult research field of long standing (see, for example, the reviews of Baranger, 1962; Griem, 1964; Aller, 1963; Cooper, 1966; Jefferies, 1968). It is thus not the purpose of this section to make a general review of the field, but rather to record established results of line-profile properties which are of direct application to the study of hot-gas opacities.

It is well known that the experimentally recorded profiles of spectrum lines in emission or absorption depend upon three effects: (a) instrumental effects, (b) intrinsic properties of emitting or absorbing particles and, (c) environmental effects. In the theoretical discussion of opacities it is often assumed that purely instrumental effects have already been allowed for. Intrinsic radiation-damping properties of the emitting or absorbing atom or molecule give rise to its "natural" width which is usually so small compared to other effects that it can be neglected. Two classes of environmental effect are important, however, and often occur together. The first depends upon the random motion of the absorbing or emitting particle, the component of whose motion in the line of sight gives rise to a Doppler broadening of the line. The second results from the influences of collisions of neighboring particles on the emitter or absorber and is thus called collision broadening. Both effects are temperature dependent, the Doppler effect strongly so. These two concurrent effects are usually treated phenomenologically by folding together the Doppler and Lorentz line shapes into a resultant profile called the Voigt function. This function is also useful in accounting approximately for some intrumental effects. Because of its wide applicability to situations of practical interest, we will review the origin and properties of the Voigt function in some detail, after a short summary of individual profile-determining effects.

The effects of geometrical and of physical optics determine the intrumental profile of a spectral feature. The gross size of the image of the entrance slit of the instrument at the detector is determined through geometrical optics, by the magnification of the optical system involved. Further, diffraction effects arising from the passage of a truncated wave front through the instrument result in diffraction "wings" at each side of the geometrical image of the entrance slit. In optimum adjustment the entrance slit is usually made as small as possible so that the major contributor to the instrumental profile is the primary diffraction maximum of the entrance slit. In what follows it will be assumed that this profile is either negligible compared to other influences or is known and has been taken into account. It can often be approximated by a triangular-shaped function. Van de Hulst and Reesinck (1947) have suggested that useful approximations to a rectangular slit profile and to the diffraction pattern of a straight slit can often be obtained by comparing the Fourier transforms of those patterns with an expansion of the Fourier transform of the Voigt function (see also Unsöld, 1955).

For the intrinsic, or natural, profile of a spectrum line, the Lorentz shape

$$\alpha_\nu = \frac{e^2 f}{mc} \frac{w}{(\nu - \nu_0)^2 + w^2} \tag{3.7.1}$$

of eqns. (3.1.10) and (3.1.15) is retained in the quantum formulation. The half half-width w is equal to $w_U + w_L$, the sum of the widths of the upper and lower levels as can be determined from the uncertainty principle, for example (Heitler, 1954, chap. v, § 18). If L is the ground state, $w_L \sim 0$ and $w_U = \tau^{-1}$, where τ is the lifetime of the upper state. The central maximum of the line profile (of width $2w$) is very narrow, often negligible compared to other broadening influences, and the wings can be fairly extensive. The algebraic form of the natural profile, eqn. (3.7.1), is often retained also in the quantitative description of the collisionally broadened line profile [see eqn. (3.1.13)], with appropriate redefinition of parameters.

The random motion in the line of sight of the absorbing or emitting atoms implies a thermal, or Doppler, profile of the form

$$I_\nu = I_0 \exp\left[-\beta \frac{(\nu - \nu_0)^2}{\nu_0^2}\right] \tag{3.7.2}$$

where $\beta = Mc^2/2kT$, M is the molecular mass, T is the absolute temperature, and c is the velocity of light. Gaussian turbulence can also be included in this motion as it leads to the same profile (Unsöld, 1955).

Collision broadening results from forces between the emitting atom and its neighbors, and their relative motion. A complete theoretical discussion of collision broadening is outside the scope of this review (see Baranger, 1962; Griem, 1964; Cooper, 1966). Line broadening by collision results from collisions with:

Atoms of same kind (Self-broadening, or "Holtsmark broadening"

Atoms of a different kind (Lorentz broadening)

Ions
Electrons } (Stark broadening, or electron impact broadening)

Two types of theories, impact (or collisional) and statistical, have been developed to explain the effects. In the former, the random truncation of emission or absorption wave trains by individual collisions is examined by classical or quantum methods, often for uncharged perturbers. In the latter, the influence of the average electric field strength of neighboring charge carriers on the emitting or absorbing atoms is taken into account statistically through the Stark effect. As implied above, the practical result of collision broadening theory is usually to predict a line profile similar in form to eqn. (3.7.1) but with width parameters w_{natural} and $w_{\text{collision}}$. In this event, the absorption coefficient becomes

$$\alpha_\nu = \frac{\pi e^2 f}{mc} \frac{(w_n + w_c)}{(\nu - \nu_0)^2 + (w_n + w_c)^2}. \tag{3.7.3}$$

The influence of w_c is nearly always much greater than w_n and gives rise to lines with very wide wings.

If external magnetic and/or electric fields are involved in addition to collision microfields, Stark and Zeeman shifts and broadening must also be taken into account. Hansen (1964) has treated the problem of combined (ionic) Stark and Doppler broadening by analytic approximation.

The Superposition of Line Shapes and Other Origins of the Voigt Function

In many typical situations of interest more than one broadening influence is present at a time. For example, natural and thermal broadening often occur in low-pressure laboratory sources. Thermal and collision broadening (illustrated in Fig. 3.4) often dominate the scene at higher

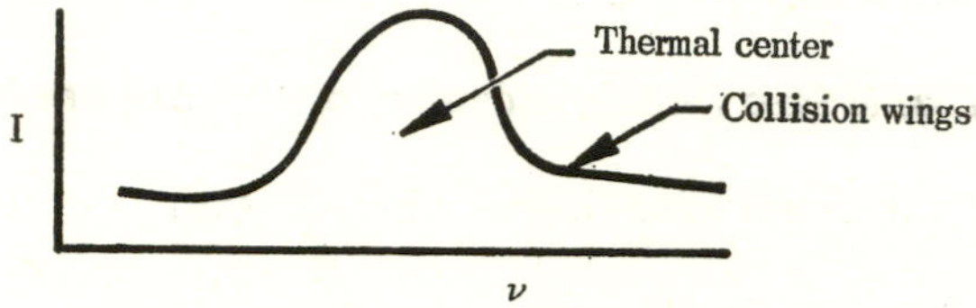

Fig. 3.4. Representative line profile showing thermal center and collision wings.

pressures. If $b_1(\nu)$ and $b_2(\nu)$ are the frequency profiles [see eqns. (2.2.4a) and (2.6.20a, b)] of the two independent effects, the combination of these effects is represented through the convolution or "folding" integral (Unsöld, 1955)

$$F(\nu) = \int_{-\infty}^{\infty} b_1(\nu - \nu')\, b_2(\nu')\, d\nu'. \tag{3.7.4}$$

A number of analytical and numerical methods have been developed to treat specific cases of this integral. For example, the combined application of natural and thermal broadening was treated by Voigt (see Mitchell and Zemansky, 1934) and the absorption coefficient so obtained is proportional to what we will call the "Voigt function", which we write in the form

$$K(x, y) = \frac{\varkappa_\nu}{\varkappa_0} = \frac{y}{\pi} \int_{-\infty}^{+\infty} \frac{\exp(-t^2)}{y^2 + (x - t)^2}\, dt. \tag{3.7.5}$$

A derivation of this formula is given by Mitchell and Zemansky (1934). In this expression, $K(x, y)$ can be considered to be a reduced absorption coefficient and the parameters are defined by

$$\varkappa_0 = \frac{S}{\alpha_D}\left(\frac{1}{\pi}\right)^{1/2}$$ (dimensional constant of the reduced absorption coefficient),

$$y = \frac{w}{\alpha_D}$$ (ratio of Lorentz to Doppler widths),

$$x = \frac{\nu - \nu_0}{\alpha_D}$$ (frequency scale in units of Doppler width), (3.7.6)

$$S = \int_{-\infty}^{+\infty} \varkappa_\nu \, d\nu$$ (integrated line strength),

$$(\ln 2)^{1/2}\alpha_D = \nu_0 \left(\frac{2kT \ln 2}{Mc^2}\right)^{1/2}$$ (Doppler half-width),

where w is the Lorentz half-width, ν_0 is the frequency at the line center, ν is the frequency at which $\varkappa_\nu$ is to be evaluated, and M is the mass of the emitter.

The function $\alpha_D^{-1} (1/\pi)^{1/2} K(x, y)$ is normalized to unity over the line profile. This can be readily demonstrated by direct integration of eqn. (3.7.5) first over x and then over t. Because the effect of collision broadening can often be expressed in terms of a Lorentzian profile as mentioned above, the combination of collision and thermal broadening, a case of great practical interest, can also be described by the Voigt function if one neglects the velocity dependence of the collision half-width. The thermal and collision-broadened line has a central core which is dominated by thermal effects, and wings which are dominated by pressure, or collision effects. In this case the natural line width is negligible. Relativistic corrections to both the natural line width and the Doppler effect produce velocity-dependent terms which couple the two profiles so that they are not completely independent. Mizushima (1967) has accounted for these corrections and shown them to be quite small under most conditions. Most collision-broadening theories produce line widths and shifts which are much more strongly velocity dependent than the corrected natural line width. Modified, or generalized Voigt functions which account for this effect have been defined by Mizushima (1967) for pressure broadening and Edmonds (1968) for quadratic Stark and Van der Waals broadening, and calculations carried out to ascertain its importance. According to Edmonds (1968) these effects are still small except possibly in the dense atmosphere of dwarf stars.

Stewart (1965) has formulated the problem of a Gaussian cluster of spectrum lines. If the transmission function of a group of lines with an exponential strength distribution is computed for lines of Lorentz shape and Gaussian distribution of centers, the result turns out to be expressible in terms of the Voigt function. Thus, we see that on a phenomenological basis, the Voigt function has a wide applicability to situations of practical interest in the study of radiation in heated gases. The function arises in other contexts as well. In neutron physics, the Lorentz (also known as the "Breit–Wigner" or "dispersion") shape combines with thermal motion to produce this profile under conditions of interest in reactor theory (Solbrig, 1961). It can be placed in a more general mathematical perspective by noting that it can be interpreted as the real part of the complex error function as was pointed out by Born (1933); we will discuss this in more detail later.

In plasma physics, the theory of electrostatic plasma oscillations leads to dispersion relations with Lorentzian-like resonances. If collisions are neglected and the particles are assigned a Maxwellian velocity distribution, an average of the dispersion relation over the Maxwell distribution arises. This, then, produces the complex error function. There is a tabulation by Fried and Conte (1961) specifically developed for the plasma application. The physics of the problem is described by Simon (1965) and Levin (1967).

The above applications arise from the physical motivation of folding together a Gaussian and a dispersion shape. It turns out that our function can also arise mathematically as an approximate generalization of the method of saddle-point integration for an integrand with a pole. This leads to the appearance of the function in electromagnetic theory and acoustics.

Sommerfeld (1909) first gave an approximate solution to one of the most basic and famous problems in electromagnetic wave propagation: the field of a dipole imbedded in an infinite conducting plane. This is a model for the practical problem of a dipole antenna radiating from a point on the surface of a conducting earth. Later on, Weyl (1919) improved on Sommerfeld's result by devising a better approximation to the integral solution for the Hertz vector potential by the use of saddle-point integration and his result produces the complex error function. We have been unable to find any earlier references to it as arising from a physical problem. (Recall that the identification of $K(x, y)$ as the real part of the complex error function was not done until 1933 by Born.) A good discussion of this application, including the contributions of Van der Pol and others, is given in Stratton's book (1941) and there is brief reference to it in Sommerfeld's book on partial differential equations (Sommerfeld, 1949). A graph and a tabulation of the complex error function for the purposes of this problem was made by K. A. Norton (1936). More recent work has been done in this area by V. Fock and others; see, for example, chap. 11 of Fock (1965).

There is an acoustical analog to the above electromagnetic wave propagation problem. This is the behavior of sound waves radiated into a semi-infinite region from a point in an infinite plane of a given acoustic impedance. This was pointed out by Rudnick (1947) who applied the Sommerfeld–Weyl–Van der Pol methods to the acoustic problem (Morse and Ingard, 1961).

A tabulation of the complex error function motivated by the electromagnetic wave propagation problem has been made by Clemmow and Mumford (1952). This is now one of the standard references. They reference an interesting paper by Pauli (1938) on light diffraction theory in which he uses saddle-point integration and gets a family of functions, one member of which appears to be the complex error function. They also reference some work by Rankin (1949) on rocket ballistics, wherein this function crops up.

There are significant problems involving the Voigt profile both in radiative transfer and in these other fields which can only be solved by very elaborate computer codes. In many of these problems a detailed account of the profile is advisable, but the codes are already so large one either cannot afford the storage of a large table for the line profiles, or the alternative of a basic recalculation each time, unless this recalculation can be done efficiently and rapidly. This places a premium on the efficient calculation of the function. Some of the representations of the function are valid only over a restricted region of the x–y-plane, and therefore require the piecing together of a number of such methods unless the problem requires only a restricted range of Doppler–Lorentz mixing. And as one might expect, the problems of current interest are gradually and inevitably increasing the required range to the whole x–y-plane. In problems of atmospheric radiative transfer this is particularly obvious: at the top of the atmosphere where the density is very low, molecular lines become almost entirely Doppler. At the bottom of the atmosphere where pressure broadening dominates, they are almost entirely Lorentzian, and in between one passes through all degrees of superposition of the two.

The above discussion gives an indication of the wide applicability of and extensive literature on this function. We, therefore, present herein a fairly detailed account of its properties and the methods that have been used to compute it. A review of the Voigt function has been presented previously by one of the authors (Armstrong, 1967), and we draw heavily on that article for this section.

Equation (3.7.5) is in the more-or-less standard form used in atomic and molecular spectroscopy. A transformed choice of variables can be made to yield the form

$$\Psi(X, \eta) = \frac{1}{\eta\sqrt{\pi}} \int_{-\infty}^{+\infty} \frac{\exp\,[-(X - t)^2/\eta^2]}{1 + t^2}\, dt \qquad (3.7.7)$$

which is more-or-less the standard form used in nuclear spectroscopy. (Sometimes $\theta \equiv 2/\eta$ is used instead of η; see, for example, Dresner, 1960, chap. 3.) Expressed in this way, $\Psi(X, \eta)$ has the form of what Erdelyi *et al.* (1953, p. 195) calls a "Gauss transform" of $1/(1 + t^2)$. The definitions of the variables are as follows:

$$X \equiv \frac{\nu - \nu_0}{w} = \frac{x}{y} \quad \text{(a)}$$

$$\eta \equiv \frac{\alpha_D}{w} = y^{-1} \quad \text{(b)} \qquad (3.7.8)$$

For a neutron application X would be defined in terms of energies $E - E_0$, and the half-widths would be measured accordingly. By comparing this definition with eqn. (3.7.5), one can readily ascertain that

$$\Psi(X, \eta) = \frac{(\pi)^{1/2}}{\eta} K\left(\frac{X}{\eta}, \frac{1}{\eta}\right).$$

The primary distinction between these two representations is that one (the "atomic") has the energy, or frequency scale measured in units of the Doppler half-width, while the other (the "nuclear") has its energy scale measured in units of the Lorentz half-width. These distinctions have probably arisen because of the tendency of the early atomic problems to be near the Doppler limit and for the early nuclear problems to be near the Lorentz limit.

Let us return now to the convolution integral, eqn. (3.7.4). It is well known that the convolution of two Gaussian functions yields a third Gaussian, and that the convolution of two Lorentz profiles is a third Lorentz profile. That is to say, we can write the convolution integral for these mechanisms as a type of product

$$b_{12} = b_1 \cdot b_2. \tag{3.7.9}$$

If we now take the convolution of two Voigt profiles, we find

$$\begin{aligned} b_{1V} \cdot b_{2V} &= b_{1D} \cdot b_{1L} \cdot b_{2D} \cdot b_{2L} \\ &= b_{1D} \cdot b_{2D} \cdot b_{1L} \cdot b_{2L} = b_{3D} \cdot b_{3L}. \end{aligned} \tag{3.7.10}$$

Thus, the fact that Gaussian and Lorentz profiles each "convolute" to a similar profile establishes the fact that the convolution of two Voigt profiles leads again to a Voigt profile. This argument is due to Van de Hulst and Reesinck (1947) who appear to be the first to point out this property in the atomic field. They were apparently unaware that Bethe and Placzek (1937) had already shown it by actual calculation of the integral.

Now the physical line profile is a function of three variables: w, α_D, and $\nu - \nu_0$. These are, of course, reducible to two by taking a ratio and this has been done in both these representations. In each case, one variable is the (frequency) distance from the line center and the other variable is a ratio of widths. Let us consider $K(x, y)$ as a function in the x–y-plane as in Fig. 3.5 (from Armstrong, 1967). We note that x is the distance from the line center, and $y = 0$ is the pure Doppler case. The pure Lorentz case does not lie on an axis of coordinates, but is reached, as shown in the figure, as both x and y get large. In spite of the fact that $y = 0$ is the pure

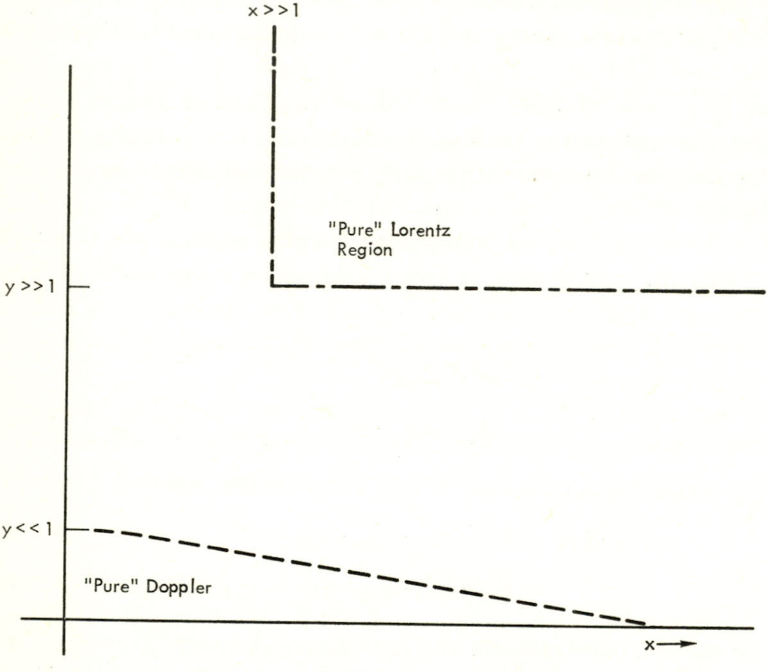

FIG. 3.5. Limiting regions of the x–y-plane.

Doppler limit, $y \ll 1$ is not a sufficient specification of the dominance of the Doppler effect. The reason lies in the extremely rapid fall-off of the Doppler wing compared to the Lorentz wing. Hence, one must require $xy \leq c$, where c is some fixed number, in addition to $y \ll 1$. The corresponding region in the x–y-plane is shown in Fig. 3.5 below the dashed line and is labelled "'pure' Doppler". An analogous effect occurs in the transition to a pure Lorentz shape. Here, sufficiently near the line center ($x = 0$)

one will never recover a pure Lorentz shape, so that mathematically speaking we must require x to be large (in terms of some insignificant but non-zero value of α_D) in addition to y. This region is indicated in Fig. 3.5 as lying above and to the right of the dash-dot line.

The structure of the line profile is as shown in Fig. 3.6 [for $K(xy)$]. The line center region as indicated in the figure corresponds to $x/y \ll 1$. For small y where the profile is near Doppler, one also needs

$$\frac{\nu - \nu_0}{(\ln 2)^{1/2} \alpha_D} \ll 1$$

but this is automatically satisfied if $x/y \ll 1$ and $y \ll 1$. The far wing lies on the lower right-hand side ($x/y \gg 1$, $x \gg 1$). Since the maximum value of $K(x, y)$ for a given value of y is at $x = 0$, the "half-width" of a line profile for a given value of y is given by the equation $K(w, y) = \frac{1}{2}K(0, y)$.

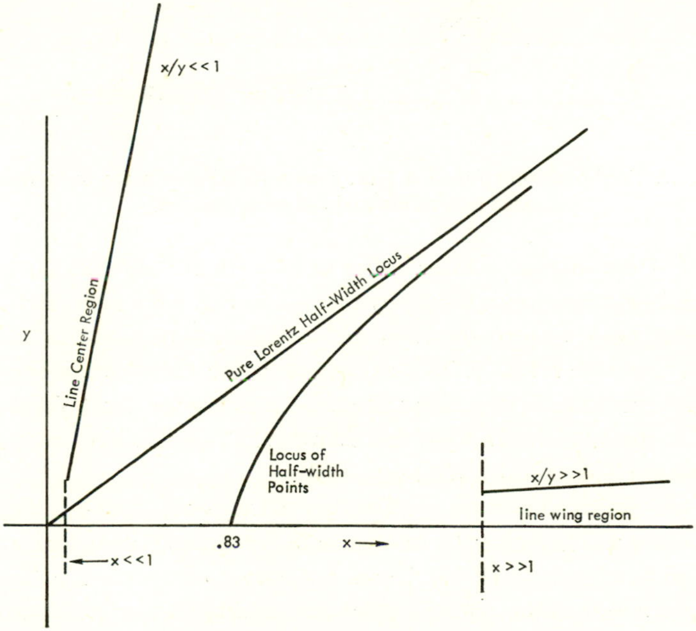

FIG. 3.6. Demarcation of the x–y-plane for the Voigt function $K(x, y)$ according to position on the line profile.

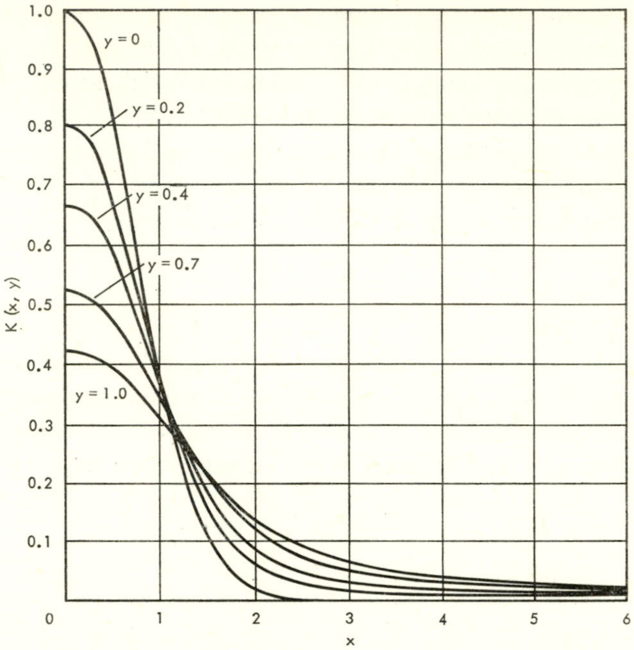

FIG. 3.7. The broadening function $K(x, y)$ as a function of x for the y-values indicated (after Finn and Mugglestone, 1965).

For $y = 0$, the value of w turns out to be $w = (\ln 2)^{1/2} = 0.83$. The locus of the half-width points is shown by the line in Fig. 3.6 which starts from this point, curves up to the right and becomes asymptotic to the line $x/y = 1$ for large x and y. This asymptote is the half-width locus for a pure Lorentz line. Simple approximate formulas for the "Voigt half-width" in terms of the Doppler and Lorentz half-widths have been given by Posener (1959) and by Whiting (1968).

The profiles produced by the function $K(x, y)$ are shown in Fig. 3.7. Such a curve was first given by Mitchell and Zemansky (1934), and later by Finn and Mugglestone (1965). These are curves of constant y. The top curve, for $y = 0$, is the pure Doppler case and the lower curves contain increasing admixtures of Lorentz broadening. The "nuclear" function $\Psi(X, \eta)$ produces a similar curve that starts from the Lorentz case for

$\eta = 0$ and proceeds through more and more Doppler admixture. A graph of this type was first given by Born (1933). Figure 3.8 shows $K(x, y)$ as a function of y for selected values of x. This gives the change at a particular point on a line profile produced by a change in the mixture of Lorentz and

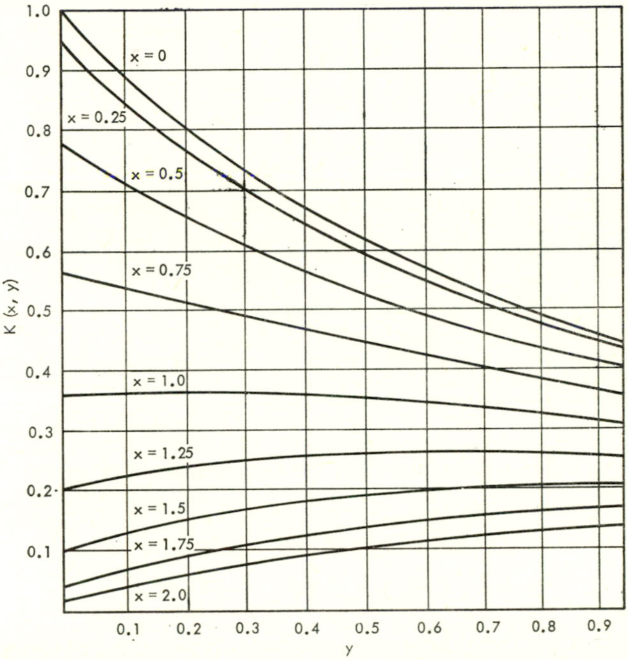

FIG. 3.8. The broadening function $K(x, y)$ as a function of y for the x-values indicated (after Finn and Mugglestone, 1965).

Doppler widths. An overall idea of the numerical behavior of the function can be obtained from Fig. 3.9 which shows a contour plot of $K(x, y)$ over the first quadrant of the x–y-plane.

In order to produce the Doppler limit from $K(x, y)$ we note that

$$\lim_{y \to 0} \frac{1}{\pi} \frac{y}{x^2 + y^2} = \delta(x) \qquad (3.7.11)$$

where $\delta(x)$ is the Dirac delta function.

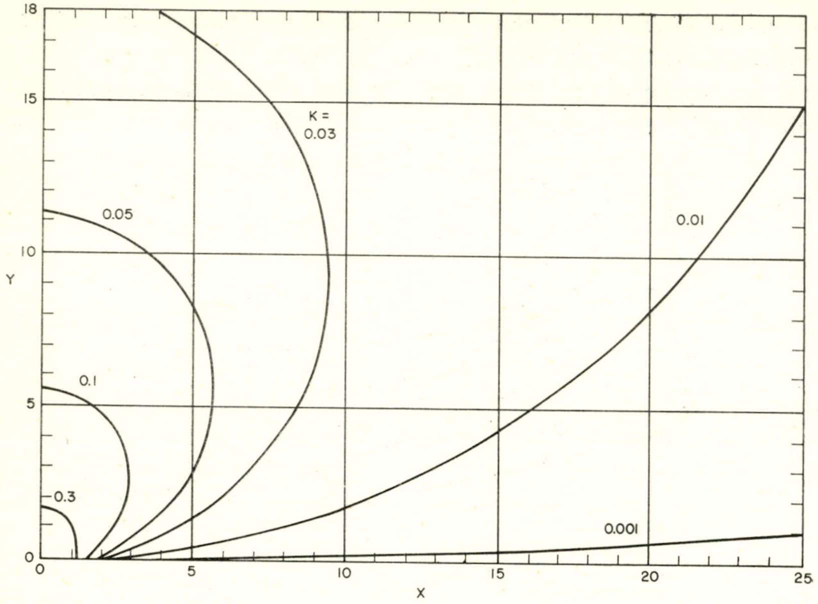

FIG. 3.9. Contour map of the Voigt function $K(x, y)$.

Inserting this into eqn. (3.7.5) we find

$$\lim_{y\to 0} K(x, y) = \int e^{-t^2} \delta(x - t)\, dt = e^{-x^2}. \tag{3.7.12}$$

The reason for pointing this out is not that it represents anything surprising or unusual, but that it shows that the representation eqn. (3.7.5) *cannot be used to calculate $K(x, y)$ for small y*. Attempting to do so would be tantamount to attempting to calculate a delta function, and this is not a very profitable procedure.

A similar situation prevails for the psi function. We note that

$$\lim_{\eta\to 0} \exp\left(-\frac{X^2}{\eta^2}\right)\bigg/\eta\sqrt{\pi} = \delta(X) \tag{3.7.13}$$

is another representation of the delta function. Hence, Ψ can be written

$$\lim_{\eta\to 0} \Psi(X, \eta) = \int_{-\infty}^{+\infty} \frac{\delta(X - t)}{1 + t^2}\, dt = \frac{1}{1 + X^2}. \tag{3.7.14}$$

So the X coordinate axis, or $\eta = 0$, is the Lorentz limit in this representation.

MATHEMATICAL PROPERTIES OF THE VOIGT FUNCTION

Looking at $K(x, y)$ as a function in the x–y-plane suggests that we attempt to define it as a function of a complex variable made up of x and y. That this can be done was first pointed out by Born (1933). If we note the fact that $K(x, 0) = e^{-x^2}$ as previously shown, then the K representation of the Voigt function can be written

$$K(x, y) = \frac{y}{\pi} \int_{-\infty}^{+\infty} \frac{K(t, 0)\, dt}{(x - t)^2 + y^2}. \qquad (3.7.15)$$

But this is precisely the form (cf. Morse and Feshbach, 1953, p. 371) that complex variable theory yields for the real part of an analytic function of $z = x + iy$ in the upper half plane in terms of its values on the real axis. This would appear then to be about the simplest complex generalization of K. The same theory that yields eqn. (3.7.15) for the real part of any analytic function of z provides an analogous formula for the imaginary part of the function in the upper half-plane.

$$L(x, y) = \frac{1}{\pi} \int_{-\infty}^{+\infty} \frac{(x - t)\, K(t, 0)\, dt}{(x - t)^2 + y^2}. \qquad (3.7.16)$$

If we put $W(z) = K(x, y) + iL(x, y)$ and factor the denominator: $(x - t)^2 + y^2 = [(x - t) + iy]\,[(x - t) - iy]$, we find that K is the real part of the function

$$W(z) = \frac{i}{\pi} \int_{-\infty}^{+\infty} \frac{e^{-t^2}}{z - t}\, dt \qquad (3.7.17)$$

in the upper half-plane, and L is its imaginary part.

It can be determined (Faddeyeva and Terent'ev, 1961) that the entire function that reduces to $W(z)$ for $y > 0$ is $w(z)$ given by

$$w(z) = e^{-z^2}\left[1 + \frac{2i}{\sqrt{\pi}} \int_0^z e^{t^2} dt\right] = e^{-z^2} \operatorname{erfc}(-iz) \qquad (3.7.18)$$

where erfc (x) is the complementary error function

$$\frac{2}{\pi^{1/2}} \int_x^{\infty} e^{-t^2} dt$$

[cf. eqn. (2.8.43)]. For the detailed mathematical properties of the error function and its relatives, see Gautschi (1964). From this result one can ascertain most readily that $K(0, y)$, the Voigt function on the y-axis is given by

$$K(0, y) = e^{y^2} \operatorname{erfc}(y). \tag{3.7.19}$$

From this equation, from eqn. (3.7.12), and Fig. 3.9, we see that $K(x, y) \leq 1$ over the entire first quadrant. The same feature is true for the imaginary part $L(x, y)$ of $w(z)$. Since the imaginary part of $w(0, y)$ vanishes, we see that L vanishes along the y-axis. For vanishing y, we find from equation (3.7.18),

$$w(x, 0) = e^{-x^2} - iL(x, 0), \tag{3.7.20}$$

where

$$L(x, 0) = \frac{2}{\pi^{1/2}} e^{-x^2} \int_0^x e^{t^2} dt$$

is known as Dawson's integral or Dawson's function (Cody, Paciorek and Thatcher, 1970). Some of the methods of computing $K(x, y)$ make use of this function (Armstrong, 1967). The complex error function of equation (3.7.18) has the properties

$$w(-z) = 2e^{-z^2} - w(z) \tag{3.7.21}$$

$$w(z) = \overline{w(-z)}$$

where the bar signifies complex conjugation. There are many representations, series expansions, asymptotic expansions, etc., for this function; it is not practical or desirable to review them all herein. But we will point out some fundamental features whose understanding is particularly relevant to computation and applications of the Voigt function.

As mentioned earlier, the Voigt function, or, more particularly, the complex error function, arises in a generalization of the method of saddle-point integration. The manner in which this occurs is as follows. This method approximates an integral

$$I = \int_C e^{f(z)} dz$$

by expansion about a saddle point where one can take $f(z) \cong f(z_0) - \{f''(z_0)\}/2\,(z - z_0)^2$ and the contour can be replaced by a straight line such that $f''(z_0)\,(z - z_0)^2$ is real. The result of the method is

$$I \approx \frac{e^{f(z_0)+i\eta}}{\sqrt{|f''(z_0)|}} \int_{-\infty}^{+\infty} e^{(-|f''|/2)\xi^2} d\xi.$$

These observations form the basis of a generalization of the saddle-point method to functions which have a pole near the saddle point, viz. integrals which are of the form

$$J = \int_C g(z)\, e^{f(z)} dz$$

where

$$g(z) = \frac{A}{z - z_1} + g_1(z).$$

A first approximation to this integral is

$$J \approx A\, e^{f(z_0)} \int_{-\infty}^{+\infty} \frac{e^{-t^2}}{t - p}\, dt = iA\, e^{f(z_0)} w(p) \tag{3.7.22}$$

where

$$p = |\tfrac{1}{2} f''(z_0)|^{1/2}\,(z_1 - z_0). \tag{3.7.23}$$

For more details, the foreword by Fock in the *Tables* of Faddeyeva and Terent'ev (1961), or the book by Clemmow (1966) is recommended.

Since we have shown the Voigt function to be the real part of the complex error function, it is interesting to enquire as to whether the imaginary part of this function, eqn. (3.7.16), also has a physical significance. The total neutron cross section, for instance, has in it a term

$$\sigma_p(E) = \sigma_0 k \frac{\Gamma_n (E - E_0)}{(E - E_0)^2 + (\Gamma/2)^2} \tag{3.7.24}$$

representing the interference between the absorption amplitude and the potential scattering amplitude. σ_0 is given by

$$\frac{4\pi}{k^2} g_J \frac{\Gamma_n}{\Gamma}$$

where g_J is a spin weight, Γ_n is the neutron width of the level, k is the neutron wave number, E is the neutron energy, E_0 is the neutron energy of the resonance, and Γ is the total width of the level. (As written, all these quantities should be evaluated in the center-of-mass coordinate system.) When this expression is averaged over the Doppler motion of the nuclei it yields the imaginary part of the function $w(z)$, viz. the function $L(x, y)$ defined by eqn. (3.7.16). An analogous expression to eqn. (3.7.24) occurs in the theory of refractivity, and is useful in parameterizing the shapes of autoionization resonances (Shore, 1967). Hence, if one wishes to account for the effects of thermal motion on these resonances, the function $L(x, y)$ can again be used.

A Maclaurin expansion (Γ now signifies the Γ-function)

$$w(z) = \sum_{n=0}^{\infty} \frac{(iz)^n}{\Gamma[(n/2) + 1]} \tag{3.7.25}$$

and an asymptotic expansion

$$\left(\frac{\pi}{i}\right) w(z) = \sum_{n=0}^{N} \frac{\Gamma(n + \frac{1}{2})}{z^{2n+1}} + R_n \tag{3.7.26}$$

where the remainder term R_n is given by

$$R_n = \int_{-\infty}^{+\infty} \frac{t^{2n+1} e^{-t^2}}{z^{2n+1}(z - t)} \, dt$$

can be readily developed from $w(z)$. This function can be seen to satisfy the differential equation

$$w'(z) + 2zw(z) = \frac{2i}{\sqrt{\pi}} \tag{3.7.27}$$

and continued fraction expansions can be developed. As noted before, eqn. (3.7.5) cannot be used for small y. In the early days of astrophysical spectroscopy, where this was the important case, a representation due to Reiche (1913)

$$K(x, y) = \frac{1}{\sqrt{\pi}} \int_0^\infty e^{-yt - t^2/4} \cos xt \, dt \qquad (3.7.28)$$

was used to develop a power series in y from which K could be calculated for small y. Unless y is quite small, however, this series is rather cumbersome. By differentiation and integration through the Cauchy–Riemann equations, the form of $L(x, y)$ analogous to eqn. (3.7.28) can be easily determined to be

$$L(x, y) = \frac{1}{\sqrt{\pi}} \int_0^\infty e^{-yt - t^2/4} \sin xt \, dt. \qquad (3.7.29)$$

By putting these last two expressions together, one obtains

$$W(x, y) = \frac{2}{\sqrt{\pi}} \int_0^\infty e^{2izt - t^2} dt \qquad (3.7.30)$$

which displays our function as a Laplace transform of the Gaussian.

It can also be expressed in terms of the confluent hypergeometric function by use of

$$\operatorname{erf}(z) = \frac{2z}{\sqrt{\pi}} e^{-z^2} M(1, \tfrac{3}{2}, z^2) \qquad (3.7.31)$$

(Gautchi, 1964) or

$$\operatorname{erfc}(z) = \frac{1}{\sqrt{(\pi z)}} e^{-z^2/2} W_{-1/4, 1/4}(z^2) \qquad (3.7.32)$$

(Whittaker and Watson, 1927, p. 341). The imaginary part on the real axis, $L(x, 0)$, is also a Hilbert transform (Fried and Conte, 1961). The representation $w(z)$ given by eqn. (3.7.17) can be expressed as the "minus one-th" Hermite polynomial,

$$W(z) = \frac{2}{\sqrt{\pi}} H_{-1}(-iz)$$

as can be verified by means of the integral representation for the Hermite polynomials (Morse and Feshbach, 1953, p. 786):

$$H_n(x) = \frac{2^n}{\sqrt{\pi}} \int_{-\infty}^{+\infty} (x + it)^n e^{-t^2} dt.$$

This emphasizes the close connection between the complex error function and Hermite polynomials, but does not appear to be of practical value. The coefficient of H_{-1} in the recurrence relation for these polynomials vanishes; hence, these recurrence relations cannot be used to compute H_{-1} in terms of the other H_n.

A continued fraction expansion was used by Fried and Conte (1961). This approach was further developed for efficient computation by Thatcher (1967).

Computation of the Voigt Function

A number of tabulations of the Voigt function and functions related to it have been made. Among the more extensive are those by Posener (1959), Fried and Conte (1961), Faddeyeva and Terent'ev (1961), Hummer (1965), Young (1965), Finn and Mugglestone (1965), and Karpov (1965). This last reference tabulates Dawson's function of complex argument [see eqn. (3.7.20)].

Much of the work done on this function prior to the mid-sixties does not reflect the influence of the modern digital computer, and has been extensively reviewed already by one of the authors (Armstrong, 1967). Therefore, we will restrict this discussion to a brief survey of efficient calculational methods by modern standards.†

For fast-calculation purposes, it is important to note that Gauss–Hermite quadrature (Abramowitz and Stegun, 1964):

$$\int_{-\infty}^{+\infty} e^{-t^2} f(t)\, dt = \sum_{n=1}^{N} A_n f(t_n) \tag{3.7.33}$$

is ideally suited to calculate $W(z)$ for large y (Faddeyeva and Terent'ev, 1961). In fact, as long as $y > 0$, this representation converges to the true

† Explicit computer codes based on some of the older methods are given by Armstrong (1967), Hummer (1964) and a more advanced one by Gautschi (1969).

value, viz.

$$\frac{i}{\pi}\int_{-\infty}^{+\infty}\frac{e^{-t^2}}{z-t}=\sum_{n=1}^{\infty}\frac{A_n}{z-t_n} \tag{3.7.34}$$

as $N\to\infty$. The importance of this method lies in the fact that it does not require the calculation of any exponentials, sines or cosines, but only rational functions of x and y for both the real and imaginary parts of $w(z)$. In practice y does not have to be very large for a relatively small number of terms (say 20 or so) to yield very high accuracy. Because of this absence of exponentials, sines, and cosines, and the fast convergence of the series, it is unlikely to be superseded for large y. Up until about 1967 the calculational methods proposed were limited in their efficacy to very specific regions in the x–y-plane. To cover the entire plane† required the piecing together of several distinct methods of computation. This situation is well illustrated by the review of Armstrong (1967). Many of these are quite simple, and for problems whose x–y values tend to be concentrated in limited regions of the plane may still be very useful in practice.‡ A further development of such methods has been presented by Reichel (1968) who gives algorithms for simple machine computation of the Voigt function in the form of eqn. (3.7.7) (and its imaginary counterpart). Reichel's methods are based on (for $x < 7$) the method originally suggested by Gronwall (Zemansky, 1930; reviewed by Penner, 1959) and (for $x > 7$) the asymptotic series in x^{-1} first given by Plass and Fivel (1953). It appears that for very large y, Reichel's methods would need to be supplemented by a third method for efficient computation.

Recently, more uniform methods have been developed (Rybicki, 1967; Chiarella and Reichel, 1968; Gautschi, 1970) which are computationally valid over the entire first quadrant of the x–y-plane, although they tend to pose a higher degree of mathematical difficulty or sophistication. The

† More specifically, the first quadrant of the x–y-plane, since values in the other quadrants can be obtained from those in the first quadrant by use of eqns. (3.7.21).

‡ Whiting (1968) has given an empirical approximation to the Voigt function that aims at up to 5 per cent maximum error. It is not clear that his approximation, which is already somewhat complicated, offers an advantage over some of the fundamental methods in the literature. For this accuracy only a few terms are required in some of the basic expansions, so that these become quite simple.

method suggested by Rybicki (1967) is similar to that of Chiarella and Reichel (1968), but much simpler, albeit at the expense of some rigor. Since its simplicity and elegance also carry over into the computational scheme, we will describe it briefly.

It is based on the sampling representation of the Gaussian, considered as a "band-limited" function (Hamming, 1962, chap. 23):

$$\exp(-t^2) = \sum_{n=-\infty}^{+\infty} \exp[(-nh)^2] \frac{\sin \pi (t - nh)/h}{\pi (t - nh)/h}. \qquad (3.7.35)$$

If we insert this expansion into eqn. (3.7.17) for the representation $W(z)$ of which $K(x, y)$ is the real part, we obtain

$$W(z) = \frac{1}{2\pi} \sum_{n=-\infty}^{+\infty} e^{-(nh)^2} \int_{-\infty}^{+\infty} \frac{[e^{-i(\pi/h)(t-nh)} - e^{i(\pi/h)(t-nh)}]\, dt}{(\pi/h)(t - nh)(t - z)} \qquad (3.7.36)$$

where we have made use of the complex exponential formula for the sine function. The integrand has poles on the real axis at $t = nh$, and off the real axis at $t = z$. By the calculus of residues, we can replace the integral along the real axis by a closed semicircular contour in the lower half plane for the first term in eqn. (3.7.36) and in the upper half plane for the second term. The pole for $t = z$ will fall in one or the other depending, of course, on the value of z. We are concerned here with $y \geq 0$ and, hence, take the pole for $t = z$ to lie in the upper quadrant. The poles on the real axis can be evaluated in the Cauchy principal value sense (Copson, 1935) in which case they are split between the upper and lower contours, or they can be shifted slightly to lie on either side of the real axis. In any case the same result is obtained:

$$w(z) = \frac{h}{\pi i} \sum_{n=-\infty}^{+\infty} e^{-(nh)^2} \left\{ \frac{1 - e^{-i(\pi/h)(z-nh)}}{nh - z} \right\}. \qquad (3.7.37)$$

We can separate off the real part of this equation in the following manner:

$$K(x, y) = \mathrm{Re}\,[W(z)] = \frac{h}{\pi} \sum_{n=-\infty}^{n=+\infty} e^{-(nh)^2}$$

$$\times \left\{ \frac{(x - nh)\,\mathrm{Im}\,(E_n) + y\,[1 - \mathrm{Re}\,(E_n)]}{(x - nh)^2 + y^2} \right\}. \qquad (3.7.38)$$

We have set $E_n \equiv (-1)^n \exp(i\pi z/h)$ in this equation and used its real and imaginary parts rather than its complete reduction since the equation can be coded in this form in a programming language (e.g. FORTRAN IV) that includes complex arithmetic. The completely reduced form is algebraically more complicated and, therefore, more tedious to program and to compute. Equation (3.7.38) can be coded in a very few steps, and is valid with some slight reorganization over the entire first quadrant in the x–y-plane. Rybicki (1967) found that terminating the series at $n = \pm 15$ (31 terms) for $h = \frac{1}{3}$ yields an accuracy of eight significant figures or better. Since the exponential coefficients $\exp(-n^2h^2)$ do not need to be recomputed for each value of $K(x, y)$ to be obtained, this method is relatively fast. For more rigorous mathematical detail, including treatment of the error involved, the reader is referred to Chiarella and Reichel (1968). If the complex exponential in eqn. (3.7.37) is written out in terms of cos + i sin, the sine term turns out to be identical, by comparison with eqn. (3.7.35), to the sampling representation expansion of $\exp(-z^2)$. Thus, we can also write eqn. (3.7.37) as

$$w(z) = \frac{h}{\pi i} \sum_{n=-\infty}^{n=+\infty} e^{-(nh)^2} \left\{ \frac{1 - \cos \pi (z - nh)/h}{nh - z} \right\} + e^{-z^2}. \tag{3.7.39}$$

This form affords a useful reorganization of eqn. (3.7.38) for computing when y is small. Since the first term in eqn. (3.7.39) vanishes as $y \to 0$ and the second produces the proper limit $\exp(-x^2)$, we see that the contour integration procedure has effectively continued $W(z)$ onto the real axis and into the form $w(z)$.

Finally, we cite the highly refined computational procedure for $w(z)$ recently developed by Gautschi (1970). This method, which basically involves the evaluation of a truncated Taylor series, is also uniformly effective over the entire first quadrant. It automatically reduces to Gauss–Hermite quadrature, in effect, for large $|z|$. The algorithm which Gautschi gives also automatically selects values of the number of terms in the series and of other parameters employed, as a function of z in order to minimize machine computation time for a given desired accuracy.

Chapter 4

Theory of Molecular Absorption

AT RELATIVELY low temperatures (up to a few thousand degrees), a heated gas contains many molecular species which significantly influence the transport of radiation. In this chapter, therefore, a review is made of the quantal description of the molecular contribution to gaseous absorption coefficients. We will be primarily concerned with diatomic molecules because of the relatively more complete theoretical description that is available. For the details of molecular structure and spectroscopic nomenclature, the reader is referred to the classic treatise of Herzberg (1950), to Nicholls (1969), and to Kovacs (1969). For an excellent introduction to the basic quantum mechanics involved, see chapter II of Landau and Lifshitz (1959). We will not, however, discuss the molecular photoelectric effect or autoionization. For these subjects the reader is referred to Ditchburn and Öpik (1962), Marr (1967), and, for example, Smith (1970) and Tuckwell (1970).

It was pointed out in Section 3.3 that the quantity which determines the probability of a U–L electric dipole transition for an atomic system is what we will call here the *transition strength matrix element* $S_{UL} = |er_{\alpha\beta}|^2$ (see eqns. (3.3.11), (3.3.17) and (3.5.2). In this chapter we shall adopt the notation

$$S_{UL} = |R_e|^2 = |\int \Psi_U^* \boldsymbol{M} \Psi_L \, d\tau|^2, \tag{4.1.1}$$

where R_e $(= er_{\alpha\beta})$ is called the *transition moment.* Ψ_U and Ψ_L are the complete wave functions of the upper and lower states, $\boldsymbol{M}$ is the electric dipole moment and $d\tau$ is the element of configuration space. Matrix elements similar to S_{UL} control higher and other-pole transition (Allen, 1963). Transition probability parameters such as absorption coefficients, oscillator strengths, Einstein A and B coefficients, are, apart from constants, the product of S_{UL} with appropriate powers of frequency (see eqns. (4.1.17), (4.1.18), (4.1.19), and also Nicholls, 1969).

In this chapter we specifically discuss the form that S_{UL} takes for *molecular* transitions. We also review the molecular transition-probability parameters which are determined by S_{UL}, and particularly emphasize absorption coefficients. We will review in a subsequent volume calculations which have been made during the past few years on the molecular contributions to the absorption coefficient of heated air.

4.1. The Born–Oppenheimer Approximation and its Consequences

The internal degrees of freedom of the molecule influence the transition strength matrix element of a molecular spectral feature. Consider the U, v', J', Λ', $M' \leftrightarrow L$, v'', J'', Λ'', M'' transition. U and L are the upper and lower electronic states, or components of multiplets. For the other quantum numbers, primes refer to the upper levels and double primes refer to the lower levels. v is the vibrational quantum number, J is the total or rotational quantum number. Λ is the quantum number of the component of the electronic angular momentum along the internuclear axis, and M, the magnetic quantum number (which should not be confused with $\boldsymbol{M}$, the electric dipole moment), refers to the component of J in the direction of an externally applied magnetic field. It can assume any of $2J + 1$ values.

Born and Oppenheimer (1927) proposed a molecular model in which it was assumed that the electronic state of the molecule was negligibly affected by the nuclear vibration and rotation. This assumption allows the complete molecular wave function of eqn. (4.1.1) to be written as a product of factors that take account of the electronic, the vibrational and the rotational motion independently:

$$\Psi_{evJ\Lambda M} = \psi_e(r_s, r)\,[\psi_v(r)/r]\,\psi_{J\Lambda M}(\theta, \chi, \varphi). \qquad (4.1.2)$$

(A discussion of vibration–rotation interaction effects is postponed until Section 4.3.) Under this approximation, the total energy E of the molecule is comparably written as the sum of the three components E_{elect}, E_{vib} and E_{rot}:

$$E = E_{\text{elect}} + E_{\text{vib}} + E_{\text{rot}}, \tag{4.1.3}$$

$\Psi_{e,v,J,\Lambda,M}$ is the complete molecular wave function of the electronic state e, vibrational level v, with J, Λ and M defined as above. $\psi_e(\boldsymbol{r}_s, r)$ is the electronic wave function written in subsequent equations as ψ_U or ψ_L for upper or lower states, respectively. The $\boldsymbol{r}_s$ represent electron coordinates relative to the internuclear axis, r is the internuclear separation, $\psi_v(r)$ is the vibrational wave function of the nuclear one-dimensional oscillator in the vth level appropriate to a particular molecular potential (centrifugal effects have been neglected). $\psi_{J\Lambda M}(\theta, \chi, \varphi)$ is the wave function of the symmetric top rotator. θ, χ, φ are the Euler angles of the molecular coordinate system relative to the fixed one. Kovacs (1969) gives a very complete discussion of these angles and coordinate systems.

The dipole moment $\boldsymbol{M}$ can be resolved into contributions $\boldsymbol{M}_e$ arising from the molecular electrons and $\boldsymbol{M}_n$ arising from the nuclei,

$$\boldsymbol{M} = \boldsymbol{M}_e + \boldsymbol{M}_n \tag{4.1.4}$$

where

$$\boldsymbol{M}_e = -\sum_s e\boldsymbol{r}'_s = \{\sum_s - e\boldsymbol{r}_s\} \cdot D(\theta, \chi, \varphi). \tag{4.1.5}$$

$\boldsymbol{r}'_s$ is the position vector of the sth electron relative to external axes, and $\boldsymbol{r}_s$ is the position vector of the sth electron relative to the figure axis of the molecule. $D(\theta, \chi, \varphi)$ is the dyadic appropriate to the transformation of axes. Its elements are the direction cosines of the angles between axes of the two coordinate systems. The configuration space element $d\tau$ of eqn. (4.1.1) may be written as

$$d\tau = d\tau_e\, dV = d\tau_e r^2 \sin\theta\, d\theta\, d\varphi\, dr \tag{4.1.6}$$

where dV is the volume element for the vibration and rotation and $d\tau_e$ is the configuration space element for electrons.

From eqns. (4.1.1), (4.1.2), (4.1.4), (4.1.6) R_e may now be written as

$$R^{Uv'J'\Lambda'M'}_{Lv''J''\Lambda''M''} = \int \psi_U^* \frac{\psi_{v'}}{r} \psi_{J'\Lambda'M'} (\boldsymbol{M}_e + \boldsymbol{M}_n) \psi_L \frac{\psi_{v''}}{r} \psi_{J''\Lambda''M''} \, d\tau_e \, dV$$

$$= \int \psi_U^* \frac{\psi_{v'}}{r} \psi_{J'\Lambda'M'} \boldsymbol{M}_e \psi_L \frac{\psi_{v''}}{r} \psi_{J''\Lambda''M''} \, d\tau_e \, dV$$

$$+ \int \psi_U^* \psi_L \, d\tau_e \int \frac{\psi_{v'} \psi_{J'\Lambda'M'}}{r} \boldsymbol{M}_n \frac{\psi_{v''} \psi_{J''\Lambda''M''}}{r} \, dV$$

for a molecular line. Orthogonality between ψ_U and ψ_L reduces the second term on the right hand side to zero.

Using eqns. (4.1.5) and (4.1.6) the first member may be rewritten as

$$|R^{Uv'J'\Lambda'M'}_{Lv''J''\Lambda''M''}| = \int \psi_{v'} \left(\int \psi_U^* \left|\sum e\boldsymbol{r}_s\right| \psi_L \, d\tau_e\right) \psi_{v''} \, d\tau \int \psi_{J'\Lambda'M'}$$
$$\times \, |\underline{D}(\theta, \chi, \varphi)| \, \psi_{J''\Lambda''M''} \, d\Omega \qquad (4.1.7)$$

where $d\Omega = \sin\theta \, d\theta \, d\varphi$.

The first integral on the right-hand side of eqn. (4.1.7) when summed, if necessary, over degenerate electronic states and squared is called the *band strength* $S^{Uv'}_{Lv''}$. The second integral determines the selection rules. It has been studied for various types of molecular transitions by Dennison (1926), Kronig and Rabi (1927), Reiche and Rademacher (1926, 1927), Schadee (1964), Kovacs (1969), Hougen (1970), and others. When summed over the degenerate quantum numbers M' and M'' and squared it becomes the Hönl–London (1925), or line intensity, factor $S^{J'\Lambda'}_{J''\Lambda''}$.

After squaring and summing over degeneracies, eqn. (4.1.7) yields

$$S^{Uv'J'\Lambda'}_{Lv''J''\Lambda''} = |R|^2 = S^{Uv'}_{Lv''} \cdot S^{J'\Lambda'}_{J''\Lambda''} \qquad (4.1.8)$$

$S^{Uv'J'\Lambda'}_{Lv''J''\Lambda''}$ is the strength matrix of the individual molecular band line. $S^{Uv'}_{Lv''}$ is the band strength defined in eqn. (4.1.9) and is a very important quantity (see the discussion of Section 4.4). It can be factored into two terms controlled, respectively, by electronic and vibrational aspects of the motion of the molecule. $S^{J'\Lambda'}_{J''\Lambda''}$ is the Hönl–London factor (see Section 4.2 for a more complete discussion) or line intensity factor mentioned above. It is often a quotient of simple polynomial functions of J and Λ and has been tabulated for branches of most allowed molecular transitions.

The band strength $S_{Lv''}^{Uv'}$ is an average, with respect to the vibrational wave functions,

$$S_{Lv''}^{Uv'} = |\int \psi_{v'} R_e(r) \psi_{v''} dr|^2 \tag{4.1.9}$$

of the electronic transition moment† $R_e(r)$ defined as:

$$R_e(r) = \int \psi_U^* M_e \psi_L^* d\tau_e = \int \psi_U^* |-\sum er_s| \psi_L d\tau_e. \tag{4.1.10}$$

The electronic transition moment is thus an average with respect to electronic wave functions of the electric dipole moment.

If $R_e(r)$ is independent of r, eqn. (4.1.9) becomes

$$S_{Lv''}^{Uv'} = R_e^2 |\int \psi_{v'} \psi_{v''} dr|^2 = R_e^2 q_{v'v''}. \tag{4.1.11}$$

On the other hand, if, as is much more realistic, $R_e(r)$ varies with r in a polynomial fashion:

$$R_e(r) = \sum_n a_n r^n,$$

it is possible (Fraser, 1954a, b) to use the *r-centroid* ($\bar{r}_{v'v''}$) approximation (Drake and Nicholls, 1970), where

$$\bar{r}_{v'v''} = \frac{\int \psi_{v'} \psi_{v''} r \, dr}{\int \psi_{v'} \psi_{v''} \, dr} \tag{4.1.12a}$$

$$(\bar{r}_{v'v''})^n = \frac{\int \psi_{v'} \psi_{v''} r^n \, dr}{\int \psi_{v'} \psi_{v''} \, dr} \tag{4.1.12b}$$

to write eqn. (4.1.9) as

$$S_{Lv''}^{Uv'} = |\int \psi_{v'} \sum_n a_n r^n \psi_{v''} dr|^2 = |\sum_n a_n \int \psi_{v'} \psi_{v''} r^n dr|^2$$

$$= |\sum_n a_n \bar{r}_{v'v''}^n|^2 |\int \psi_{v'} \psi_{v''} dr|^2 = R_e^2 (r_{v'v''}) q_{v'v''}. \tag{4.1.13}$$

† In the case of infrared vibration–rotation transitions where $\psi_U = \psi_L$, and where $\psi_{v'}$, and $\psi_{v''}$ both refer to levels of one molecular potential, eqn. (4.1.9) becomes

$$S_{Lv''}^{Uv'} = |\int \psi_{v'} M \psi_{v''} dr|^2. \tag{4.1.9a}$$

See discussion in Section 4.4.

The vibrational overlap integral square $q_{v'v''}$ in eqns. (4.1.11) and (4.1.13) is called the *Franck–Condon factor*:

$$q_{v'v''} = |\int \psi_{v'}\psi_{v''}\, dr|^2. \tag{4.1.14}$$

It is responsible for the operation of the Franck–Condon Principle in the determination of relative band intensities. The r-centroid is a characteristic internuclear separation associated with the band. Franck–Condon factors and r-centroids are discussed more fully in Section 4.3.

Finally, using eqns. (4.1.8) and (4.1.13), the total line strength is written in eqn. (4.1.15) as the product of the square of the electronic transition moment, the Franck–Condon factor, and the Hönl–London factor:

$$S_{Lv''J''\Lambda''}^{Uv'J'\Lambda} = R_e^2(\bar{r}_{v'v''}) \cdot q_{v'v''} \cdot S_{J''\Lambda''}^{J'\Lambda'}. \tag{4.1.15}$$

The sum rules for the Hönl–London factors and Franck–Condon factors are:

$$\sum_{J'} S_{J''\Lambda''}^{J'\Lambda'} = 2J'' + 1; \quad \sum_{J''} S_{J''\Lambda''}^{J'\Lambda'} = 2J' + 1, \tag{4.1.16a}$$

$$\sum_{v''} q_{v'v''} = 1 = \sum_{v'} q_{v'v''}. \tag{4.1.16b}$$

A useful interpretation of eqn. (4.1.15) is that $R_e^2(r)$ controls the magnitude of the transition strength for the whole band system, $q_{v'v''}$ determines its distribution from band to band in the system and $S_{J''\Lambda''}^{J'\Lambda'}$ influences the distribution from line to line in a band. This is an approximate point of view, because both $R_e\,(\bar{r}_{v'v''})$ and $q_{v'v''}$ influence the distribution of transition strength from band to band. However, $q_{v'v''}$ can vary by many orders of magnitude from band to band in a system and $R_e\,(\bar{r}_{v'v''})$ varies relatively slowly across a system. Nicholls and Stewart (1962) and Tatum (1967) have discussed these topics more fully.

Any application of eqn. (4.1.15) to the study of molecular intensities requires a good knowledge of $R_e(r)$, $q_{v'v''}$, and $S_{J''\Lambda''}^{J'\Lambda'}$ for all components of the contributing transitions. These factors are, thus, discussed separately in following sections.

A number of common molecular transition–probability parameters may now be defined in terms of $S_{Lv''J''\Lambda''}^{Uv'J'\Lambda'}$. It is well known (Allen, 1963) that the general expressions for oscillator strength f, and the Einstein A and B coefficients are [see also eqns. (3.3.11), (3.3.17), (3.3.58)]

$$|gf| = |g_U f_{UL}| = |g_L f_{LU}| = \frac{8\pi^2 \nu_{UL} S_{UL}}{3he^2}, \tag{4.1.17}$$

$$g_U A_{UL} = \frac{64\pi^4 \nu_{UL}^3 S_{UL}}{3hc^3}, \tag{4.1.18}$$

$$|g_U B_{UL}| = |g_L B_{LU}| = \frac{8\pi^2 S_{UL}}{3h^2} \tag{4.1.19}$$

where g is the statistical weight. In the case of a molecular line, we make the following substitutions:

$$g_U = d_U(2J' + 1); \quad g_L = d_L(2J'' + 1),$$

$$S_{UL} = R_e^2(\bar{r}_{v'v''})\, q_{v'v''} S_{J''\Lambda''}^{J'\Lambda'};$$

d_U and d_L are the degeneracies of the electronic states U and L and are $(2S + 1)$ for $\sum$ states ($\Lambda = 0$) and $2(2S + 1)$ for other electronic states (Davidson, 1962). The spectroscopic notation is standard (Herzberg, 1950).

Equations (4.1.17), (4.1.18) and (4.1.19) are thus rewritten as follows for a molecular line:

$$d_U(2J'+1) f_{Lv''J''}^{Uv'J'}\downarrow = d_L(2J''+1) f_{Lv''J''}^{Uv'J'}\uparrow = \frac{8\pi^2 m}{3he^2} \nu_{Lv''J''}^{Uv'J'} R_e^2(\bar{r}_{v'v''}) q_{v'v''} S_{J''\Lambda''}^{J'\Lambda'}, \tag{4.1.20}$$

$$d_U(2J' + 1)\, A_{Lv''J''}^{Uv'J'}\downarrow = \frac{64\pi^4}{3hc^3} (\nu_{Lv''J''}^{Uv'J'})^3\, R_e^2(\bar{r}_{v'v''})\, q_{v'v''} S_{J''\Lambda''}^{J'\Lambda'}, \tag{4.1.21}$$

$$d_U(2J' + 1)\, B_{Lv''J''}^{Uv'J'}\downarrow = d_L(2J'' + 1)\, B_{Lv''J''}^{Uv'J'}\uparrow = \frac{8\pi^2}{3h^2} R_e^2(\bar{r}_{v'v''})\, q_{v'v''} S_{J''\Lambda''}^{J'\Lambda'}. \tag{4.1.22}$$

The arrows indicate whether a transition is upward ($L \to U$) or downward ($U \to L$).

For the purposes of this discussion perhaps the most important transition–probability parameter is the absorption coefficient. As pointed out in Chapter 2, eqns. (2.1.1a) and (2.1.6), the three most commonly defined absorption coefficients, when integrated over a spectral feature are:

$$\text{Atomic or molecular} \quad \int \alpha_\nu \, d\nu = \frac{\pi e^2}{mc} f_{LU} \tag{4.1.23}$$

$$\text{Linear or volume} \quad \int \mu_\nu \, d\nu = \frac{\pi e^2}{mc} N_v(L) f_{LU} \tag{4.1.24}$$

$$\text{Mass} \quad \int \varkappa_\nu \, d\nu = \frac{\pi e^2}{mc\varrho} N_v(L) f_{LU} \tag{4.1.25}$$

α_ν is the optical absorption cross section and μ_ν is the reciprocal of the photon mean free path. Equation (4.1.24) is rewritten for a molecular line as

$$\int \mu_\nu \, d\nu = \frac{8\pi^3}{3hc} \frac{N_v(Lv''J'')}{d_L(2J''+1)} \nu_{Lv''J''}^{Uv'J'} R_e^2(\bar{r}_{v'v''}) \, q_{v'v''} S_{J''\Lambda''}^{J'\Lambda'}. \tag{4.1.26}$$

This equation may be modified to take account of the normalized line profile $b(\nu)$ introduced in eqn. (2.2.4a) [see also eqns. (3.3.42) and (3.3.43)] which obeys $\int b(\nu) \, d\nu = 1$. The effect is simply to multiply the right-hand side above by $b(\nu)$. Thus,

$$\mu_\nu = \frac{8\pi^3}{3hc} \frac{N_v(Lv''J'')}{d_L(2J''+1)} \nu_{Lv''J''}^{Uv'J'} R_e^2(\bar{r}_{v'v''}) \, q_{v'v''} S_{J''\Lambda''}^{J'\Lambda'} b(\nu). \tag{4.1.27}$$

As was pointed out in Section 3.7, various instrumental, thermal and environmental influences control $b(\nu)$.† It is very difficult to propose a completely satisfactory theoretical form for it which will satisfy all circumstances. A Voigt profile is often used to take account of the combined effects of thermal and collision broadening (see Section 3.7). The last influence produces a line shape which is similar in general form to, although much broader than, the Lorentz shape produced by natural broadening. A Lorentz profile [eqn. (2.6.20a)], in which the half width at half height w is a variable parameter that can be modified as circumstances warrant, is

† An often unconsidered source of apparent broadening is unresolved or only partly resolved rotation structure as in the case of components of Λ-doubling (Herzberg, 1950).

thus often used:

$$b(\nu) = \frac{1}{\pi} \frac{w}{w^2 + (\nu - \nu_{Lv''J''}^{Uv'J'})^2}. \tag{4.1.28}$$

Equations (4.1.26), (4.1.27), and equations derived from them have been the basis of a number of calculations of the molecular contributions to absorption coefficients of hot air (e.g. Churchill *et al.*, 1966).

A formal summation over all the lines of band is possible in each equation from (4.1.20) to (4.1.26) to produce similar equations which apply to the entire $v'v''$ band. In such summations, an approximation of the form

$$\sum \nu^n R_e^2 qS = \bar{\nu}^n R_e^2 q \sum S = \bar{\nu}^n R_e^2 q\,(2J + 1) \tag{4.1.29}$$

is involved where n is 0, 1 or 3, depending upon which transition probability parameter is involved. It is assumed in such an approximation that a characteristic frequency $\bar{\nu}$ can be assigned to a whole band. This is a fair approximation for a band whose rotational structure is not too greatly developed and which does not extend over too extensive a frequency range. Care must be taken not to force such approximations on cases where they do not reasonably apply.

The following equations for integrated bands result:

$$d_U f_{Lv''}^{Uv'}\downarrow = d_L f_{Lv''}^{Uv'}\uparrow = \frac{8\pi^2 m}{3he^2}\, \nu_{v'v''} R_e^2(\bar{r}_{v'v''})\, q_{v'v''}, \tag{4.1.30}$$

$$d_U A_{Lv''}^{Uv'}\downarrow = \frac{64\pi^4}{3hc^3}\, \nu_{v'v''}^3 R_e^2(\bar{r}_{v'v''})\, q_{v'v''}, \tag{4.1.31}$$

$$d_U B_{Lv''}^{Uv'}\downarrow = d_L B_{Lv''}^{Uv'}\uparrow = \frac{8\pi^2}{3h^2}\, R_e^2(\bar{r}_{v'v''})\, q_{v'v''}, \tag{4.1.32}$$

$$\int_{\text{band}} \mu_\nu \, d\nu = \frac{8\pi^3}{3hc} \frac{N_v(Lv'')}{d_L}\, \nu_{v'v''} R_e^2(\bar{r}_{v'v''})\, q_{v'v''}. \tag{4.1.33}$$

It is often convenient to rewrite eqn. (4.1.33) so as to take account of the contributions of bands (or parts of bands) which fall within the frequency increment $\Delta\nu$. Thus,

$$\bar{\mu}_\nu = \frac{1}{\Delta\nu} \sum_{\substack{\text{bands}\\ \Delta\nu}} \int_{\text{bands}} \mu_\nu \, d\nu. \tag{4.1.34}$$

Equations (4.1.30) to (4.1.33) have been used to interpret band intensity measurements in absorption or emission in terms of transition-probability parameters. Equation (4.1.34) has been used in preliminary "broad band" estimates of opacities of hot gases.

In some experiments, vibrational lifetimes $\tau_{v'}$, have been measured. They are related to Einstein A coefficients by

$$\tau_{v'} = \frac{1}{\sum\limits_{v''} A_{v'v''}}. \tag{4.1.35}$$

A formal summation procedure, similar in principle to that of eqn. (4.1.29). has been used by some to define the effective oscillator strength f_{LU} of whole band systems in analogy to the oscillator strength of an atomic line. Summation is over all v' which combine with a given v'':

$$d_L f_{LU} = d_L \sum_{v'} f_{v''v'} = \frac{8\pi^2 m}{3he^2} \sum_{v'} \nu_{v'v''} R_e^2(\bar{r}_{v'v''})\, q_{v'v''}. \tag{4.1.36}$$

If an effective $\bar{\nu}$ for the system can be defined, and if $R_e(r)$ is a constant, eqn. (4.1.36) can be rewritten [using eqn. (4.1.16b)] as

$$d_L f_{LU} = \frac{8\pi^2 m}{3he^2} \bar{\nu} \bar{R}_e^2 \sum_{v'} q_{v'v''} = \frac{8\pi^2 m}{3he^2} \bar{R}_e^2 \bar{\nu}. \tag{4.1.37}$$

Dividing eqn. (4.1.37) by eqn. (4.1.30) we obtain

$$f_{LU} = \frac{f_{v'v''}\bar{\nu}}{q_{v'v''}\nu_{v'v''}}. \tag{4.1.38}$$

Equation (4.1.38) should be used with extreme care, if at all, in view of the somewhat gross assumptions involved in the definition of $\bar{\nu}$ and $\bar{R}_e$. Schadee (1967) has suggested that eqn. (4.1.36) provides a useful definition of f-number if the $\bar{r}_{v'v''}$ dependence of R_e is retained. Because of the (approximately) unique relationship between $\bar{r}_{v'v''}$ and $\nu_{v'v''}$ (which will be discussed in Section 4.3), eqn. (4.1.36) can be interpreted as a frequency-dependent f-number definition. Schadee (1967) performed calculations of this f-number as a function of wavelength for a number of systems (C_2 Swan bands, CN red and violet systems, CO^+ comet tail and first negative systems, N_2 first and second positive systems, N_2^+ first negative, NO β and

γ, and the O_2 Schumann–Runge systems) and verified that under the approximations used, the f-number so obtained was a smooth function of wavelength. While this does yield a simple interpretation of the data, it is questionable that is provides sufficient generality and advantage over the calculation of arrays of smoothed band strengths (as will be discussed in Section 4.4).

It is clear from the discussion of eqns. (4.1.20) to (4.1.22), (4.1.26), (4.1.27), (4.1.30) to (4.1.34) and (4.1.38), that a good knowledge of ν, $R_e(r)$, $q_{v'v''}$, $S_{J''\Lambda''}^{J'\Lambda'}$ and $N_v(L)$ is necessary before they can be used in theoretical studies of opacities. The factors which control $N_v(L)$ are discussed by Gilmore (1967). Although it might appear that ν is always known in principle from previous spectroscopic research, this is not always the case, particularly when one is dealing with $\nu_{Lv''J''}^{Uv'J'}$ in very hot gases where high values of J' will be excited. Unambiguous band analyses have been carried out for a relatively small number of bands of each system and those analyses which have been made usually involve only those J-values excited in common laboratory sources running at temperatures under 1000 °K. Extrapolation of the low-temperature analyses to the high degrees of rotational development involved in hot gases, using the usual equations (Herzberg, 1950) for $\nu_{Lv''J''}^{Uv'J'}$ can be a somewhat approximate procedure, for no account is taken of possible perturbations at high J-numbers. The same argument holds to a lesser extent for extrapolation of vibrational analyses.

A review is made in subsequent sections of Hönl–London factors, Franck–London factors, r-centroids, electronic transition moments and band strengths, their properties and the methods by which they may be obtained.

4.2. The Hönl–London Angular Factors $S_{J''\Lambda''}^{J'\Lambda'}$

The line strengths for a symmetric top model of a diatomic molecule were first studied with the old quantum theory by Hönl and London (1925). They were reinvestigated by Dennison (1926), Reiche and Rademacher (1927), Kronig and Rabi (1927), Hill and Van Vleck (1928), Budo (1935; 1936; 1937), and Earls (1935), using quantum mechanics, for branches of bands of allowed transitions and for most common coupling cases. Herzberg (1950), Johnson (1949), Jevons (1932), Mulliken (1931) and others list tables of the Hönl–London factors for these cases. Schlapp (1932), Kovacs (1960) and others have studied the factors for some forbidden transitions. Schadee (1964) and Tatum (1967) have compiled tables for important transitions. Tatum also provides an extensive pedagogical discussion aimed at clarifying some of the complications involved in the calculation of intensities. Two important reviews (Kovacs, 1969; Hougen, 1970) which contain new results have recently appeared, and others have been written (Whiting and Nicholls 1970). It is clear from these commentaries that Hönl–London factors do not yet exist for all electronic transitions and coupling cases, and that some ambiguity in that definition remains. The sum rules for Hönl–London factors were given in eqn. (4.1.16a). As a simple example, the Hönl–London expressions for allowed singlet transitions are given for the *P*, *Q* and *R* branches by the following equations (Herzberg, 1950) for $\Delta\Lambda = 0$:

$$S_J^R = \frac{(J'' + 1 + \Lambda'')(J'' + 1 - \Lambda'')}{J'' + 1} = \frac{(J' + \Lambda')(J' - \Lambda')}{J'}$$

$$S_J^Q = \frac{(2J'' + 1)\Lambda''^2}{J''(J'' + 1)} = \frac{(2J' + 1)\Lambda'^2}{J'(J' + 1)} \qquad (4.2.1)$$

$$S_J^P = \frac{(J'' + \Lambda'')(J'' - \Lambda'')}{J''} = \frac{(J' + 1 + \Lambda')(J' + 1 - \Lambda')}{J' + 1},$$

for $\Delta\Lambda = +1$

$$S_J^R = \frac{(J'' + 2 + \Lambda'')(J'' + 1 + \Lambda'')}{4(J'' + 1)} = \frac{(J' + \Lambda')(J' - 1 + \Lambda')}{4J'}$$

$$S_J^Q = \frac{(J'' + 1 + \Lambda'')(J'' - \Lambda'')(2J'' + 1)}{4J''(J'' + 1)} = \frac{(J' + \Lambda')(J' + 1 - \Lambda')(2J' + 1)}{4J'(J' + 1)} \qquad (4.2.2)$$

$$S_J^P = \frac{(J'' - 1 - \Lambda'')(J'' - \Lambda'')}{4J''} = \frac{(J' + 1 - \Lambda')(J' + 2 - \Lambda')}{4(J' + 1)},$$

and for $\Delta\Lambda = -1$

$$S_J^R = \frac{(J'' + 2 - \Lambda'')(J'' + 1 - \Lambda'')}{4(J'' + 1)} = \frac{(J' - \Lambda')(J' - 1 - \Lambda')}{4J'}$$

$$S_J^Q = \frac{(J'' + 1 - \Lambda'')(J'' + \Lambda'')(2J'' + 1)}{4J''(J'' + 1)} = \frac{(J' - \Lambda')(J' + 1 + \Lambda')(2J' + 1)}{4J'(J' + 1)} \tag{4.2.3}$$

$$S_J^P = \frac{(J'' - 1 + \Lambda'')(J'' + \Lambda'')}{4J''} = \frac{(J' + 1 + \Lambda')(J' + 2 + \Lambda')}{4(J' + 1)}.$$

Since the selection rule $\Delta J = 0, \pm 1$ determines the P, Q or R branch concerned, it is only necessary to specify the branch and the J-value in the superscripts and subscripts to the factors. For branches of bands of higher multiplicity the K-value has to be specified (Johnson, 1949). The intensity profile in a branch is determined from line to line by $S_J \exp(-E_J/kT)$ for thermally maintained populations. In the early literature S_J was written as i_J and called the intensity factor (Jevons, 1932). Here $J = J'$ for emission and $J = J''$ for absorption transitions. These formulas are valid for Hund's coupling case (a), which is the one that occurs most frequently, since the eigenfunctions used to define R_e and $S_{J''\Lambda''}^{J'\Lambda'}$ are eigenfunctions for that case. These Hönl–London factors for singlet–singlet transitions may also be applied to all other transitions provided that both states are Hund's case (a). The quantum number Λ in these formulas should be replaced by Ω, and the normalization, eqn. (4.1.16a), changes from $2J + 1$ to $(2J + 1)(2S + 1)$ (Schadee, 1967). For the other coupling cases, appropriate transformations of eigenfunctions must be invoked and the situation becomes complicated. The most complete tabulation of these factors is that of Kovacs (1969), mentioned above, who gives a great many tables of explicit formulas as well as extensive discussion of the theory. Hougen (1970) does not give tables, but rather, describes procedures for calculating them and illustrates these procedures with sample calculations.

4.3. The Franck–Condon Principle and Vibrational Features of Molecular Absorption

A strict vibrational selection rule ($\Delta v = \pm 1$) applies to *infrared vibrational–rotation* transitions between levels of a parabolic (simple harmonic oscillator) molecular potential. In this case, the potential (electronic state) in common to the upper and the lower vibrational levels. The vibrational wave functions of eqn. (4.1.9a) are Hermite functions and are mutually orthogonal. $\boldsymbol{M}$ is assumed to be linear in r. Under these circumstances (see Herzberg, 1950, p. 80) $S_{Lv''}^{Uv'}$ vanishes unless the above selection rule applies. Realistic potentials are, however, somewhat anharmonic and thus some weak infrared vibration–rotation bands are observed for transitions which break the selection rule. Infrared vibration–rotation transitions are only observed in heteronuclear molecules. Homonuclear molecules, because of their symmetry, have no permanent dipole moment and are thus infrared inactive.

The Franck–Condon principle takes the place of a strict selection rule for vibrational transitions between *different electronic states* by specifying their relative probabilities. In the qualitative discussion of the photodissociation of diatomic molecules, Franck (1925) pointed out that a spontaneous electronic transition affects neither the instantaneous position nor momentum of the nuclei. Thus, the most preferred molecular transitions are "vertical" ones in which r does not change on the molecular energy vs. internuclear separation diagrams. Further, the molecule is, on a time average basis, most likely to be found near the turning points (r_{min} or r_{max}) of its oscillation, and, thus, those v'–v'' transitions are most likely to occur where one of the turning points of the upper level v' is nearly equal to one of the turning points of the lower level v''.

Condon (1926) placed this suggestion on a quantitative basis and showed (Condon, 1928) that it was the square of the overlap integral of vibrational wave functions (or "Franck–Condon factor" as it was called by Bates, 1952) which was a measure of the relative probability of the $v'v''$ transition [see eqn. (4.1.14)]. This is consistent with Franck's suggestion, as the vibrational wave functions have large antinodes in the region of r_{min} and r_{max} of the classic oscillator. When the overlap between these regions of the two wave functions is large, the Franck–Condon factor will be large. Thus, "vertical" transitions at the turning points of molecular oscillations will be most probable. Other transitions are not forbidden. Their relative strength will be determined by the amount of non-cancelled overlap between their wave functions. Condon (1947) has given a very good review of the circumstances leading to the development of the principle.

Franck–Condon Factors and Molecular Vibrational Potentials

In this section, we review briefly our knowledge of Franck–Condon factor arrays and related quantities for important electronic band systems of diatomic molecules. These factors, denoted by $q_{v'v''}$, were defined in eqn. (4.1.14). The sum rule which they obey has been given already by eqn. (4.1.16b); it can be obtained by expansion of $\psi_{v''}$ for the lower state in terms of the set of functions, including $\psi_{v'}$ for the upper state or vice versa and subsequent application of Parseval's theorem. Considerable effort has been expended, particularly over the past two decades, to provide tables of Franck–Condon factors for important band systems. Spindler (1965) lists some of the more recent calculations.

The first requirement for calculation of q-arrays is, of course, a knowledge of the vibrational wave functions appropriate to the molecule in the upper and lower states of the transitions. Once these are known, a number of derived quantities of the wave functions including q-values can be calculated. The wave functions are solutions of the one-dimensional Schroedinger equation for the linear single-particle oscillator of reduced mass $\mu = M_1M_2/(M_1 + M_2)$ moving under the action of the "rotationless" molecular potential $U(r)$. (We discuss the effect of vibration–rotation interaction later.) This Schroedinger equation for the nuclear vibrational motion is (Schiff, 1955, p. 305)

$$\frac{d^2\psi_v}{dr^2} + \frac{8\pi^2\mu}{h^2}[E_v - U(r)]\psi_v = 0. \qquad (4.3.1)$$

The specification of a realistic molecular potential provides the first problem. No completely realistic (non-numerical) analytic potential is available. All analytic potentials are empirical, and constitute representations of what has been thought to be a reasonable approximation to molecular behavior. The parabolic or simple harmonic potential was the first to be used. Equation (4.3.1) has closed solutions (Hermite functions) for such potentials. The model is only realistic at very low values of vibrational quantum number. Manneback (1951), Condon (1928) and others have produced limited arrays of "harmonic oscillator" q-values for a number of transitions. Aiken (1951) produced a large number of arrays of them for many band systems on the Harvard Mark I Computer. An attempt was made by Gaydon and Pearse (1939) to overcome some of the lack of realism of the parabolic potential by linearly distorting its wave functions to fit an equivalent Morse potential. Although the resulting functions are non-orthogonal, they were used with some success by a number of workers. Pillow (1949), Nicholls (1950) and others extended and applied the method to many transitions (see list of references given by Nicholls and Stewart, 1962).

The Morse (1929) potential,

$$U(r) = D_e\,[1 - \exp\{-\beta\,(r - r_e)\}]^2 \qquad (4.3.2)$$

where

$$\beta = 2\left(\frac{2\pi^2 c}{h}\,\mu\omega_e x_e\right)^{1/2}$$

has had the greatest popularity in molecular spectroscopy for many years, because, among other properties, it does predict the correct vibrational energy level array E_v for many cases. Although it is an empirical suggestion, it does have some physical plausibility in the adequate representation of the mechanical anharmonicity of many molecular states. There are nevertheless relatively severe theoretical objections to it (e.g. it often predicts the incorrect dissociation energy $D_e = \omega_e^2/4\omega_e x_e$, $U(0)$ is non-infinite, etc.). Many other empirical analytic functions have been suggested; these have been reviewed by Varshni (1957) and by Steele, Lippincott and Vanderslice (1962). The vibrational wave functions for the Morse potential are also known in closed form and have been the basis of a number of calculations of Franck–Condon factors. Some of these calculations were performed on desk calculators (Jarmain and Nicholls, 1954) before it was common to have access to digital computers. A number of approximate methods for evaluation of "Morse" Franck–Condon factors were thus developed (Fraser and Jarmain, 1953; Jarmain and Fraser, 1953; Fraser, 1954a; Wu, 1952). Franck–Condon factor arrays were computed for a large number of astrophysically and aeronomically important arrays by these methods (Jarmain, Fraser and Nicholls, 1953, 1955; Fraser, Jarmain and Nicholls, 1954; Nicholls, Fraser and Jarmain, 1959; Nicholls, Fraser, Jarmain and McEachran, 1960). Access to computers permitted the direct computation of Morse Franck–Condon Factor arrays by means of a computer program which first computes the wave functions (Nicholls, 1961). Many Franck–Condon factor arrays have since been computed to high quantum numbers by this method, some of which have been published (e.g. Nicholls, 1960; 1961; 1962a, b, c, d; 1963a; 1964a, b; 1965a, b, c, d; Ory, 1964a, b; Halmann and Laulicht, 1965). Not only radiative transitions have been studied, but excitation and ionization transitions whose cross sections are to some extent controlled by the Franck–Condon factor (Craggs and Massey, 1959) have also been studied in a number of important cases (Nicholls, 1961; 1962a; 1968a; Wacks and Krauss, 1961; Wacks, 1964). The suggestion (Nicholls, 1968b) that the matrix multiplication property of vibrational overlap integrals could be used to evaluate Franck–Condon factors has been confirmed by Bennett and Sommerville (1969).

One major limitation of "Morse" Franck–Condon factor arrays is the inability of the Morse potential to represent with realism the behavior of all molecular potentials. Rydberg (1931, 1933), Klein (1932), and Dunham (1932) developed methods of constructing molecular potentials numerically from the location of the classical turning points of the oscillator at each value. Rees (1946) placed this method on a sound analytic basis. The starting point for such calculations was a set of measured molecular constants, particularly the B_v values derived from band analysis. Jarmain (1959, 1960), Vanderslice, Mason, Maisch and Lippincott (1959, 1960) and Hurley (1962) have demonstrated the equivalence of Rydberg and Klein's WKB method (which would be expected to hold at high v-values) and Dunham's representation of the potential at low v-values. The basis of the method and the equations involved have been reviewed by Gaydon (1953, 1968), Gilmore (1965b) and Spindler (1965). Jarmain (1959, 1960), Vanderslice and co-workers (see Steele, Lippincott and Vanderslice, 1962, and references therein) and others have computed realistic Rydberg–Klein–Dunham–Rees potentials for many molecular electronic states. Gilmore (1965a, b; 1967) has made a definitive and exhaustive study of the potential energy curves by this method for all the known states of N_2, O_2, NO and their ions.

The end results of such work is a list of the classical turning points r_{min} and r_{max} for each vibrational level for which measured values of G_v and B_v (Herzberg, 1950) were used. The rectification of such a list into $U(r)$ at equally spaced intervals of r is necessary and has been performed numerically in different ways by different workers.

Computer-oriented methods have been developed by Jarmain (1961, 1963a, b), Cooley (1961a, b), and Cashion (1963) for the numerical solution of eqn. (4.3.1) appropriate to a numerical potential produced in the manner described above. These methods depend on the analysis of Numerov (1933). The resulting wave functions have then been used to compute Franck–Condon factors by straightforward numerical integration of their products. Jarmain (1963b) has discussed criteria which wave functions arising from these methods must satisfy. A number of other workers (Zare, Larson and Berg, 1965; Spindler, 1965; Wentink and Spindler, 1970) have used Cooley's program to obtain wave functions for the computation of Franck–Condon factor arrays for a number of important band systems. In particular, Hubisz (1968) has made a systematic and extended study, including vibration and rotation effects. A number of laboratories (e.g. Generosa at the Air Force Weapons Laboratory, Kirtland Air Force Base, New Mexico; Albriton and Zare at the Environmental Sciences Services Administration Laboratory, Boulder, Colorado, McCallum, Nicholls and Jarmain, 1970, at York University) have extensive compilations of realistic Franck–Condon factors in report form which are available on application.

A comparison between arrays of Franck–Condon factors computed from realistic molecular potentials and from Morse potentials indicates that in general there is good agreement between the two arrays at low quantum numbers and less good agreement at high quantum numbers. However, in some cases the agreement at high quantum numbers is much better than might have been *a priori* expected. See, for example, Spindler,

Isaacson and Wentink (1970). If there is a very great divergence between the potentials, there will be divergence between the arrays. In fact Hubisz (1968) found that reanalysis of spectra, including high quantum number data often led to new vibrational constants ω_e and $\omega_e x_e$ with negligible values of $\omega_e y_e$ and $\omega_e z_e$. The new Morse potential so derived often produces very good agreement with the RKR potential and relevant wave functions and derived quantities. Analogous results have been obtained by Cunio and Jansson (1968). Much of the above discussion of Franck–Condon factors relates to a rotationless model of the molecular vibrator (in the $J = 0$ state). It is tacitly assumed that the coupling between vibration and rotation is small and that the centrifugal energy

$$\frac{h^2}{8\pi^2\mu} \frac{[J(J+1) - \Lambda^2]}{r^2}$$

of molecular rotation which should really be added to $U(r)$ in eqn. (4.3.1) has a negligible effect on the molecular potential. This assumption is not always valid, particularly for light, rapidly rotating molecules such as hydrides, and in principle the eqn. (4.3.1) of which ψ_v is the solution should have the centrifugal energy term added to $U(r)$. Pekeris (1934) solved eqn. (4.3.1) for a Morse potential to which the rotational energy barrier has been added. Herman and Rubin (1955) and Herman, Rothery and Rubin (1958) have examined vibration–rotation interaction, using Morse–Pekeris wave functions, for vibration–rotation spectra. Fraser (1958) examined the effect of vibration–rotation interaction using the analytic approximation with perturbation correction (Fraser, 1954b). Learner (1962) made a similar study of hydrides (particularly OH) and computed overlap integrals by numerical integration of Morse–Pekeris wave functions and James (1960, 1961) made a similar study using harmonic and anharmonic oscillators and perturbation methods. Haycock (1963) extended Fraser's work with particular emphasis on hydrides. Nearly all of the early work was for transitions in light molecules where the effect is expected to be large, and the results of the studies tends to confirm this expectation, particularly for hydrides. For heavier molecules the effect appears to be small. Halmann and Laulicht (1968) solved the Schroedinger eqn. (4.3.1) numerically with an RKR potential including a rotational contribution, for the (5, 0) band of the Schumann–Runge system of O_2 and for the (0, 0) to (2,0) bands of the H_2 Lyman system. Although their primary interest was to explore the sensitivity to isotopic substituion in the diatomic molecule, their results provide a quantitative measure of the rotational effects as well for these systems. The isotope ratio of Franck–Condon factors $q(^{16}O_2)/q(^{18}O_2)$ was $\sim$2.5 and along with the r-centroids remained virtually constant out to rotational quantum number $J'' = 21$. The Franck–Condon factors themselves decreased by about 12 per cent from $J'' = 1$ to 21. (The rotational effect in H_2 and D_2 was large, in agreement with the findings noted above.) An effect of this size is not necessarily significant to high-temperature opacity studies such as will be discussed in our next volume of this series. Hubisz (1968) has carried out further numerical studies that have confirmed that the effect is small for most common molecular transitions not involving light molecules. Shumaker (1969) made calculations of Franck–Condon factors, r-centroids and band strengths for several band systems of nitrogen out to rotational quantum number 100. The effect of rotation in all these cases was extremely small.

So far we have limited the discussion of Franck–Condon factors to transitions between discrete vibrational levels, for which the sum rule eqn. (4.1.16b) holds. There is an important class of molecular transitions which, however, involve unbound continuum states in one of the levels. For example, the dissociation continuum of the O_2–Schumann–Runge and the O_2 Herzberg I systems, involve transitions between the $v'' = 0$ level of the $X^3\Sigma_g^-$ ground state and levels above the dissociation limit of the $B^3\Sigma_u^-$ and the $A^3\Sigma_u^+$ states respectively in cold O_2. For hot O_2 other values of v'' are incorporated. The wave function $\psi_{\text{cont}}(r, \nu)$ of the unbound state is a function of energy as well as internuclear separation and cannot be normalized in the same way as bound state wave functions. Normalization is carried out with respect to energy and involves the asymptotic amplitude (see Section 3.4, eqns. (3.4.54ff.) and, for example, Jarmain and Nicholls, 1964). A Franck–Condon density $q_{v''\text{cont}}(\nu)$ is defined by

$$q_{v'',\text{cont}}(\nu) = \left|\int \psi_{v''}(r)\, \psi_{\text{cont}}(r, \nu)\, dr\right|^2 \tag{4.3.3}$$

and the Franck–Condon factor for the entire continuum is then

$$q_{v'',\text{cont}} = \int q_{v'',\text{cont}}(\nu)\, d\nu . \tag{4.3.4}$$

Under these circumstances the sum rule [eqn. (4.1.16b)] is replaced by

$$\Sigma_{v'} q_{v'v''} + \int q_{v''\text{cont}}(\nu)\, d\nu = 1\,. \tag{4.3.5}$$

The element of the volume absorption coefficient in the continuum is

$$d \int \mu_\nu \, d\nu = \mu_\nu \, d\nu = \frac{\pi e^2}{mc} \frac{N_v(Lv'')}{dL} \frac{df}{d\nu} d\nu = \frac{\pi e^2}{mc} N_v(Lv'')\, R_e^2(r)\, dq_{\text{cont}}\,. \tag{4.3.6}$$

Jarmain and Nicholls (1964, 1967) have recently computed Franck–Condon densities for the Schumann–Runge continuum from $v'' = 0$, using wave functions derived from realistic potentials and also for the Herzberg I continuum from $v'' = 0, 1, 2$. The densities were compared with measured absorption coefficients, and by use of eqn. (4.3.6), $R_e(r)$ was determined. Their calculations have been confirmed by recent experimental measurements in the photodissociation continuum by Hasson, Nicholls and Degen (1970) and Hasson Hébert, and Nicholls (1970).

An earlier approach to this sort of problem derives from the suggestion of Winans and Stueckelberg (1928), that from a Franck–Condon point of view it will be the terminal antinode of the continuous wave function which makes an important contribution to the overlap integral. They thus suggest that in the absence of firm knowledge of the form of the continuum wave function it should be replaced by a δ-function at the classical turning point r_1 of the oscillator. The Franck–Condon density is then $|\psi_{v''}(r_1')|^2$. This ingenious idea was tested by Coolidge, James and Present (1936) in an H_2 transition using a Dunham potential and a differential analyzer to obtain realistic wave functions. They obtained good agreement between the two results. However, a similar test carried out by Jarmain and Nicholls (1964) for the O_2 Schumann–Runge transition led to no agreement between the two methods. Ditchburn and Young (1962) used the δ-function method for comparison with experimental measurements on the O_2 Herzberg I continuum. Doyle (1968) has recently made a more accurate calculation on H_2 which has confirmed the earlier results of Coolidge, James and Present on the accuracy of the δ-function approximation for H_2. In addition, he found that the neglect of rotation was a larger source of error than the δ-function approximation for temperatures at which several rotational states are populated.

A sufficiently large number of Franck–Condon Factor arrays for discrete bands have now been calculated under similar circumstances that the systematic properties of these arrays can be studied. Two related aspects of these systematics have been investigated: the geometry of the Condon loci, and the geometry of the Franck–Condon factor surface (Nicholls, 1969). It was a very early observation in molecular spectroscopy that the most commonly observed bands of a system lie in well-defined regions of the v'–v''-plane on and around an open limbed, quasiparabolic curve, which is roughly symmetric about $v' = v''$ and whose vertex is in the region of (0, 0). Condon (1926) in a semiclassical discussion, was able to show that for a simple harmonic oscillator potential the strong bands would indeed lie on a parabolic locus in the v'–v''-plane. The parabolae have, thus, been called "Condon parabolae". For band systems, where Δr_e is very small, the parabola collapses into a diagonal line. As Δr_e increases, the parabola widens, its vertex moves away from (0, 0), and subsidiary "parabolae" nested within the primary outer parabola develop. Detailed examination of these loci of the stronger bands show that they are not parabolic, but lie on open curves, which remind one of parabolae. They are by no means all symmetric about $v' = v''$. They are, in fact, the projection on the v'–v''-plane of the three-dimensional surface of $q_{v'v''}$ vs. v' and v''.

The definition of $q_{v'v''}$ (eqn. (4.1.14)) suggests that the shape of the Condon loci can be predicted by requiring equality in r between one antinode of one wave function and one of the other. This requirement enhances the magnitude of the overlap integral. The primary locus runs through those v' and v'' values for which there is overlap between one terminal antinode of one wave function with one terminal antinode of the other (Nicholls, 1962e). The subsidiary loci run through those v' and v'' for which there is overlap between a terminal antinode of one wave function and a subsidiary antinode of the other (Nicholls, 1963b; Murty and Nicholls, 1967). This approach has been very fully investigated by Murty (1964), who has shown that each locus has six possible segments of which three usually occur.

The systematic variation in shape of the Condron loci with Δr_e is a reflection of a similar behavior of the surfaces. It has been described elsewhere (Nicholls, 1964d, 1965f, 1969) how with increase in Δr_e the surfaces change from a diagonal ridge to a series of nested horseshoe-shaped ridges whose apex moves steadily away from (0, 0). It has been possible to represent this behavior by a series of plots (one for each v', v'') of log $q_{v'v''}$ vs. log $\beta'\Delta r_e$. β' is the harmonic mean of the β's in the exponents of the two Morse potentials [eqn. (4.3.2)] and $\beta'\Delta r_e$ is thus a transition parameter characteristic of the whole system (Nicholls, 1964c, 1965f). The curves have a systematic undulatory shape as would be expected, and can be used for the rough interpolation of Franck–Condon factors.

The *r*-centroid Approximation

The r-centroid approximation defined in eqn. (4.1.12a, b) is another important derived quantity of vibrational wave functions used in the interpretation of intensities of molecular spectra. Fraser (1954a, b) has shown that as a result of the compact nature of the wave function product $\psi_{v'}\psi_{v''}$ [see eqn. (4.3.7)] eqn. (4.1.12b) is a good approximation for band systems for which $\mu_A\omega_e \sim 10^4$ (μ_A is the reduced mass in atomic mass units and ω_e is in cm^{-1}), 0.01 Å $< \Delta r_e <$ 0.25 Å, and for which v' and v'' do not greatly exceed 10.

This approximate property allows one to resolve the integral of the product of $\psi_{v'}$ and $\psi_{v''}$ and a polynomial function of r into the product of the vibrational overlap integral and the polynomial taken at an argument $\bar{r}_{v'v''}$ [see eqn. (4.1.13)]. The approximation has been studied systematically by Drake and Nicholls (1970) in a synthetic-molecule-oriented, computer-based study. They confirmed that for a wide variety of transitions, for bands for which $q_{v'v''}$ was not extremely small, the approximation was valid, by establishing that the quantity

$$Y^n_{v'v''} = \frac{\int \psi_{v'} r^n \psi_{v''}\, dr}{(\bar{r}_{v'v''})^n \int \psi_{v'}\psi_{v''}\, dr}$$

is nearly unity for these bands.

The r-centroid is a characteristic r-value associated with the v'–v'' transition and recent work has shown that it may be identified with the "Franck–Condon principle" value of r for the v'–v'' transition in question as is demonstrated below. Equations (4.1.12a, b) imply that the wave function product $\psi_{v'}\psi_{v''}$ approximates a high order member of the sequence defining a generalized function (Temple, 1955; Lighthill, 1959), such as a δ-function or damped oscillatory function, strongly peaked at $\bar{r}_{v'v''}$. Approximately one may thus rewrite equations (4.1.12a, b) as

$$\frac{\psi_{v'}\psi_{v''}}{\int \psi_{v'}\psi_{v''}\, dr} \cong \delta\,(r - \bar{r}_{v'v''}). \tag{4.3.7}$$

A recent study using tabulated Morse wave functions for a number of systems (Nicholls, 1966) has confirmed this view for many bands on the primary Condon Locus. James (1966) has given a critique of the r-centroid approximation and the use of the condition established by eqn. (4.1.12b). However, his discussion is more or less superseded by that of Halevi (1965) who established a rigorous mathematical basis for the validity of this approximation. The band-strength equation in the r-centroid approximation, eqn. (4.1.15), can be interpreted as a Taylor expansion about the position of the r-centroid. Thus, the transition moment $R_e(\bar{r}_{v'v''})$ is to be modified by the factor $[1 + \varrho\,(v'v'')]^2$ where

$$\varrho\,(v', v'') = \frac{R_e''(\bar{r}_{v'v''})}{2R_e(\bar{r}_{v'v''})}\left[\frac{(v', \bar{r}^2 v'')}{(v', v'')} - \frac{(v', rv'')^2}{(v', v'')^2}\right] \tag{4.3.8}$$

since the first derivative term in the Taylor expansion vanishes by definition of the r-centroid†. The quantity $\varrho\,(v', v'')$ yields a measure of the accuracy of the approximation, or, alternatively, it can be used as a correction. The scalar product notation $(v', r^n v'')$ was introduced in Section 3.6 and is, of course, a short-hand notation for $\int \psi_{v'} r^n \psi_{v''}\, dr$ in this case. Jansson and Cunio (1968) have applied Halevi's correction to the N_2 first positive system.

The r-centroid also satisfies the relationship (Nicholls and Jarmain, 1956)

$$U'\,(\bar{r}_{v'v''}) - U''\,(\bar{r}_{v'v''}) = E_{v'} - E_{v''} \tag{4.3.9}$$

which has been the basis for many early calculations of r-centroid arrays and also for a theoretical demonstration that the r-centroid is the Franck–Condon principle value of r (Nicholls, 1966). James (1966) has given a

† The double prime on R_e in this equation signifies differentiation.

detailed discussion of the graphical method of solving the above equation originally suggested by Jarmain and Fraser.

Arrays of r-centroids have been computed for many band systems by a number of methods—some of which are based on eqn. (4.3.9) and some of which involve direct calculation of the integral quotient of eqn. (4.1.12). The properties, methods of calculation and arrays of r-centroids have been fully discussed in a number of papers (Nicholls and Jarmain, 1955, 1956; Nicholls, Parkinson, Robinson and Jarmain, 1956; Nicholls and Stewart, 1962; Nicholls, 1965g, 1966). The r-centroids are related to the r_e's and the turning values $r'_{1,2}$ of the oscillators in the v' and v'' levels (Nicholls and Jarmain, 1959). It is also possible to relate r-centroids and Franck–Condon factors to the expectation value $\bar{r}_{v'}$ or $\bar{r}_{v''}$ of r in the level v' or v'' through the equation:

$$\bar{r}_{v' \text{ or } v''} = \sum_{v'' \text{ or } v'} \bar{r}_{v'v''} q_{v'v''} \tag{4.3.10}$$

(Nicholls and Jarmain, 1955). That is, the expectation value is a weighted average of r-centroids and Franck–Condon factors for all transitions out of the level concerned.

One of the most useful properties for applications of the r-centroid is its monotonic dependence on wavelength $\lambda_{v'v''}$ or frequency $\nu_{v'v''}$ for the bands of a system. When $r'_e > r''_e$, $\bar{r}_{v'v''}$ increases monotonically with $\lambda_{v'v''}$. When $r'_e < r''_e$, $\bar{r}_{v'v''}$ decreases monotonically with $\lambda_{v'v''}$. This behavior, and the two possible types of $\bar{r}_{v'v''}$ surfaces (the three-dimensional representation of $\bar{r}_{v'v''}$ as a function of v' and v''), which it implies, is directly related to the two types of $\lambda_{v'v''}$ surface. In both, $\lambda_{v',0}$ is less than $\lambda_{0,v''}$. When $r'_e > r''_e$, $\lambda_{v'v''} > \lambda_{00}$ and conversely. The monotonic relationship allows simple interpolation of uncalculated $\bar{r}_{v'v''}$ values and also allows $\lambda_{v'v''}$ or $\nu_{v'v''}$ to be used as independent variables in place of $\bar{r}_{v'v''}$ in some applications in the rescaling of measured intensities to provide a table of band strengths (see Section 4.4).

After a number of $\bar{r}_{v'v''}$ arrays had been calculated, it was observed that $\bar{r}_{v'+1,\, v''+1} - \bar{r}_{v'v''}$ was approximately constant for each of the v'–v'' sequences. The constant was different for each sequence. The theoretical justification for this was provided and it gave a simple means of making up an $\bar{r}_{v'v''}$ array once the leading entries $\bar{r}_{v',v}$, $\bar{r}_{v',v''}$ for the sequences and the constant differences were known. The constant differences imply that the $\bar{r}_{v'v''}$ surface is approximately plane. As expected from the definition of the *r*-centroid, and as confirmed by Drake and Nicholls (1970), the approximation is valid for bands of non-vanishing Franck–Condon factor for most band systems.

4.4. The Electronic Transition Moment and Band Strengths

The electronic transition moment $R_e(r)$ is the final of the three factors which control a line strength. It is formally defined in eqn. (4.1.10) and because of our poor detailed knowledge of molecular electronic wave functions (recent *ab initio* studies notwithstanding), we are forced to study it experimentally. The electronic transition moment enters into a discussion of molecular transition probabilities from a number of standpoints. It is, of course, interesting in its own right, as a knowledge of its magnitude and variation with *r* for a large number of different systems gives greater understanding of problems of molecular structure. It has been pointed out a number of times in the preceding discussion that from a transition probability point of view, it is the band strength $S_{v'v''}$ or the product

$$R_e^2\,(\bar{r}_{v'v''})\,q_{v'v''}$$

which is a more important quantity than either of its factors. In the applications discussed below, we shall thus look on $R_e(r)$, $q_{v'v''}$ and $\bar{r}_{v'v''}$ as means to the end of determination of reliable and smoothed band strengths.

$R_e(r)$ has been studied experimentally in three main ways:

(a) from intensities of emission spectra,

(b) from intensities of absorption spectra,

(c) from measurement of lifetimes of states.

The consequences of these will be discussed briefly in turn.

The emission intensity $I_{v'v''}$ of an optically then molecular band is given by (Nicholls and Stewart, 1962)

$$I_{v'v''} = KN_v\,(v')\,\nu^4_{v'v''}R_e^2\,(\bar{r}_{v'v''})\,q_{v'v''} = KN_v\,(v')\,\nu^4_{v'v''}S_{v'v''} \qquad (4.4.1)$$

where K is a constant which allows for units and geometry. In emission, geometrical considerations make absolute measurements very difficult, and thus, this equation has been used in many *relative measurements* of band intensities. There are a number of other practical difficulties associated with overlap in structure between adjacent bands which can add significantly to possible errors in measured $I_{v'v''}$. The application of computer-generaled synthetic spectra can materially assist in overcoming these difficulties.

One obvious application of eqn. (4.4.1) is to measure $I_{v'v''}$ band by band by photographic or photoelectric means, and to interpret relative $S_{v'v''}$ band by band from

$$S_{v'v''} = I_{v'v''}/KN_v(v')\nu^4_{v'v''}. \tag{4.4.2}$$

That approach requires either a knowledge of N_v (e.g. from demonstrated thermal excitation of spectra in a furnace or shock-tube) or the determination of relative $S_{v'v''}$ along progressions of bands for which $N_{v'}$ = const. The method has been applied to furnace-excited C_2 spectra for example, by King (1948), Phillips (1954, 1957) and Hagan (1963) and others. The method, while straightforward in application, has the disadvantage that the relative error varies from band to band. Strong bands have profiles of large area and are, in general, more accurately measured than weak bands.

An alternative application of eqn. (4.4.1) (Fraser, 1954a) is to plot $(I_{v'v''}/q_{v'v''}\nu^4_{v'v''})^{1/2}$ vs. $\bar{r}_{v'v''}$ for progressions of bands in which v' is common in each. This plot is equivalent to $N_{v'}^{1/2}(v')\,R_e(r)$ vs. r. It has been applied to many band systems (see Nicholls and Stewart, 1962; Nicholls, 1964c). The result is a set of segments (one for each v = const. progression of bands) which delineate the relative variation of $R_e(r)$ with r. The segments are displaced in ordinate from each other by an amount controlled by $N_v^{1/2}(v')$. Rescaling procedures allow all of the segments to be placed on the same ordinate scale and provide a knowledge of N_v. All the measured intensities have then played a role in the delineation of $R_e(r)$, and a smooth empirical curve can be fitted by least squares methods to the final set of points.

If the aim of such work is to study $R_e(r)$ *per se*, then the procedure outlined above can be used with the reservation that $\bar{r}_{v'v''}$ is an approximately defined quantity. If the aim is, however, to provide a set of smoothed $S_{v'v''}$ ($= R_e^2 q$) values from the smoothed $R_e(r)$ curve and the Franck–Condon factors, the discussion of the preceding section on r-centroids indicates that $(I/q\nu^4)^{1/2}$ could equally well have been plotted against ν or λ in progressions because of the monotonic relationship between $\bar{r}_{v'v''}$ and wavelength or frequency. Further, provided that no large extrapolation is made outside the range of $\bar{r}_{v'v''}$ (or λ or ν) for the system this procedure will also allow estimates to be made of relative band strengths for bands not measured, providing that a consistent estimate of $q_{v'v''}$ and $\bar{r}_{v'v''}$, λ or ν is available for each of them so that the correct ordinate of the $R_e(r)$ plot can be read off.

Another point should also be made in this connection, obviously, different forms of $R_e(r)$ will result from the use of each $q_{v'v''}$ (and $\bar{r}_{v'v''}$) array calculated on the basis of a different model of $U(r)$. However, the compensation achieved between division by $q_{v'v''}$ in the determination of $R_e(r)$, and the multiplication by $q_{v'v''}$ in the determination of $S_{v'v''}$ will make the eventual smoothed $S_{v'v''}$ arrays less sensitive to the model of potential used than has often been thought.

The relative $S_{v'v''}$ arrays (often expressed on a scale where $S_{0,0}$ is 100) can be, and have been in some cases, placed on an absolute scale by comparison with absolute absorption coefficients for a few bands or by use of lifetime measurements (see, for example, Nicholls, 1964d; Hasson, Nicholls and Degen, 1970). One advantage of emission measurements, in spite of their relative nature, is that, in general, more bands of a system are excited in emission than are observed in cold absorption (see, however, the remarks below).

If the aim or the work is the provision of data on $R_e(r)$, the realism of the potential model used to compute $q_{v'v''}$ arrays is of great importance as is the adequacy of the r-centroid approximation for the band system under consideration. There has been relatively little discussion of any possible systematic difference in behavior of $R_e(r)$ with r from band system to band system. Bates (1949) proposed that the dependence upon r would be most strong for perpendicular ($\Delta\Lambda = \pm 1$) transitions than for parallel ($\Delta\Lambda = 0$) transitions. The change of the electronic structure is more severe in the former than the latter. In one survey which was made of the meagre experimental data (Nicholls, 1962a), support was found for Bates' proposal.

The band absorption coefficient (eqn. (4.1.33)) has been used by a number of authors in recent years to provide information on band oscillator strengths, and in some cases, on behavior of $R_e(r)$ with r from a plot of $(\mu_\nu \text{ (band)}/\nu_{v'v''}q_{v'v''})^{1/2}$ vs. $\bar{r}_{v'v''}$. See, for example, Bethke (1959a, b), Treanor and Wurster (1960), Hasson, Hébert, and Nicholls (1970). One advantage of absorption work is that absolute absorption coefficients can be measured directly in most cases (when $N_v(L)$ is known). While much absorption work has been done on the $v'' = 0$ progression in cold gas samples, the technically more difficult time-resolved shock-tube spectroscopy of hot gases is being increasingly used. Band oscillator strengths are a commonly measured parameter in such work and more bands per system are in principle accessible than when using cold gas samples.

Equation (4.1.35) has been used during the past decade to interpret lifetimes $\tau_{v'}$ of vibrational levels in terms of $R_e(r)$. The lifetime is the reciprocal of the sum of Einstein A-coefficients of all (v', v'') transitions which combine with v'. Consequently, care has to be exercised when interpreting lifetime measurements that the sum is realistic, that no processes competing with radiation depopulate the level, and that it is not being populated by unsuspected cascade transitions.

Three techniques have been developed for the measurement of molecular lifetimes:

(a) The delayed coincidence method (Heron, McWhirter and Rhoderick 1954, 1956; Brannen *et al.*, 1955; Bennett and Dalby, 1959, 1960a, b; Dalby, 1964; Schwenker, 1965).

(b) The decay of discharge method (Jeunehomme, 1965; Wentink, 1964).

(c) The phase shift method (Demtroder, 1962; Brewer *et al.*, 1962; Fink and Welge, 1964; Hesser and Dressler, 1965).

The above references are typical, but by no means extensive. In the delayed coincidence technique the spectrum feature is repetitively excited by a chopped electron beam and detected through a monochromator with a photomultiplier. Electrical delays are inserted in the detector circuit and the lifetime is inferred from the slope of a semilogarithmic plot of coincidence (with respect to the exciting pulse) counting rate vs. delay time. This is a modification of a commonly used method in nuclear physics. Provided cascading and other secondary effects do not occur, it is a very accurate method and has been exploited to the full by Bennett and co-workers. The results are usually extrapolated to zero pressure to avoid the influence of secondary processes, and a dependence of $\tau_{v'}$ upon v' is often found, which is not surprising.

In the decay of radiation from a discharge, the decay of the radiation of a spectral feature from a pulsed discharge is studied. Care must be taken in interpretation of the results that a single atomic process (radiation) is being observed and not the cumulative effect of a number of overlapping processes in the discharge. The method is in principle quite simple.

In the phase-shift method, the spectral feature of interest is excited by a sinusoidally modulated electron beam of controllable frequency. The lifetime can be inferred from either a measurement of amplitude or phase (relative to the electron beam) of the output signal. The phase measurements are used in practical application. This method is being used with great success on molecules of astrophysical interest at Princeton observatory (Lawrence, 1965).

In the interpretation of all of these methods, some prior knowledge of the number of bands making a significant contribution to the sum over v'' of the $A_{v'v''}$ is required before $\tau_{v'}$ can be interpreted in terms of $R_e^2(\bar{r}_{v'v''})$. In simple cases (e.g. (0, 0) bands) of systems where no other contributors to the $v' = 0$ progression make a contribution,

$$\tau_{v'} = \frac{1}{A_{00}} \tag{4.4.3}$$

from which $R_e(\bar{r}_{v'v''})$ can be immediately inferred. In cases where there is more than one contributor the "branching ratios" have to be estimated, often approximately by use of the Franck–Condon factor ratios.

Lifetime measurements have been used in a number of cases to place relative band strengths on an absolute basis (Nicholls, 1964c).

Conclusion

IN THIS book we have attempted as far as possible to lay a foundation for the multifaceted subject of thermal radiation phenomena. Although the central ideas of the subject were orginally developed to describe the physics of stellar atmospheres, the concepts of thermal radiation phenomena have come to have a wide application to hot-gas phenomenology in general. In the past two decades they have been energetically applied (as we indicated in the introduction) to a very diverse range of phenomena related to terrestrial hot gases. They have been used to treat, for example, radiative properties of rocket engine exhausts, radiative heat load on re-entering satellites, radiation from, and absorption in many shock-wave situations, and the growth and decay phenomenology of nuclear fireballs. In many of these applications there is also strong coupling between the flow of radiation and the flow of gas. The methods of radiation gas-dynamics, which are closely related to thermal radiation phenomena, are used in these cases. In many of the problems of terrestrial hot-gas phenomena there is, in contrast to most astrophysical applications, lack of spatial homogeneity and variation of the parameters with time. The problems are thus often more complicated to formulate and to solve than the more idealized astrophysical ones. There are, of course, exceptions to this statement.

In all applications of thermal radiation phenomena *the* important parameter which controls much of the physics of the situation is the absorption coefficient of the material through which the radiation passes. In astrophysical applications usually only a few simple species (often atomic or ionic) and processes contribute to the total absorption coefficient. Its dominant spectral properties can thus often be relatively easily determined and realistic mean values of the coefficient are readily developed. In contrast, in most of the recently studied applications to terrestrially produced hot gases the situation is much more complex. The atomic and molecular species which contribute to the spectral absorption coefficient are nearly always more numerous in any particular problem. The number densities of the important species are often temperature- and time-dependent. The processes which contribute to the total spectral absorption coefficient are also often quite numerous. To determine realistic mean absorption coefficients in these cases implies that a complex total spectral absorption coefficient has first to be determined with some accuracy.

This is a fairly difficult problem, particularly in the range of temperatures (say 2000–25,000 °K) in which molecules can exist. The number of spectral transitions available to a molecular absorber is much greater than for an atomic absorber, and often less is known about each transition. Three types of information are needed to estimate the spectral absorption coefficient: (a) thermochemical and kinetic data to allow for accurate prediction of the abundances of absorbing species; (b) firm knowledge of the spectral features which will contribute to the absorption coefficient; (c) transition probability data for each spectral feature. Such data are used in the construction and use of extensive computer programs for generation of "synthetic" spectra. The need for such data has led to the critical evaluation of what data are available and their reliability. It came as a great surprise to many workers in the field to discover how much apparently "standard" information, even on location of spectral lines, was not available. In fact, the highly developed spectra of thermally excited gases, which are not necessarily similar to many commonly excited spectra of the conventional spectroscopic research laboratory, form an important aspect of study in plasma physics. Our spectroscopic knowledge of highly ionized species, of spectral line profiles and of molecular spectra under high-temperature conditions of high rotational development is somewhat meagre.

In a subsequent volume of this series we will examine in some detail specific applications of the material of this book to terrestrial thermal radiation phenomena, and in particular to thermal radiation in heated air. As will become clear from that volume, perhaps one of the most valuable contributions made by research activity in thermal radiation phenomena in recent years has been the great stimulus that it has given to the development of contemporary atomic and molecular physics, both experimental and theoretical. That is perhaps one of its major justifications.

References

The numbers in brackets after each reference refer to the pages on which the work is cited.

ABRAMOWITZ, M. and STEGUN, I.A., Editors (1964) *Handbook of Mathematical Functions*, National Bureau of Standards, Washington, D.C. [76, 90, 234.]

AIKEN, H.H. (1951) *Computation of the Intensities of Vibration Spectra of Electronic Bands in Diatomic Molecules*, Harvard Comp. Lab. Report No. 27, Cambridge, Mass. [253.]

ALDER, K., BOHR, A., HUUS, T., MOTTELSON, B. and WINTHER, A. (1956) *Rev. Mod. Phys.* **28**, 432. [201, 202–3, 205, 207, 209.]

ALLEN, C.W. (1963) *Astrophysical Quantities*, 2nd ed., The Athlone Press, Univ. of London. [27, 164, 166, 170–1, 173–4, 238, 243.]

ALLER, L.H. (1963) *Astrophysics. The Atmospheres of the Sun and Stars*, 2nd ed., Ronald Press Co., New York. [13, 18, 23, 27, 107, 109, 215.]

ALTICK, P.L. and GLASSGOLD, A.E. (1964) *Phys. Rev.* **133**, A632. [137.]

ARMSTRONG, B.H. (1959) *Proc. Phys. Soc. (London)* **74**, 136. [137.]

ARMSTRONG, B.H. (1962) *Astrophys. J.* **136**, 309. [53.]

ARMSTRONG, B.H. (1965c) *J. Quant. Spectrosc. and Rad. Trans.* **5**, 55 [*JQSRT* from here on]. [57.]

ARMSTRONG, B.H. (1967) *JQSRT* **7**, 61. [222, 224, 230, 234–5.]

ARMSTRONG, B.H. (1968a) *J. Atm. Sci.* **25**, 312. [61.]

ARMSTRONG, B.H. (1968b) *JQSRT* **8**, 1577. [83, 85–6, 88, 91–5, 101–2.]

ARMSTRONG, B.H. (1969a) *JQSRT* **9**, 1039. [35.]

ARMSTRONG, B.H. (1969b) *J. Atm. Sci.* **26**, 741. [102.]

ARMSTRONG, B.H., JOHNSTON, R.R. and KELLY, P.S. (1965) *JQSRT* **5**, 55. [10.]

ARMSTRONG, B.H., JOHNSTON, R.R., KELLY, P.S., DEWITT, H.E. and BRUSH, S.G. (1967) "Opacity of high-temperature air", in *Progress in High Temperature Physics and Chemistry*, Vol. I, ed. by CARL ROUSE, Pergamon Press, Oxford. [10, 11, 51, 54, 137, 163.]

ARMSTRONG, B.H., SOKOLOFF, J., NICHOLLS, R.W., HOLLAND, D.H. and MEYEROTT, R.E. (1961) *JQSRT* **1**, 143. [10, 11.]

ASHKIN, M. (1966) *Phys. Rev.* **141**, 41. [185, 196, 199, 213.]

BARANGER, M. (1962) *Spectral Line Broadening in Plasmas*, in *Atomic and Molecular Processes*, ed. by D.R. BATES, Academic Press, New York. [31, 215, 217.]

BATES, D.R. (1946) *Monthly Not. Roy. Astron. Soc.* **106**, 423, 432. [127, 137, 147, 153.]

BATES, D.R. (1949) *Proc. Roy. Soc. (London)* A **196**, 217. [265.]

BATES, D.R. (1952) *Monthly Not. Roy. Astron. Soc.* **112**, 614. [251.]

BATES, D.R. and DAMGAARD, A. (1949) *Phil. Trans. Roy. Soc. (London)* A **242**, 101. [10, 115, 121, 166, 208.]

BATES, D.R. and MASSEY, H.S.W. (1943) *Phil. Trans. Roy. Soc. (London)* A **239**, 269. [137.]

BATES, D.R. and SEATON, M.J. (1949) *Monthly Not. Roy. Astron. Soc.* **109**, 698. [137.]

BATES, D.R., BUCKINGHAM, R.A., MASSEY, H.S.W. and UNWIN, J.J. (1939) *Proc. Roy. Soc. (London)* A **170**, 322. [137.]

BEKEFI, G. (1966) *Radiation Processes in Plasmas*, John Wiley & Sons, Inc., New York. [20–1, 184, 206.]

BENNETT, R.G. and DALBY, F.W. (1959) *J. Chem. Phys.* **31**, 434. [266.]

BENNETT, R.G. and DALBY, F.W. (1960a) *J. Chem. Phys.* **32**, 1111. [266.]

BENNETT, R.G. and DALBY, F.W. (1960b) *J. Chem. Phys.* **32**, 1716. [266.]

BENNETT, R.J.M. and SOMERVILLE, W.B. (1969) *Nature* **223**, 489. [254.]

BERGER, J.M. (1956) *Astrophys. J.* **124**, 550. [184–5.]

BERGER, J.M. (1957) *Phys. Rev.* **105**, 35. [184–5.]

BERNSTEIN, J. and DYSON, F.J. (1959) *The Continuous Opacity and Equations of State of Light Elements at Low Densities*, General Atomic Report, BA-848, General Atomic Division of General Dynamics Corp., San Diego, Calif. [52.]

BETHE, H.A. and JACKIW, R.W. (1968) *Intermediate Quantum Mechanics*, W.A. Benjamin, Inc., New York. [136.]

BETHE, H. and PLACZEK, G. (1937) *Phys. Rev.* **51**, 450. [223.]

BETHE, H.A. and SALPETER, E.E. (1957) *Quantum Mechanics of One- and Two-electron Systems*, Academic Press, New York. [52, 56, 112, 121, 126, 133, 135–7, 143, 149, 157–8, 161.]

BETHKE, G.W. (1959a) *J. Chem. Phys.* **31**, 669. [265.]

BETHKE, G.W. (1959b) *J. Chem. Phys.* **31**, 662. [265.]

BIBERMAN, L.M. and NORMAN, G.E. (1963) *JQSRT* **3**, 221. [10.]

BIEDENHARN, L.C. (1956) *Phys. Rev.* **102**, 262. [203.]

BIEDENHARN, L.C. and VAN DAM, H. (1965) *Quantum Theory of Angular Momentum*, Academic Press, New York. [183.]

BLATT, J.M. and WEISSKOPF, V.F. (1952) *Theoretical Nuclear Physics*, John Wiley and Sons, Inc., New York. [113, 155.]

BLUMENTHAL, G.R. and GOULD, R.J. (1970) *Rev. Mod. Phys.* **42**, 237. [184.]

BOND, J.W., Jr., WATSON, K.M. and WELCH, J.A., Jr. (1965) *Atomic Theory of Gas Dynamics*, Addison-Wesley, Reading, Mass. [12, 18, 52, 81.]

BORN, M. (1933) *Optik*, Julius Springer, Berlin. Photo-lithoprint reproduction by Edwards Brothers, Inc., Ann Arbor, Mich., 1943. [220–1, 227, 229.]

BORN, M. (1946) *Atomic Physics*, Hafner Publishing Co., Inc., New York. [43.]

BORN, M. and OPPENHEIMER, J.R. (1927) *Annalen der Physik* **84**, 457. [8, 239.]

BORN, M. and WOLF, E. (1959) *Principles of Optics*, Pergamon Press, London. [20.]

BRANNEN, E., HUNT, F.R., ADLINGTON, R.H. and NICHOLLS, R.W. (1955) *Nature* **175**, 810. [266.]

BREENE, R.G., Jr., and NARDONE, M.C. (1960) *J. Opt. Soc. Am.* **50**, 1111. [184–5.]

BREENE, R. G., Jr., and NARDONE, M. C. (1961) *J. Opt. Soc. Am.* **51**, 692. [184–5.]

BREENE, R. G. and NARDONE, M. C. (1962) *JQSRT* **2**, 273. [10.]

BREENE, R. G., Jr., and NARDONE, M. C. (1963) *J. Opt. Soc. Am.* **53**, 924–928. [184–5.]

BREWER, L., JAMES, C. G., BREWER, R. G., STAFFORD, F. E., BERG, R. H. and ROSEN, G. M. (1962) *Rev. Sci. Instr.* **33**, 1450. [266.]

BRUSSARD, P. J. and VAN DE HULST, H. C. (1962) *Rev. Modern Phys.* **34**, 507. [184–5, 204, 206.]

BUDO, A. (1935) *Z. Physik* **96**, 219. [248.]

BUDO, A. (1936) *Z. Physik* **98**, 437. [248.]

BUDO, A. (1937) *Z. Physik* **105**, 579. [248.]

BURGESS, A. (1958) *Monthly Not. Roy. Astron. Soc.* **118**, 477. [204, 207.]

BURGESS, A. (1964) *Memoirs of the Roy. Astron. Soc.* **69**, Part I. [137.]

BURGESS, A. and SEATON, M. J. (1960) *Monthly Not. Roy. Astron. Soc.* **120**, 121. [10, 127, 137, 158, 163, 171, 208.]

BURKE, P. G. and SMITH, K. (1962) *Rev. Mod. Phys.* **34**, 458. [200.]

CARSON, T. R., MAYERS, D. F. and STIBBS, D. W. N. (1968) *Monthly Not. Roy. Astron. Soc.* **140**, 483. [11.]

CASE, K. M., DE HOFFMANN, F. and PLACZEK, G. (1953) *Introduction to the Theory of Neutron Diffusion*, Los Alamos Sci. Lab. Report, Los Alamos, New Mexico (June, 1953). [72, 74, 75.]

CASHION, J. K. (1963) *J. Chem. Phys.* **39**, 1872. [255.]

CHANDRASEKHAR, S. (1939) *An Introduction to the Study of Stellar Structure*, Univ. of Chicago Press, Chicago. [13, 18, 23–4, 26, 44.]

CHANDRASEKHAR, S. (1950) *Radiative Transfer*, Oxford Univ. Press. [13, 17–8, 23, 30, 35, 60, 63.]

CHANDRASEKHAR, S. and BREEN, F. (1946) *Astrophys. J.* **104**, 430. [185, 212.]

CHIARELLA, C. and REICHEL, A. (1968) *Math. Comp.* **22**, 137. [235–7.]

CHURCHILL, D. R., ARMSTRONG, B. H., JOHNSTON, R. R. and MULLER, K. G. (1966) *JQSRT* **6**, 371. [10, 246.]

CHURCHILL, D. R. and MEYEROTT, R. E. (1965) *JQSRT* **5**, 69. [10.]

CLEMMOW, P. C. (1966) *The Plane Wave Representation of Electromagnetic Fields*, Pergamon Press, Oxford. [231.]

CLEMMOW, P. C. and MUMFORD, C. M. (1952) *Phil. Trans. Roy. Soc.* A **245**, 189. [221.]

CODY, W. J., PACIOCEK, K. A. and THACHER, H. C., Jr., (1970) *Math. Comp.* **24**, 171. [230.]

COGLEY, A. C. (1970) *JQSRT* **10**, 1065. [79.]

COMPTON, A. H. (1926) *X-Rays and Electrons*, D. Van Nostrand Co., New York. [112.]

CONDON, E. U. (1926) *Phys. Rev.* **28**, 1182. [251, 259.]

CONDON, E. U. (1928) *Phys. Rev.* **32**, 858. [251, 253.]

CONDON, E. U. (1947) *Am. J. Phys.* **15**, 365. [252.]

CONDON, E. U. (1968) *JQSRT* **8**, 369. [21.]

CONDON, E. U. and SHORTLEY, G. H. (1935) *The Theory of Atomic Spectra*, Cambridge Univ. Press (reprinted, 1957). [126, 135, 163–4, 168, 171–2, 177.]

COOLEY, J. W. (1961a) *Maths. Comput.* **15**, 363. [255.]

COOLEY, J. W. (1961b) AEC Report No. NYO-9490 (May 1, 1961). [255.]

COOLIDGE, A. S., JAMES, H. M. and PRESENT, R. D. (1936) *J. Chem. Phys.* **4**, 193. [258.]

COOPER, J.W. (1962) *Phys. Rev.* **128**, 681. [137.]

COOPER, J. (1966) "Plasma spectroscopy", in *Reports on Progress in Physics* **29**, 35–130. [12, 215, 217.]

COOPER, JOHN W. and MARTIN, J.B. (1962) *Phys. Rev.* **126**, 1482. [137.]

COPSON, E.T. (1935) *Theory of Functions of a Complex Variable*, Oxford Univ. Press, London. [236.]

COX, A.N. (1964) *JQSRT* **4**, 737. [10.]

COX, A.N. (1965) "Stellar absorption coefficients and opacities", Chap. 3 in *Stellar Structure* (Vol. VIII of *Stars and Stellar Systems*), ed. by L.H. ALLER and D.B. MCLAUGHLIN, Univ. of Chicago Press, Chicago. [10.]

COX, A.N. and STEWART, J.N. (1970) *Astrophys. J. Suppl.* **19**, 243. [11.]

CRAGGS, J.D. and MASSEY, H.S.W. (1959) The collision of electrons with molecules, *Handbuch der Physik* 37/1, 314, S. FLÜGGE, Ed., Springer, Berlin. [254.]

CUNIO, B.E. and JANSSON, R.E.W. (1968) *JQSRT* **8**, 1763. [256.]

DALBY, F.W. (1964) Transition probabilities of molecular spectra, *Handbuch der Physik* **27**, 464, S. FLÜGGE, Ed., Springer, Berlin. [266.]

DALGARNO, A. (1969) "Radiative transitions", in *Atomic Physics*, ed. by V.W. HUGHES, B. BEDERSON, V.W. COHEN and F.M.J. PICHANICK, Plenum Press, New York. [163.]

DALGARNO, A. and LANE, N. (1966) *Astrophys. J.* **145**, 623. [214.]

DALGARNO, A. and PARKINSON, D. (1960) *J. Atmos. and Terr. Phys.* **18**, 335. [137.]

DALGARNO, A., HENRY, R.J.W. and STEWART, A.L. (1964) *Planetary and Space Science* **12**, 235. [137.]

DAVIDSON, N. (1962) *Statistical Mechanics*, McGraw-Hill, New York. [111, 244.]

DAVIS, H.J. (1964) *Thermal Radiation from High-temperature Air*, Harry Diamond Laboratories Technical Report TR-1225 (May 1964). [62–3.]

DEMTRODER, W. (1962) *Z. Physik* **166**, 42. [266.]

DENNISON, D.M. (1926) *Phys. Rev.* **28**, 318. [241, 248.]

DEVORE, R.V. (1964) *Phys. Rev.* **136**, A666; with errata (1965) *ibid.* **140**, AB3. [185, 192, 207, 211, 213.]

DIRAC, P.A.M. (1958) *The Principles of Quantum Mechanics*, 4th ed., Clarendon Press, Oxford. [112, 135.]

DITCHBURN, R.W. and ÖPIK, U. (1962) "Photoionization processes", in *Atomic and Molecular Processes*, ed. by D.R. BATES, Academic Press, New York and London. [137, 238.]

DITCHBURN, R.W. and YOUNG, P.A. (1962) *J. Atmos. and Terr. Phys.* **24**, 127. [258.]

DOLGINOV, A.Z., GNEDIN, YU.N. and SILANT'EV, N.A. (1970) *JQSRT* **10**, 707. [22.]

DOYLE, R.O. (1968) *JQSRT* **8**, 1555. [258.]

DRAKE, J. and NICHOLLS, R.W. (1970) *Chem. Phys. Letters* **3**, 457. [242, 260, 262.]

DRESNER, L. (1960) *Resonance Absorption in Nuclear Reactors*, Pergamon Press, Oxford. [222.]

DUNHAM, J.L. (1932) *Phys. Rev.* **41**, 713, 721. [254.]

EARLS, L.T. (1935) *Phys. Rev.* **48**, 423. [248.]

EDDINGTON, A.S. (1926) *The Internal Constitution of the Stars*, Cambridge Univ. Press (Dover Reprint, 1959). [7, 18.]

EDMONDS, A.R. (1957) *Angular Momentum in Quantum Mechanics*, Princeton Univ. Press. [171, 176–7, 179, 183, 193, 195.]

EDMONDS, F.N., Jr. (1968) *JQSRT* **8**, 1447. [220.]

ELSASSER, W.M. (1938) *Phys. Rev.* **54**, 126. [9, 81.]

ELSASSER, W. (1942) *Heat Transfer by Infrared Radiation in the Atmosphere*, Harvard Meteorological Studies No.6, Harvard Univ. Blue Hill Meteor. Obs., Milton, Mass. [18, 63, 83, 85, 99.]

ELWERT, G. (1939) *Ann. Physik* **34**, 178. [205.]

ÉNOMÉ, S. (1969) *Publications of the Astronomical Society of Japan* **21**, 367. [20.]

EPSTEIN, P.S. (1930) *Proc. Natl. Acad. Science (U.S.)* **16**, 37. [20.]

ERDELYI, A., MAGNUS, W., OBERBETTINGER, F. and TRICOMI, F.G. (1953) *Higher Transcendental Functions*, Vol.II, McGraw-Hill Book Co., Inc., New York. [222.]

FADDEYEVA, V.N. and TERENT'EV, N.M. (1961) *Tables of the Probability Integral for Complex Argument*, Pergamon Press, Oxford. [229, 231, 234.]

FERMI, E. (1924) *Z. Physik* **26**, 54. [8.]

FERMI, E. (1950) *Nuclear Physics*, Notes compiled by J. OREAR, A.H. ROSENFELD and R.A. SCHLUTER, Univ. of Chicago Press, Chicago. [113.]

FEYNMAN, R.P. (1962) *Quantum Electrodynamics*, W.A.Benjamin, Inc., New York, second printing with corrections (first printing, Dec. 1961). [113.]

FINK, E. VON and WELGE, K.H. (1964) *Z. Naturf.* **19A**, 1193. [266.]

FINN, G.D. and MUGGLESTONE, D. (1965) *Monthly Not. Roy. Astr. Soc.* **129**, 221. [226–7, 234.]

FIRSOV, O.B. and CHIBISOV, M.I. (1961) *Soviet Physics JETP* **12**, 1235. [196, 211.]

FOCK, V.A. (1965) *Electromagnetic Diffraction and Propagation Problems*, Pergamon Press, Oxford. [221.]

FOLDY, L.L. (1945) *Phys. Rev.* **67**, 107. [21.]

FOWLER, R.H. (1936) *Statistical Mechanics*, 2nd ed., Cambridge Univ. Press. [8.]

FRANCK, J. (1925) *Trans. Faraday Soc.* **21**, 536. [251.]

FRANKEN, P.A. (1969) "Collisions of light with atoms", in *Atomic Physics*, ed. by V.W. HUGHES, B. BEDERSON, V.W. COHEN and F.M.J. PICHANICK, Plenum Press, New York. [131.]

FRANKLIN, PHILIP (1940) *A Treatise on Advanced Calculus*, John Wiley & Sons, Inc., New York. [91.]

FRASER, P.A. (1954a) *Can. J. Phys.* **32**, 515. [242, 254, 260, 264.]

FRASER, P.A. (1954b) *Proc. Phys. Soc. (London)* A **67**, 939. [242, 256, 260.]

FRASER, P.A. (1958) *The influence on Vibration–rotation Interaction on the Intensities of Diatomic Molecular Lines and Bands*, Sci. Rep. No.6, Contract AF 19(604)-1718, Univ. of Western Ontario, London, Ontario. [256.]

FRASER, P.A. and JARMAIN, W.R. (1953) *Proc. Phys. Soc. (London)* A **66**, 1145. [254.]

FRASER, P.A., JARMAIN, W.R. and NICHOLLS, R.W. (1954) *Astrophys. J.* **119**, 286. [254.]

FREEMAN, B.E. (1963) *Opacity and Absorption Coefficients for Ionic Air*, General Atomic Report, GAMD-4566, General Atomic Division, General Dynamics Corp., San Diego, Calif. [51.]

FRIED, B.D. and CONTE, S.D. (1961) *The Plasma Dispersion Function*, Academic Press, New York. [220, 233, 234.]

GAUNT, J.A. (1928) *Trans. Roy. Soc. (London)* A **228**, 195. [163.]

GAUNT, J.A. (1930) *Phil. Trans. Roy. Soc. (London)* A **229**, 163 (see also *Proc. Roy. Soc.* A **126**, 654, 1930). [7, 137, 184–5, 196, 202.]

GAUTSCHI, W. (1964) "Error function and Fresnel integrals", in *Handbook of Mathematical Functions*, ed. by M. ABRAMOWITZ and I. STEGUN, National Bureau of Standards, Applied Math. Series No. 55, Washington, D.C. [230, 233.]

GAUTSCHI, WALTER (1969) *Comm. ACM* **12**, 635. [234–5.]

GAUTSCHI, WALTER (1970) *SIAM J. Numer. Anal.* **7**, 187. [235, 237.]

GAYDON, A. G. (1968) *Dissociation Energies*, Chapman & Hall, 3rd edition (2nd edition, 1953). [254.]

GAYDON, A. G. and PEARSE, R. W. B. (1939) *Proc. Roy. Soc. (London)* A **173**, 37. [253.]

GILMORE, F. R. (1965a) *JQSRT* **5**, 125. [255.]

GILMORE, F. R. (1965b) *JQSRT* **5**, 369. [254–5.]

GILMORE, F. R. (1967) *The Equilibrium Thermodynamics Properties of High Temperature Air*, DASA Report 1971-1, Vol. 1 (May 1967), Lockheed Research Laboratories, Palo Alto, Calif. [4, 248, 255.]

GODSON, W. L. (1955) *J. Meteor.* **12**, 272 and 533. [87.]

GOLDBERGER, M. L. and WATSON, K. M. (1964) *Collision Theory*, John Wiley & Sons, Inc., New York. [22, 114, 155, 185, 190, 197.]

GOLDEN, S. A. (1967) *JQSRT* **7**, 483. [82.]

GOLDEN, S. A. (1968) *JQSRT* **8**, 877. [82.]

GOLDEN, S. A. (1969) *JQSRT* **9**, 1067. [82.]

GOODY, R. M. (1952) *Quart. J. Roy. Met. Soc.* **78**, 165. [9, 81.]

GOODY, R. M. (1964) *Atmospheric Radiation, Theoretical Basis*, Oxford Univ. Press. [1–2, 11, 13, 18–9, 44, 61, 63, 65, 77, 80, 81, 83, 87–8, 90, 92, 94, 97.]

GORDON, W. (1929) *Ann. Physik* (5) **2**, 1031. [207.]

GOULD, R. J. (1970) *Am. J. Phys.* **38**, 189. [184–5.]

GRANT, I. P. (1958) *Monthly Not. Roy. Astron. Soc.* **118**, 241. [184–5, 201, 203–5.]

GREEN, J. M. (1958) *The Free–Free Gaunt Factor in an Ionized Medium*, Rand Research Memorandum RM-2223-AEC (July, 1958). The Rand Corp., Santa Monica, Calif. [210.]

GREENE, J. (1959) *Astrophys. J.* **130**, 693. [185.]

GRIEM, H. R. (1964) *Plasma Spectroscopy*, McGraw-Hill Book Co., New York. [12, 112, 122, 215, 217.]

HAGAN, L. (1963) Ph.D. Thesis, University of California, UCRL 19620. [264.]

HALEVI, P. (1965) *Proc. Phys. Soc. (London)* **86**, 1051. [261.]

HALL, H. (1936) *Rev. Mod. Phys.* **8**, 358. [137.]

HALMANN, M. and LAULICHT, I. (1965) *J. Chem. Phys.* **42**, 137. [254.]

HALMANN, M. and LAULICHT, I. (1968) *JQSRT* **8**, 935. [256.]

HAMMING, R. W. (1962) *Numerical Methods for Scientists and Engineers*, McGraw-Hill Book Co., Inc., New York. [236.]

HANSEN, C. F. (1964) *J. Opt. Soc. Am.* **54**, 1198. [217.]

HARTREE, D. R. (1957) *The Calculation of Atomic Structures*, John Wiley & Sons, Inc., New York. [11, 115, 163.]

HASSON, V., HÉBERT, G. R., and NICHOLLS. R. W. (1970) *J. Phys. B. Ser. 2* **3**, 1188. [258, 265.]

HASSON, V., NICHOLLS, R. W. and DEGEN, V. (1970) *J. Phys. B. Ser. 2* **3**, 1192. [258, 265.]

HAYCOCK, Miss S. (1963) *Line Intensities in Diatomic Electronic Spectra. The Effect of*

Vibration–Rotation Interaction, M.Sc. Thesis, Univ. of Western Ontario, London, Ontario. [256.]

HAZLEHURST, J. and SARGENT, W.L.W. (1959) *Astrophys. J.* **130**, 276. [44.]

HEITLER, W. (1954) *The Quantum Theory of Radiation*, 3rd ed., Clarendon Press, Oxford. [23, 30, 103–6, 111–3, 117, 131, 133, 137, 139, 188, 216.]

HERMAN, R.C. and RUBIN, R. (1955) *Astrophys. J.* **121**, 2. [256.]

HERMAN, R.C., ROTHERY, R. and RUBIN, R. (1958) *J. Molec. Spectrosc.* **2**, 369. [256.]

HERON, S., MCWHIRTER, R.W.P. and RHODERICK, E.H. (1954) *Nature* **174**, 564. [266.]

HERON, S., MCWHIRTER, R.W.P. and RHODERICK, E.H. (1956) *Proc. Roy. Soc. (London)* **A 234**, 565. [266.]

HERZBERG, G. (1950) *Molecular Spectra and Molecular Structure, I. Spectra of Diatomic Molecules*, D.Van Nostrand, Princeton, New Jersey. [238, 244, 248–9, 251, 255.]

HESSER J.E. and DRESSLER, K. (1965) *Astrophys. J.* **142**, 389. [266.]

HILL, E.L. and VAN VLECK, J.H. (1928) *Phys. Rev.* **32**, 250. [248.]

HILLENDAHL, R.W. (1969) *Astrophys. Letters* **4**, 179. [60.]

HIRSCHFELDER, J.O. and MAGEE, J.L. (1945) *Opacity and Thermodynamic Properties of Air at High Temperatures*, Los Alamos Scientific Laboratory Report 296, Los Alamos, New Mexico. [9.]

HITSCHFELD, W. and HOUGHTON, J.T. (1961) *Quart. J. Roy. Met. Soc.* **87**, 569. [83.]

HOLSTEIN, T. (1965) Westinghouse Scientific Paper, 65-1E2-GASES-P2 (unpublished). [213.]

HÖNL, H. and LONDON, F. (1925) *Z. Physik* **33**, 803. [241, 248.]

HOUGEN, J.T. (1970) *The Calculation of Rotational Energy Levels and Rotational Line Intensities in Diatomic Molecules*, NBS Monograph No.115, National Bureau of Standards, Washington, D.C. (issued June 1970). [241, 249.]

HOWELL, K.M. (1959) *Revised Tables of 6-j Symbols*, Research Report 59-1, University of Southampton. [179.]

HUBISZ, J. (1968) Ph.D. Thesis, York Univ., Toronto. [255–7.]

HUEBNER, W.F. (1964) *JQSRT* **4**, 753–60. [51.]

HUMMER, D.G. (1964) *The Voigt Function: an Eight Significant-Figure Table and Generating Procedure*, Joint Institute for Laboratory Astrophysics Report No.24 (Nov. 1964), Univ. of Colorado, Boulder, Colo. [234.]

HUMMER, D.G. (1965) *Mem. Roy. Astr. Soc.* **70**, 1. [234.]

HUNDLEY, R.O. (1962) *Bremsstrahlung During the Collision of Low-Energy Electrons with Neutral Atoms and Molecules*, Rand Research Memorandum RM-3334-ARPA (Oct. 1962). The Rand Corp., Santa Monica, Calif. [196, 214.]

HURLEY, A.C. (1962) *J. Chem. Phys.* **36**, 1117. [254.]

JAMES, T.C. (1960) *J. Chem. Phys.* **32**, 1770. [256.]

JAMES, T.C. (1961) *J. Chem. Phys.* **35**, 767. [256.]

JAMES, T.C. (1966) *J. Molec. Spectrosc.* **20**, 77. [260–1.]

JANSSON, P.A. and KORB, C.L. (1968) *JQSRT* **8**, 1399. [79.]

JANSSON, R.E.W. and CUNIO, B.E. (1968) *JQSRT* **8**, 1747. [261.]

JARMAIN, W.R. (1959) *J. Chem. Phys.* **31**, 1137. [254.]

JARMAIN, W.R. (1960) *Can. J. Phys.* **38**, 217. [254.]

JARMAIN, W.R. (1961) *Numerical Solution of the Schroedinger Equation*, Scientific

Report 2, Contract AF 19(604)-4560. Univ. of Western Ontario, London, Ontario. [255.]
JARMAIN, W.R. (1963a) *Can. J. Phys.* **41**, 414. [255.]
JARMAIN, W.R. (1963b) *Can. J. Phys.* **41**, 1926. [255.]
JARMAIN, W.R. and FRASER, P.A. (1953) *Proc. Phys. Soc. (London)* A **66**, 1153. [254.]
JARMAIN, W.R. and NICHOLLS, R.W. (1954) *Can. J. Phys.* **32**, 201. [253.]
JARMAIN, W.R. and NICHOLLS, R.W. (1964) *Proc. Phys. Soc. (London)* **84**, 417. [257–8.]
JARMAIN, W.R. and NICHOLLS, R.W. (1967) *Proc. Phys. Soc. (London)* **90**, 545. [258.]
JARMAIN, W.R., FRASER, P.A. and NICHOLLS, R.W. (1953) *Astrophys. J.* **118**, 228. [254.]
JARMAIN, W.R., FRASER, P.A. and NICHOLLS, R.W. (1955) *Astrophys. J.* **122**, 55. [254.]
JEANS, J.H. (1926a) *Monthly Not. Roy. Astron. Soc.* **86**, 328. [44.]
JEANS, J.H. (1926b) *Monthly Not. Roy. Astron. Soc.* **86**, 444. [44.]
JEFFERIES, J.T. (1968) *Spectral Line Formation*, Ginn-Blaisdell, New York. [215.]
JEUNEHOMME, M. (1965) *J. Chem. Phys.* **42**, 4086. [266.]
JEVONS, W. (1932) *Band Spectra of Diatomic Molecules*, The Physical Society, London. [249–50.]
JOHN, T.L. (1964) *Monthly Not. Roy. Astron. Soc.* **128**, 93. [209, 712.]
JOHN, T.L. (1966) *Monthly Not. Roy. Astron. Soc.* **131**, 315. [185, 197, 199, 209, 212.]
JOHNSON, R.C. (1949) *An Introduction to Molecular Spectra*, Methuen, London. [249.]
JOHNSTON, R.R. (1964) *Phys. Rev.* **136**, A958. [137.]
JOHNSTON, R.R. (1967) *JQSRT* **7**, 815. [183.]
JOHNSTON, R.R. and PLATAS, O.R. (1969) *Atomic Lines and Radiation From High Temperature Air*, Lockheed Missiles and Space Co. Technical Report N-3L-69-1, Palo Alto, Calif. (July 15, 1969). [6, 72.]
JOHNSTON, R.R., ARMSTRONG, B.H. and PLATAS, O.R. (1965) *JQSRT* **5**, 49. [10.]
JUDD, B.R. (1963) *Operator Techniques in Atomic Spectra*, McGraw-Hill Book Co. Inc., New York. [163.]
KAPLAN, L.D. (1952) *J. Meteor.* **9**, 139. [83.]
KARPOV, K.A. (1965) *Tables of the Function* $w(z) = e^{-z^2} \int_0^t e^{x^2} dx$ *in the Complex Domain*, Pergamon Press, Oxford. [234.]
KARZAS, W.J. and LATTER, R. (1961) *Astrophys. J. Suppl.* **6**, 167. [137, 184–5, 201, 203, 206.]
KELLER, G. and MEYEROTT, R.E. (1955) *Astrophys. J.* **122**, 32. [10.]
KELLY, P.S. (1959) *Techniques for Calculation of F.P.C. and Nuclear Matrix Elements*, Technical Report, Dept. of Physics, Univ. of Calif. at Los Angeles. [163.]
KELLY, P.S. (1964a) *Astrophys. J.* **140**, 1247. [115, 182.]
KELLY, P.S. (1964b) *JQSRT* **4**, 117. [115.]
KENNARD, E.H. (1938) *Kinetic Theory of Gases*, McGraw-Hill Book Co., Inc., New York. [41, 43.]
KERKER, M. (1969) *The Scattering of Light and Other Electromagnetic Radiation*, Academic Press, New York and London. [30.]
KHARE, S. and RUDGE, M. (1965) *Proc. Phys. Soc. (London)* **86**, 355. [213.]
KING, R.B. (1948) *Astrophys. J.* **108**, 87. [264.]
KIVEL, B. (1967) *JQSRT* **7**, 27. [213.]
KIVEL, B. and MAYER, H. (1965/1954) *JQSRT* **5**, 13. [9.]
KLEIN, O. (1932) *Z. Physik* **76**, 221. [254.]

KONDRAT'YEV, K. YA. (1965) *Radiative Heat Exchange in the Atmosphere*, Pergamon Press, Oxford and New York. [11, 83, 92.]

KONDRAT'YEV, K. YA. (1969) *Radiation in the Atmosphere*, Academic Press, New York and London. [11.]

KOURGANOFF, V. (1952) *Basic Methods in Transfer Problems*, Oxford Univ. Press. [13, 18, 23, 39.]

KOVACS, I. (1960) *Can. J. Phys.* **38**, 955. [249.]

KOVACS, I. (1969) *Rotational Structure in the Spectra of Diatomic Molecules*, Adam Hilger, London. American ed., American Elsevier Publishing Co., Inc., New York. [238, 240–1, 249–50.]

KRAMERS, H.A. (1923) *Phil. Mag.* **46**, 836. [7, 137, 184–5, 202.]

KRAMERS, H.A. (1958) *Quantum Mechanics*, North-Holland Publishing Co., Amsterdam. Translated by D. TER HAAR. [103.]

KRONIG, R.L. and RABI, I.I. (1927) *Phys. Rev.* **29**, 262. [241, 248.]

KYLE, T.G. (1967) *Astrophys. J.* **148**, 845. [82.]

KYLE, T.G. (1968) *JQSRT* **8**, 1455. [79.]

LADENBURG, R. and REICHE, F. (1913) *Ann. Phys.* **42**, 181. [79.]

LANDAU, L.D. and LIFSHITZ, E.M. (1959) *Quantum Mechanics Non-relativistic Theory*, Pergamon Press, London. [238.]

LANDAU, L.D. and LIFSHITZ, E.M. (1962) *The Classical Theory of Fields*, 2nd ed., Section 70, Addison-Wesley, Reading, Mass. [202.]

LANDSHOFF, R.K.M. and MAGEE, J.L., editors (1969) *Thermal Radiation Phenomena*, Vol. I, *Radiative Properties of Air*, IFI/Plenum Press, New York and Washington. [10.]

LAWRENCE, G.M. (1965) *JQSRT* **5**, 359. [267.]

LAX, M. (1951) *Rev. Mod. Phys.* **23**, 287. [21–2, 30.]

LEARNER, R.C. M. (1962) *Proc. Roy. Soc. (London)* **A 269**, 311. [256.]

LEVIN, J.M. (1967) *Phys. Fluids* **10**, 1298 and 1307. [220.]

LEVINSON, I.B. and NIKITIN, A.A. (1965) *Handbook for Theoretical Computation of Line Intensities in Atomic Spectra* (Russian translation by Israel Program for Scientific Translations), Daniel Davey & Co., Inc., 257 Park Ave. South, New York. [163, 164, 173, 175.]

LIBERMAN, D. (1962) *Upper Limits on the Rosseland Mean Opacity*, Los Alamos Report LA-2700. [56.]

LIGHTHILL, M. (1959) *Introduction to Fourier Analysis and Generalized Functions* (Cambridge University Press). [260.]

LOW, F.E. (1958) *Phys. Rev.* **110**, 974. [196.]

MAGEE, J.L. and HIRSCHFELDER, J.O. (1958/1947) *Thermal Radiation Phenomena*, LA-1020 Chapter IV (unclassified). Reproduced in Los Alamos Report LA-2000, *Blast Wave*, by H.A. BETHE, K. FUCHS, J.O. HIRSCHFELDER, J.L. MAGEE, R.E. PIERLS and J. VON NEUMANN. Los Alamos Scientific Laboratory, Los Alamos, New Mexico. [9.]

MALKMUS, W. (1967) *J. Opt. Soc. Am.* **57**, 323. [82.]

MANNEBACK, C. (1951) *Physica* **17**, 1001. [253.]

MARR, G.V. (1967) *Photoionization Processes in Gases*, Academic Press, New York and London. [136–7, 238.]

MARSHAK and BETHE (1940) *Astrophys. J.* **91**, 239. [8.]
MASSEY, H.S.W. and BURHOP, E.H.S. (1969) *Electronic and Ionic Impact Phenomena*, Vol. 1, 2nd ed., Oxford Univ. Press. [32, 198.]
MASSEY, H.S.W. and SMITH, R. (1936) *Proc. Roy. Soc. (London)* A **155**, 472. [137.]
MAUE, A. (1932) *Ann. Phys.* **13**, 161. [137.]
MAYER, H. (1964) *JQSRT* **4**, 585 (1st Opacity Conf. Issue). [10, 66.]
MAYER, H. (1947) Methods of Opacity Calculations, Los Alamos Scientific Laboratory Report LA 647, unpublished. [8–9, 13, 18, 25, 31–2, 47, 49, 50, 69, 81, 112, 114, 131, 185.]
McCALLUM, J.C., NICHOLLS, R.W. and JARMAIN, W.R. (1970) *Franck–Condon Factors and Related Quantities for Diatomic Molecular Band Systems, CRESS* Spectroscopic Report No. 1, York Univ., Toronto. [255.]
MENZEL, D. and PEKERIS, C. (1935) *Monthly Not. Roy. Astron. Soc.* **96**, 77. [8, 137, 184–5, 204.]
MEYEROTT, R.E. (1956) *The Threshold of Space*, M. ZELIKOFF, ed., Pergamon Press, New York and London, p. 259. [10.]
MIHALAS, D.M. and MORTON, D.C. (1965) *Astrophys. J.* **142**, 253. [11.]
MILNE, E.A. (1924) *Phil. Mag.* **47**, 209. [7, 137, 146.]
MILNE, E.A. (1930) Thermodynamics of the stars, in *Handbuch der Astrophysik*, Band III, Erste Hälfte, herausgegeben von G. EBERHARD, A. KOHLSCHUTTER and H. LUDENDORFF, Springer Verlag, Berlin. Reprinted in *Selected Papers on the Transfer of Radiation*, ed. by D.H. MENZEL, Dover Publications, New York, 1966. [18, 44, 106.]
MITCHELL, A.C.G. and ZEMANSKY, M.W. (1934) *Resonance Radiation and Excited Atoms*, Cambridge Univ. Press, Cambridge (reprinted, 1961). [218, 226.]
MIZUSHIMA, M. (1967) *JQSRT* **7**, 505. [219–20.]
MIZUSHIMA, M. (1970) *Quantum Mechanics of Atomic Spectra and Atomic Structure*, W.A. Benjamin, Inc., New York. [163.]
MJOLSNESS, R.C. and RUPPEL, H. (1967) *Phys. Rev.* **154**, 98. [200.]
MOISEIWITSCH, B. (1963) *Proc. Phys. Soc. (London)* **81**, 35. [214.]
MORSE, P.M. (1929) *Phys. Rev.* **34**, 57. [253.]
MORSE, P.M. (1940) *Astrophys. J.* **92**, 27. [8.]
MORSE, P.M. and FESHBACH, H. (1953) *Methods of Theoretical Physics*, McGraw-Hill Book Co., Inc., New York. [13, 43, 229, 234.]
MORSE, P.M. and INGARD, K.V. (1961) Linear Acoustic Theory in *Akustik I*, Vol. XI/1 of *Handbuch der Physik* (S. FLÜGGE, ed.), Springer-Verlag, Berlin. [221.]
MOTT, N.F. and MASSEY, H.S.W. (1965) *The Theory of Atomic Collisions*, Clarendon Press, Oxford, 3rd ed. [137, 151.]
MULLIKEN, R.S. (1931) *Rev. Mod. Phys.* **3**, 89. [249.]
MURTY, M. (1964) *Condon Loci of Diatomic Molecular Spectra*, M. Sc. Thesis, University of Western Ontario, London, Ontario. [259.]
MURTY, M. and NICHOLLS, R.W. (1967) *Nature* **213**, 1009. [259.]
NICHOLLS, R.W. (1950) *Phys. Rev.* **77**, 44. [253.]
NICHOLLS, R.W. (1960) *Can. J. Phys.* **38**, 1705. [254.]
NICHOLLS, R.W. (1961) *J. Res. Nat. Bur. Standards* **65A**, 451. [254.]
NICHOLLS, R.W. (1962a) *JQSRT* **2**, 433. [254, 265.]

NICHOLLS, R.W. (1962b) *Can. J. Phys.* **40**, 523. [254.]
NICHOLLS, R.W. (1962c) *Can. J. Phys.* **40**, 1772. [254.]
NICHOLLS, R.W. (1962d) *J. Res. Nat. Bur. Standards* **66A**, 227. [254.]
NICHOLLS, R.W. (1962e) *Nature* **193**, 966. [259.]
NICHOLLS, R.W. (1963a) *J. Chem. Phys.* **38**, 1029. [254.]
NICHOLLS, R.W. (1963b) *Nature* **199**, 794. [259.]
NICHOLLS, R.W. (1964a) *J. Res. Nat. Bur. Standards* **68A**, 75. [254.]
NICHOLLS, R.W. (1964b) *J. Res. Nat. Bur. Standards* **68A**, 535. [254.]
NICHOLLS, R.W. (1964c) *Nature* **204**, 373. [260, 264, 267.]
NICHOLLS, R.W. (1964d) *Annales de Geophysique* **20**, 144. [260, 265.]
NICHOLLS, R.W. (1965a) *J. Chem. Phys.* **42**, 804. [254.]
NICHOLLS, R.W. (1965b) *Astrophys. J.* **141**, 819. [254.]
NICHOLLS, R.W. (1965c) *J. Res. Nat. Bur. Standards* **69A**, 369. [254.]
NICHOLLS, R.W. (1965d) *J. Res. Nat. Bur. Standards* **69A**, 397. [254.]
NICHOLLS, R.W. (1965f) *JQSRT* **5**, 647. [260.]
NICHOLLS, R.W. (1965g) *Proc. Phys. Soc. (London)* **85**, 159. [262.]
NICHOLLS, R.W. (1966) *Proc. Phys. Soc. (London)* **89**, 181. [260–2.]
NICHOLLS, R.W. (1968a) *J. Phys. B Ser.* 2 **1**, 1192. [254.]
NICHOLLS, R.W. (1968b) *Nature* **219**, 151. [254.]
NICHOLLS, R.W. (1969) "Electronic spectra of diatomic molecules", Chap. 6 of *Electronic Structure of Atoms and Molecules:* Vol. III of *Physical Chemistry*, ed. by H. EYRING, D. HENDERSON and W. JOST, Academic Press, Inc., New York. [238–9, 259–60.]
NICHOLLS, R.W. and JARMAIN, W.R. (1955) *J. Chem. Phys.* **23**, 1561. [262.]
NICHOLLS, R.W. and JARMAIN, W.R. (1956) *Proc. Phys. Soc. (London)* A **69**, 253. [261–2.]
NICHOLLS, R.W. and JARMAIN, W.R. (1959) *Proc. Phys. Soc.* **74**, 133. [262.]
NICHOLLS, R.W. and STEWART, A. (1962) "Allowed transitions", Chap. 2, *Atomic and Molecular Processes*, ed. D.R. BATES, Academic Press, New York and London. [243, 253, 262–4.]
NICHOLLS, R.W., FRASER, P.A. and JARMAIN, W.R. (1959) *Combustion and Flame* **3**, 13. [254.]
NICHOLLS, R.W., FRASER, P.A., JARMAIN, W.R. and MCEACHRAN, R.P. (1960) *Astrophys. J.* **131**, 399. [254.]
NICHOLLS, R.W., PARKINSON, W.H., ROBINSON, D. and JARMAIN, W.R. (1956) *Proc. Phys. Soc. (London)* A **69**, 713. [262.]
NIELSEN, J.R., THORNTON, V. and DALE, E.B. (1944) *Rev. Mod. Phys.* **16**, 307. [15.]
NORRIS, J. and BASCHEK, B. (1970) *Astrophys. J. Suppl.* **19**, 305, 327, 337. [11.]
NORTON, K.A., (1936) *Proc. I. R. E.* **24**, 1367. [221.]
NUMEROV, B., (1933) *Pub. Obs. Cont. Astrophys. Russ.* **2**, 188. [255.]
OHMURA, H. and OHMURA, T. (1960) *Astrophys. J.* **131**, 8. [196, 212.]
OHMURA, H. and OHMURA, T. (1961) *Phys. Rev.* **121**, 513. [212.]
OHMURA, T. (1964) *Astrophys. J.* **140**, 282. [212.]
O'MALLEY, T., SPRUCH, L. and ROSENBERG, L. (1961) *J. Math. Phys.* **2**, 491. [214.]
OPPENHEIMER, J.A. (1928) *Phys. Rev.* **31**, 349. [7, 137.]
OPPENHEIMER, J.A. (1929) *Z. Physik* **55**, 725. [7, 137.]

ORY, H.A. (1964a) *Astrophys. J.* **139**, 346. [254.]
ORY, H.A. (1964b) *Astrophys. J.* **139**, 557. [254.]
PAULI, W. (1938) *Phys. Rev.* **54**, 924. [221.]
PEACH, G. (1962) *Monthly Not. Roy. Astron. Soc.* **124**, 371. [10, 137.]
PEACH, G. (1965) *Monthly Not. Roy. Astron. Soc.* **130**, 361–77. [11, 137, 184–5, 208–9, 214.]
PEACH, G. (1967a) *Memoirs Roy. Astron. Soc.* **71**, 1. [11, 184.]
PEACH, G. (1967b) *Memoirs Roy. Astron. Soc.* **71**, 13. [11, 137.]
PEACH, G. (1967c) *Memoirs Roy. Astron. Soc.* **71**, 29. [11, 137.]
PEACH, G. (1970) *Memoirs Roy. Astron. Soc.* **73**, 1. [11, 137.]
PEKERIS, C.L. (1934) *Phys. Rev.* **45**, 98. [256.]
PENNER, S.S. (1959) *Quantitative Molecular Spectroscopy and Gas Emissivities*, Addison-Wesley, Reading, Mass. [37, 61–2, 235.]
PENNER, S.S. and OLFE, D.B. (1968) *Radiation and Reentry*, Academic Press, New York and London. [12, 18, 64.]
PHILLIPS, J.G. (1954) *Proceedings of The Royal Society of Liege*. [264.]
PHILLIPS, J.G. (1957) *Astrophys. J.* **125**, 153. [264.]
PILLOW, M.E. (1949) *Proc. Phys. Soc. (London)* A **62**, 237. [253.]
PLASS, G.N. (1952) *J. Meteor.* **9**, 429. [83, 85–6, 90, 99, 101.]
PLASS, G.N. and FIVEL, D.I. (1953) *Astrophys. J.* **117**, 225. [235.]
POSENER, D.W. (1959) *Austral. J. Phys.* **12**, 184. [226, 234.]
POWER, E.A. (1964) *Introductory Quantum Electrodynamics*, Longmans, Green & Co., Ltd., London. [117.]
RACAH, G. (1942a) *Phys. Rev.* **61**, 186. [163.]
RACAH, G. (1942b) *Phys. Rev.* **62**, 438. [163, 167, 170.]
RACAH, G. (1943) *Phys. Rev.* **63**, 367. [163, 165, 179, 181.]
RACAH, G. (1949) *Phys. Rev.* **76**, 1352. [163.]
RANKIN, R.A. (1949) *Phil. Trans. Roy. Soc.* A **241**, 457. [221.]
REES, A.L.G. (1946) *Proc. Phys. Soc. (London)* **59**, 998. [254.]
REICHE, F. (1913) *Verh. Phys. Med. Ges.* **15**, 3. [233.]
REICHE, F. and RADEMACHER, H. (1926) *Z. Physik* **39**, 444. [241.]
REICHE, F. and RADEMACHER, H. (1927) *Z. Physik* **41**, 453. [241, 248.]
REICHEL, A. (1968) *JQSRT* **8**, 1601. [235.]
ROBERTS, O.F.T. (1930) *Proceedings of the Royal Society of Edinburgh* **50**, 225. [83.]
RODGERS, C.D. (1968) *J. Roy. Meteor. Soc.* **94**, 99. [82.]
RODGERS, C.D. and WALSHAW, C.D. (1966) *Quart. J. Roy. Met. Soc.* **92**, 67. [83, 87.]
ROHRLICH, F. (1959) *Astrophys. J.* **129**, 441, 449. [163–4, 166–7, 169.]
ROSE, M.E. (1957) *Elementary Theory of Angular Momentum*, John Wiley & Sons, New York. [124–5, 127, 129.]
ROSSELAND, S. (1924) *Monthly Not. Roy. Astron. Soc.* **84**, 525. [7.]
ROSSELAND, S. (1935) *Theoretical Astrophysics*, Oxford Univ. Press. [2.]
ROTENBERG, M., BIVINS, R., METROPOLIS, N. and WOOTOD, K. (1959) *The 3-j Symbols*, The Technology Press of the Massachusetts Institute of Technology, Cambridge, Mass. [179.]

RUDKJØBING, M. (1947) *Pub. Copenhagen Observatory* **145**. [31.]
RUDNICK, I. (1947) *J. Acous. Soc. Am.* **19**, 348. [221.]
RYBICKI, G. (1967) Private communication. [235–7.]
RYDBERG, R. (1931) *Z. Physik* **73**, 376. [254.]
RYDBERG, R. (1933) *Z. Physik* **80**, 514. [254.]
SAMPSON, D.H. (1965a) *JQSRT* **5**, 211. [19.]
SAMPSON, D.H. (1965b) *Radiative Contributions to Energy and Momentum Transport in a Gas*, Interscience Publishers (John Wiley & Sons, Inc.), New York. [71.]
SCHADEE, A. (1964) *Bull. Astron. Inst. Neth.* **17**, 341. [241, 249.]
SCHADEE, A. (1967) *JQSRT* **7**, 169. [112, 247, 250.]
SCHIFF, L.I. (1955) *Quantum Mechanics*, 2nd ed., McGraw-Hill Book Co., Inc., New York. [113, 252.]
SCHIFF, L.I. (1968) *Quantum Mechanics*, 3rd ed., McGraw-Hill Book Co., Inc., New York. [23, 32, 113, 116, 119, 122, 130–1, 137, 142–3, 149, 155, 201, 207.]
SCHLAPP, R. (1932) *Phys. Rev.* **39**, 806. [249.]
SCHUSTER, A. (1905) *Astrophys. J.* **21**, 1. [18, 19]
SCHWARZSCHILD, K. (1906) *Göttinger Nachrichten*, p.41. [7, 18, 19.]
SCHWENKER, R.P. (1965) *J. Chem. Phys.* **42**, 1895. [266.]
SEATON, M.J. (1951) *Proc. Roy. Soc. (London)* A **208**, 408. [137.]
SEATON, M.J. (1958) *Monthly Not. Roy. Astron. Soc.* **118**, 504. [137.]
SEITZ, W.S. and LUNDHOLM, D.V. (1964) *J. Opt. Soc. Am.* **54**, 315. [81.]
SEKERA, Z. (1968) *JQSRT* **8**, 17. [30.]
SHORE, B.W. (1967) *Rev. Mod. Phys.* **39**, 439. [114, 185, 232.]
SHORE, B.W. and MENZEL, D.H. (1965) *Astrophys. J. Suppl. Series* **12**, 187. [183.]
SHORE, B.W. and MENZEL, D.H. (1968) *Principles of Atomic Spectra*, John Wiley and Sons, Inc., New York. [163, 183.]
SHUMAKER, J.B., Jr. (1969) *JQSRT* **9**, 153. [257.]
SIMMONS, F.S. (1967) *JQSRT* **7**, 111. [79.]
SIMON, A. (1965) "Linear oscillations of a collisionless plasma", in *Plasma Physics* (pp.163–95), International Atomic Energy Agency, Vienna (Trieste Symposium, Oct. 1964). [220.]
SIMON, A., VAN DER SLUIS, J.H. and BIEDENHARN, L.C. (1954) *Tables of the Racah Coefficients*, Oak Ridge National Laboratory Report ORNL 1679, Oak Ridge, Tenn. [167, 171, 179.]
SLATER, J.C. (1960) *Quantum Theory of Atomic Structure*, Vol. I, McGraw-Hill Book Co., Inc., New York. [112, 122–3.]
SLATER, J.C. and KRUTTER, H.M. (1935) *Phys. Rev.* **47**, 559. [8.]
SMITH, A.L. (1970) *JQSRT* **10**, 1129. [238.]
SOLBRIG, A.W., Jr. (1961) *Am. J. Phys.* **29**, 257. [220.]
SOMMERFELD, A. (1909) *Ann. d. Physik* **28**, 665. [220.]
SOMMERFELD, A. (1939) (reprinted 1951): *Atombau und Spektrallinien*, 2. Aufl., Bd. 2, Vieweg & Sohn, Braunschweig. [137, 185, 203.]
SOMMERFELD, A. (1949) *Partial Differential Equations in Physics*, translated by E. G. STRAUS, Academic Press, Inc., New York. [221.]
SPINDLER, R.J. (1965) *JQSRT* **5**, 165. [252, 254–5.]
SPINDLER, R.J., ISAACSON, L. and WENTINK, T. (1970) *JQSRT* **10**, 621. [256.]

STEELE, D., LIPPINCOTT, E. R. and VANDERSLICE, J. T. (1962) *Rev. Mod. Phys.* **34,** 239. [253–4.]

STEWART, J. C. (1964) *JQSRT* **4**, 723. [10.]

STEWART, J. C. (1965) *JQSRT* **5**, 489–93. [220.]

STEWART, J. C. and PYATT, K.D., Jr. (1961) *Theoretical Studies of Optical Properties*, Air Force Special Weapons Center Report, AFSWC-TR-61-71, Kirtland Air Force Base, New Mexico. [51, 59.]

STEWART, J. C. and PYATT, K.D., Jr. (1966) *Astrophys. J.* **144**, 1203. [10, 11.]

STEWART, J. C. and ROTENBERG, M. (1965) *Phys. Rev.* **140**, A1508. [10, 11, 115.]

STOBBE, M. (1930) *Ann. d. Physik* **7**, 661. [7, 137.]

STONE, J. (1963) *Radiation and Optics*, McGraw-Hill Book Co., Inc., New York. [20, 103.]

STRACK, S. (1962) *A. R. S. Journal* **32**, 1404. [63.]

STRATTON, J. (1941) *Electromagnetic Theory*, McGraw-Hill Book Co., Inc., New York. [221.]

STRÖMGREN, B. (1932) *Zs. f. Astrophy.* **4**, 118. [8.]

STRÖMGREN, B. (1933) *Zs. f. Astrophy.* **7**, 222. [8.]

TATUM, J. B. (1967) *Astrophys. J. Suppl.* **14**, No. 124, 21. [243, 249.]

TAYLOR, R. (1963) *J. Chem. Phys.* **39**, 2354. [213.]

TEMPLE, G. (1955) *Proc. Roy. Soc.* A **228,** 125. [260.]

THATCHER, H. C., Jr. (1967) "Computation of the complex error function by continued fractions", *Blanch Anniversary Volume*, Aerospace Research Laboratory, U.S. Air Force, Washington, D.C., pp. 315–37. [234.]

TOLMAN, R. C. (1938) *The Principles of Statistical Mechanics*, Clarendon Press, Oxford. [114.]

TRAVIS, L. D. and MATSUSHIMA, S. (1968) *Astrophys. J.* **154**, 689. [11, 137.]

TREANOR, C. E. and WURSTER, W. H. (1960) *J. Chem. Phys.* **32**, 758. [265.]

TUCKWELL, H. C. (1970) *JQSRT* **10**, 653. [238.]

UNSÖLD, A. (1955) *Physik der Sternatmosphären*, 2nd ed., Springer-Verlag, Berlin. [18, 216–8.]

UREY, H. C. (1924) *Astrophys. J.* **59**, 1. [8.]

VAN DE HULST, H. C. (1957) *Light Scattering by Small Particles*, John Wiley, New York. [30.]

VAN DE HULST, H. C. and REESINCK, J. J. M. (1947) *Astrophys. J.* **106**, 121. [216, 223.]

VANDERSLICE, J. T., MASON, E. A., MAISCH, W. G. and LIPPINCOTT, E. R. (1959) *J. Molec. Spectrosc.* **3**, 17. [254.]

VANDERSLICE, J. T., MASON, E. A., MAISCH, W. G. and LIPPINCOTT, E. R. (1960) *J. Molec. Spectrosc.* **5**, 83. [254.]

VARSHNI, Y. P. (1957) *Rev. Mod. Phys.* **29**, 664. (See Errata, Vol. **31**, 839, 1959.) [253.]

VOROBYOV, V. S. and NORMAN, G. E. (1964) *Optics and Spectroscopy* **17**, 96. [10.]

WACKS, M. E. (1964) *J. Chem. Phys.* **41**, 930. [254.]

WACKS, M. E. and KRAUS, M. (1961) *J. Chem. Phys.* **35**, 1902. [254.]

WALSHAW, C. D. and RODGERS, C. D. (1963) *Quart. J. Roy. Met. Soc.* **89**, 122. [89.]

WATSON, W. D. (1969) *Astrophys. J.* **157**, 375. [11.]

WATSON, W. D. (1970) *Astrophys. J. Suppl.* **19**, 235. [11.]

WEINBERG, A.M. and WIGNER, E.P. (1958) *The Physical Theory of Neutron Chain Reactors*, Univ. of Chicago Press, Chicago. [74–5.]

WEISSLER, G.L. (1956) In *Handbuch der Physik* **21**, Ed. S. FLÜGGE, Springer, Berlin. [137.]

WENTINK, T. and SPINDLER, R.J. (1970) *JQSRT* **10**, 609. [255.]

WENTZEL, G. (1926) *Zeit. f. Phys.* **40**, 574. [137.]

WEYL, H. (1919) *Ann. d. Physik* **60**, 481. [221.]

WHEELER, J. and WILDT, R. (1942) *Astrophys. J.* **95**, 281. [185, 212.]

WHITE, H.E. (1934) *Introduction to Atomic Spectra*, McGraw-Hill Book Co., Inc., New York. [106.]

WHITING, E.E. (1968) *JQSRT* **8**, 1379. [226.]

WHITING, E.E. and NICHOLLS, R.W. (1971) to be published. [249.]

WHITTAKER, E.T. and WATSON, G.N. (1927) *A Course of Modern Analysis*, 4th ed., Cambridge Univ. Press, Cambridge. [233.]

WINANS, J.G. and STUECKELBERG, E.C.G. (1928) *Proceedings of the National Academy of Sciences* **14**, 867. [258.]

WU, T.Y. (1952) *Proc. Phys. Soc. (London)* A **65**, 965–72. [254.]

WYBOURNE, B.G. (1970) *Symmetry Principles and Atomic Spectroscopy*, Wiley-Interscience, New York. [163.]

YAMADA, H.Y. (1967) *JQSRT* **7**, 997. [79.]

YAMADA, H.Y. (1968) *JQSRT* **8**, 1463. [79.]

YAMAMOTO, G. (1951) *Sci. Reports of Tohoku Univ.*, Ser. 5, **3**, 130. [83, 85, 91, 99, 101.]

YAMAMOTO, G. and AIDA, M. (1970) *JQSRT* **10**, 593. [80.]

YOUNG, C. (1965) *JQSRT* **5**, 549. This paper describes the computation method but does not include the computer program which is stated to be available from the author. The tables are given in a report: *Tables for Calculating the Voigt Profile*, Univ. of Michigan, College of Engineering Technical Report 05863-7-T, July 1965 (O.R.A. Project 05863), Ann Arbor, Mich. [234.]

ZARE, R.N., LARSON, E.O. and BERG, R.A. (1965) *J. Molec. Spectrosc.* **15**, 117. [255.]

ZEL'DOVICH, YA.B. and RAIZER, YU.P. (1966) *Physics of Shock Waves and High-Temperature Hydrodynamic Phenomena*, Academic Press, New York and London. [10.]

ZEMANSKY, M.W. (1930) *Phys. Rev.* **36**, 219. [235.]

ZHELEZNYAKOV, V.V. (1967) *Astrophys. J.* **148**, 849. [20.]

Index